The East Asian High-Tech Drive

The East Asian High-Tech Drive

Edited by

Yun-Peng Chu
Professor of Economics and Director,
Research Center for Taiwan Economic Development,
National Central University, Taiwan

Hal Hill
H.W. Arndt Professor of Southeast Asian Economies,
Division of Economics, Research School of Pacific and Asian Studies,
Australian National University, Australia

Edward Elgar
Cheltenham, UK • Northampton, MA, USA

Published by
Edward Elgar Publishing Limited
Glensanda House
Montpellier Parade
Cheltenham
Glos GL50 1UA
UK

Edward Elgar Publishing, Inc.
136 West Street
Suite 202
Northampton
Massachusetts 01060
USA

A catalogue record for this book
is available from the British Library

Library of Congress Cataloguing in Publication Data
The East Asian high-tech drive / edited by Yun-Peng Chu, Hal Hill.
p. cm.
Includes bibliographical references and index.
1. Technological innovations–Economic aspects–East Asia. 2. Technological innovations–Government policy–East Asia. 3. High technology–Economic aspects–East Asia. 4. High technology industries–East Asia. 5. Industries–Technological innovations–East Asia. 6. East Asia–Economic policy. I. Chu, Yun-Peng. II. Hill, Hal, 1948-

HC460.5.Z9T432 2005
338'.064095—dc22 2005054311

ISBN-13: 978 1 84542 618 7
ISBN-10: 1 84542 618 5

Printed and bound in Great Britain by MPG Books Ltd, Bodmin, Cornwall

Contents

Contents

Contributors

Kelly Bird

Dr Bird is a consultant with Indonesia's National Development Planning Agency, Bappenas. He received his doctoral degree at the Australian National University in 1999. Since then, his works have dealt mainly with different aspects of the recovering Indonesian economy in the aftermath of the 1997 Asia financial crisis including investment, trade and labour analysis and policies. He has worked as a consultant for USAID, the Asian Development Bank and the World Bank.

Yun-Peng Chu

Dr Yun-Peng Chu is Professor of Economics and Director of the Research Centre for Taiwan Economic Development (RCTED) at the National Central University, and an Adjunct Professor at the Department of Economics, National Taiwan University. He was President of the Jin-Wen Institute of Technology, Taipei County, Taiwan, and Research Fellow and former Director of the Sun Yat-Sen Institute for Social Sciences and Philosophy (ISSP), Academia Sinica, Taiwan. He also served as Commissioner of the Fair Trade Commission of the Government of Taiwan. His published papers can be found in such journals as the *American Economic Review*, the *Journal of Development Economics* and *Developing Economies*, covering such areas as economic development, income distribution, international economics, competition laws and innovation studies. He has also co edited many books published by Edward Elgar and Routledge. He received his PhD degree in economics from the University of Maryland.

Hal Hill

Dr Hill is the H.W. Arndt Professor of Southeast Asian Economies at the Australian National University. His main research interests are the economies of ASEAN, especially Indonesia and the Philippines; industrialization and foreign investment in East Asia; and Australia's economic relations with the Asia-Pacific region. He has authored/edited many books on these subjects, in addition to about 110 academic papers and book chapters. He is also an occasional invited contributor to several Australian and Asian newspapers and magazines, and has worked as a consultant for the Australian government, the Indonesian government, the World Bank, the Asian Development Bank, and several United Nations agencies. He has held visiting appointments at Gadjah Mada University, the University of the Philippines, the Institute of Southeast Asian Studies, the University of Oxford and Columbia University.

Erh-Cheng Hwa
Dr Hwa is Professor of Economics at Shih-Hsin University with specialties ranging from economic development and international finance to East Asian studies. In his earlier career, he devoted himself to econometric research of the US economy at Cornell University, the National Bureau of Economic Research and SRI International. He later joined the International Monetary Fund and the World Bank where he spent most of his career, ultimately becoming Lead Economist in the bank's Department of Economic Management in the East Asia Region. He has served as head of the World Bank delegation to the PRC. Dr Hwa has published extensively on developing economies including China and Taiwan, and international economics, as well as on the US economy. He received his doctoral degree in Economics from Cornell University and his BA from National Taiwan University.

Medhi Krongkaew
Dr Medhi Krongkaew is Professor of Economics at the School of Development Economics, National Institute of Development Administration, Bangkok, Coordinator of the Community Building Forum at the Pacific Economic Cooperation Council (PECC), and a member of Thailand's Board of the National Statistical Office. Formerly he was Professor of Economics, Faculty of Economics, Thammasat University, Chair of the APEC Subcommittee on Economic and Technical Cooperation, a member of Thailand's National Budgetary Reform Committee, Director of the Institute of East Asian Studies at Thammasat University, Founding Director of the Thai APEC Study Centre at Thammasat University, a member of the Board of Directors of the East Asian Economic Association (EAEA), President of the Economic Society of Thailand and Economic Advisor to the Deputy Prime Minister (Dr Trairong Suwankiri). He has been granted numerous academic honours and has written extensively on international economics, income distribution, and East Asian economies, as well as financial and industrial studies.

Teeraya Krongkaew
Teeraya Krongkaew is a Senior Investment Officer, Financial Markets Operations Group, Bank of Thailand, Bangkok, Thailand.

Xielin Liu
Dr Liu presently serves as the Deputy Director, Research Centre for the Promotion of Science and Technology Development, Ministry of Science and Technology, PRC. He completed his doctoral study at Tsing-Hua University, China. His academic works are mainly concerned with technology policy and technology management.

Haji Mat Zin Ragayah
Dr Ragayah is Professor of Development Economics and currently Director of IKMAS, Malaysia. She studies and writes extensively in areas of income distribution and poverty as well as industry (particularly small and medium-sized industries) and trade. She has been appointed as a researcher and consultant to a number of national and international organisations including the East Asian Development Network (EADN), ESCAP, the Food and Agriculture Organization of the United Nations (FAO), the Thailand Research Development Institute (TDRI), and the World Bank Institute (WBI), as well as a number of state governments in Malaysia.

Jang-Sup Shin
Dr Shin is Associate Professor at the Department of Economics at the National University of Singapore. Before his career as an academic, he was the Deputy Economic Editor of the Maeil Business Newspaper in Korea and a part-time lecturer on Korean economic policy issues at Yonsei University. In 2003, he participated in a Joint Study Group for the Singapore–Korea Free Trade Agreement (FTA) negotiations. His main research interests are technology and economy, economic development, economic history, globalisation and financial crisis and catch-up strategy.

Gwendolyn R. Tecson
Dr Tecson is currently Professor of Economics at the School of Economics, University of the Philippines. She obtained her doctoral degree at Hitotsubashi University, Japan. Her field of research includes international economics and industrial economics with Japan being her specialized area.

Tham Siew-Yean
Dr Tham Siew-Yean is Professor and Deputy Director of IKMAS (Institute of Malaysian and International Studies), University Kebangsaan

Malaysia. Her research focuses mainly on foreign direct investment, industry and trade, including industrial competitiveness. She has been appointed as a researcher and consultant to a number of organisations including the Economic and Social Commission for Asia and the Pacific (ESCAP) of the United Nations, the Thailand Research Development Institute (TDRI), Thammasat University, the National Productivity Centre (NPS) and the Institute of Strategic and International Studies (ISIS), Malaysia.

Preface

East Asia has been an area of high economic growth for several decades. With growth comes economic transformation and changing comparative advantage. The faster the growth, the more rapid are these changes. This simple proposition sowed the seeds for change: as economic growth pushed up labour and other costs, the existing comparative advantage could no longer be sustained. To maintain high or even moderate growth, most East Asian economies resort to 'high-tech development' in their attempts to 'upgrade' their industries. Economies that already belong to the 'high-income' category will have to confront the advanced countries in technological and other developments. Middle- and low-income economies then try to catch up and quickly fill the positions previously occupied by the forerunners.

Many policy challenges and academically interesting questions arise in the process (see the Overview chapter for details). In response to this, a group of scholars with strong interests in East Asian economies got together and obtained funding from the Chiang Ching-Kuo Foundation in Taiwan for a research project on High-Tech Industrialization and Economic Development in East Asia. Participants included, in alphabetical order, Kelly Bird, Yun-Peng Chu, Hal Hill, Erh-Cheng Hwa, Medhi Krongkaew, X. Liu, Haji Mat Zin Ragayah, Jang-Sup Shin, Gwendolyn R. Tecson and Tham Siew-Yean. There were two roundtable discussions in Taipei (one in 2002 and another in 2003) organized by Yun-Peng Chu. The participants then wrote their final drafts on eight East Asian economies, including the People's Republic of China, Indonesia, the Republic of Korea, Malaysia, the Philippines, Singapore (in a comparative setting), Taiwan and Thailand, either individually or as joint authors. This book is a compilation of the research results, reviewed and edited.

There are several options regarding the order in which the eight economies appear in the text. It could be alphabetical, or based on per capita income. To highlight the theme of the book, we decided to arrange the order according to a measure of skill intensity of exports of the eight economies (see the Overview chapter for details). The resultant order is Singapore, South Korea, Taiwan, Malaysia, the Philippines, Thailand, China and Indonesia. The Overview chapter, which precedes these country chapters, summarizes the main issues involved.

We viewed the diversity of these eight important economies as both an opportunity and a challenge. The common element has been rapid growth and structural change in most of the economies for most of the period. But there are also significant differences, in growth rates, levels of development, institutional quality, factor endowments, development histories and much else. We therefore encouraged the authors to explore a range of country-specific issues and challenges, around the unifying themes of growth and technological upgrading. In the process, we hope that the volume is analytically richer, and that it will appeal to a broad readership beyond that with an immediate interest in East Asia.

In the process of preparation for the book, we received a lot of help from many people and institutions. We thank them all. In particular, we would like to thank all the authors for their participation in the project and for their various inputs and efforts. We also thank the Chiang Ching-Kuo Foundation for financial support. The assistance of Yi-Chun Chen, Jinso Chau, Shiu-Chi Chin and Nida Wang is also greatly appreciated.

Yun-Peng Chu and Hal Hill
March 2005

1. An Overview of the Issues

Hal Hill and Yun-Peng Chu*

1. INTRODUCTION

The chapters in this volume address one of the most important policy and analytical issues of our time. In the long sweep of economic development, several major 'turning points' are discernible. These are typically associated with the process of structural change, involving the shift of productive resources from agriculture to industry and services. Urbanization, and the demographic transition from high to low levels of mortality and fertility, typically accompany this transformation. Improved infrastructure and information flows result in better-functioning markets and more-integrated national economies.

These processes also lead to the well-documented transition from 'labour surplus' to 'labour scarce' economies (see for examples Oshima, 1987, and Chu, 2001). A combination of supply and demand factors begins to tighten the labour market and push up real wages. Labour supply rises more slowly owing to falling population growth rates and delayed entry into the workforce, the latter owing principally to longer periods of schooling. Employment in low productivity activities, including much of agriculture and the informal sector, declines owing to mechanization and falling relative demand for their products.

Several factors affect the pace and patterns of these processes. First and most important is the rate of economic growth. Faster rates of economic growth translate into accelerating demand for labour and hasten the end of this surplus labour phase of development. Second, the type of growth is important. In particular, 'East Asian' style growth – labour intensive, internationally oriented and socially inclusive – has major implications for how growth impacts on the labour market. Third, initial resource endowments influence the speed of structural change. For

* In preparing this chapter we have drawn freely on the contributors to this volume, in addition to a wide range of literature on the subject, including some by the authors. We are also most grateful to Yi-Chun Chen and Shiu-Chi Chin for assistance with the statistics and to Yi-Chun Chen for editorial work.

example, resource-poor economies will, *ceteris paribus*, shift more quickly out of agricultural activities than well-endowed economies. Finally, government policies are central to this process, both at the general level in their pursuit of economic growth, and in a wide range of specific initiatives ranging from educational priorities to promoting the development of modern commercial infrastructure.

This volume is concerned with how economies adjust to one of these major turning points, namely the transition to what for convenience may be referred to as 'high-tech' industrialization. This broadly refers to the process of transition from simple, labour-intensive, low value-added industrial activities to those that embody the more intensive use of human capital and technology. Such a transition is central to the process of economic development, as countries lose comparative advantage in labour-intensive activities, and are forced to compete on the basis of skills and scientific know-how.

How is this process to be managed and what are the public policy challenges? In a perfect neoclassical world of instantaneous and costless adjustment to changing relative prices, and of perfect foresight and information flows, no special importance attaches to this question. Private agents would anticipate and respond to the price signals. They would shift out of the now uncompetitive labour-intensive activities and technologies. The relevant new technologies would be adopted 'off the shelf', and modified as required, in analogous fashion to the neoclassical growth models where technology falls like 'manna from heaven'. In such an idealized and abstract construct, the role of government is restricted essentially to the provision of essential public goods (defence, security, the legal system), to ensuring a stable macroeconomic environment and an efficient financial sector, and to eliminating market failures and distortions.

At the other end of the spectrum is the view that market failure is pervasive and that markets will tend to underinvest in some of the 'public goods' that are central to the upgrading process, such as education and research and development (R&D).[1] Innovation is a slow, costly and uncertain process according to this argument. There are collective action and coordination failures. Both individual firms and capital markets tend to be myopic. They typically do not foresee and anticipate the evolution of dynamic comparative advantage. This is especially the case in high-growth, smaller, resource-poor economies, where comparative advantage can change very quickly. Accordingly, so the argument goes, there is a danger of growth being suddenly choked off, as firms are unable to adapt to an environment of rapidly rising real wages and appreciating exchange rates. Hence, far-sighted economic planners need to develop a commercial environment that facilitates structural adjustment, and enables firms to shift to more technology-intensive activities.

In practice, the approach of most governments falls somewhere between these two extremes. Yet this 'middle zone' typically remains ill-defined in both an analytical and policy sense. How much intervention is justified? Which particular forms of intervention appear to work? What should be in the domain of governments, and what should be left to markets? How does the role of government evolve over the course of economic development?

In addressing these questions, the policy-making context matters critically, in at least three respects. One is the quality of institutions and the extent to which policy makers can operate in an 'insulated' environment, at least in the sense that bureaucracies and legislatures are relatively immune from 'capture' by vested interests, and are able to implement 'first-best' policies. Clearly, the less this is the case, the weaker is the argument for micro, interventionist approaches, whatever their theoretical desirability. For example, if most trade policy is driven by rent-seeking, the case for industry-specific protection on infant industry or distributional grounds is thereby weaker. The implication is that alternative policy instruments need to be developed, such as time-bound and transparent subsidies.[2]

Second, openness to the global economy is a key consideration. For tradable goods, prices reflect international supply and demand conditions. Firms in these industries that operate in competitive markets will also exert pressure on suppliers of non-tradables, to ensure that they align with international benchmarks. Hence, competitive pressures act as a discipline on both firms and governments. Policy mistakes are more quickly corrected, as firms have to meet some sort of market test.

Institutional quality and openness together reinforce each other, in the sense that the calculus for internationally oriented firms is international standards of efficiency in the provision of all relevant goods and services. These firms therefore have different expectations from governments: they spend less time lobbying for rents, but they expect governments to deliver high standards of prudential supervision of the banking sector, stable macroeconomic management, judicial independence and effective contract enforcement, and high-quality physical infrastructure. The coincidence of institutional quality and openness also has important implications for industry policy in general and industrial upgrading in particular, a point to which we shall return below.

Third, history matters. Countries' development trajectories are path dependent, certainly over the short to medium term. Once established, a particular trajectory is hard to change. For example, in the East Asian context:

1. In the transition from a command to a market economy, some (notably China) have large SOE (state-owned enterprise) sectors which include historically developed, if inefficient, R&D expertise.
2. Korea felt compelled to embark on a highly interventionist strategy, in part owing to the specifics of its history (Japanese occupation and the subsequent intense rivalry) and geo-strategic considerations (cold war threats in the Korean peninsula).
3. For a variety of reasons, some have historically entrenched openness, for example, Singapore, Malaysia, and generally Thailand.
4. There is a long established foreign presence in some countries (e.g., Singapore and Malaysia).
5. There is a long history of import substitution in some cases (e.g., the Philippines).
6. A number of countries (perhaps most especially Thailand) exhibit a pronounced 'bifurcation' in state policy capacity, between macro- and microeconomic interventions.

This volume comprises rich and detailed studies of industrialization in eight diverse East Asian economies: (People's Republic of) China, (Republic of) Korea, Indonesia, Malaysia, the Philippines, Singapore (in comparative NIE context), Taiwan (Republic of China) and Thailand. Indeed, the sample of countries is sufficiently varied as to permit generalization beyond their immediate region. For example, the group includes three high-income economies (one an OECD member), three with per capita income clustered around $1,000, and two middle-income developing countries. There is also great diversity in size. The sample includes the world's most populous nation, and the fourth most populous; five have populations between 20 and 90 million; and there is one island city state. There is one major transition economy (China), another has experienced a major change in economic policy (Indonesia in the 1960s), while others – Singapore, Malaysia, Taiwan, Thailand – have been broadly consistent in their economic policy settings for at least four decades. With the exception of the Philippines, all have grown really quickly for extended periods. Six belong to the World Bank's (1993) 'miracle' club; China would also certainly be accorded membership on the basis of its performance over the past quarter-century. Conversely, the group includes the four economies most affected by the Asian economic crisis of 1997–98. There is also great variation in geographic size and natural resource endowments. The three NIEs (newly industrializing economies) in the sample are quintessentially resource-poor, densely settled economies. By contrast, China and Indonesia are huge countries, and the latter in particular is resource rich.

Importantly, the group differs very significantly on upgrading issues. The three rich economies, Singapore, Korea and Taiwan, have clearly lost their comparative advantage in labour-intensive activities, and must compete against OECD industrial economies.[3] Another two, Malaysia and Thailand, are in the process of a major structural adjustment towards advanced industrial status. This process has been somewhat delayed by very open labour markets, enabling them to hang on to labour-intensive processes for longer than would otherwise be the case. Their weaker supply-side capabilities and the economic crisis of 1997–98 are also factors. Their governments are now recognizing the need for a major technological upgrading, and that the past recipe for high growth will have to be modified. The two lower-income Southeast Asian economies, Indonesia and the Philippines, have grown more slowly on average, and more erratically (though Indonesia grew very fast between 1967 and 1997). The major policy challenge for governments is to restore high growth. Science and technology issues are second-order priorities, and are likely to remain so for a decade or more. China is the hardest country in the sample to characterize. Its income per capita is closer to the lower Southeast Asian norms, while its growth rate for the past quarter-century is 'Northeast Asian'. Its technological base is much more advanced than any other low-income country, with the partial exception of India, owing to scale, to an unusually high R&D effort, and to an earlier (and inefficient) experiment with SOE-led heavy industry.

The country studies which follow address a common set of analytical and policy issues related to industrialization and the transition to high-tech industry, while also exploring a range of country-specific issues and challenges which reflect the diversity among the eight. Such a framework is therefore designed to provide coherence but flexibility. The authors write with great authority, which springs from a deep familiarity with the countries about which they write. Beyond the requirements of analytical and thematic consistency, however, there has been no attempt to impose a doctrinaire approach to the issues at hand. For example, some authors are obviously a good deal more sympathetic to the case for an interventionist industry policy than are others.

In this overview chapter, we assemble some of the salient lessons and conclusions from the country studies. We commence with a brief review of concepts and frameworks. This is followed by an overview of the eight economies, and a discussion of key cross-cutting issues and challenges associated with innovation and upgrading. A concluding section sums up some of our key arguments.

2. CONCEPTS AND FRAMEWORKS

Some Definitions

There is no universally accepted definition of technology in general and 'high-tech' in particular. The most common approaches define technology as 'a collection of physical processes that transform inputs into outputs and knowledge and skills that structure the activities involved in carrying out these transformations' (Kim, 1997, p. 4). Technology progress may be regarded as a 'better way of doing things' or of 'producing more from less', by employing new technologies and generating new products and processes. As a corollary, high-tech refers to activities that employ or embody inputs that in some sense are technology intensive.

At every point, however, one encounters blurred definitional boundaries among 'technology', 'innovation' and 'science' policies. There is a tendency to equate technology and innovation with pathbreaking scientific discovery and research at the frontiers of knowledge. In reality, much R&D involves not the 'R' of fundamental research, but rather the 'D' of acquiring, adapting and modifying frontier technology. Rosenberg (1994, p. 13), for example, cites estimates for the US that just 8 per cent of R&D expenditure is devoted to 'basic research', compared to 25 and 67 per cent for 'applied research' and 'development' activities, respectively. For countries within the technology frontiers, the proportion devoted to basic or applied research is obviously much lower. In the latter cases, 'development' and 'diffusion' are much more important than basic research, and it is these areas to which public policy needs to be directed. The studies in this volume provide numerous illustrations of the proposition that governments think differently about technology and adopt different policy approaches over the course of economic development.

Studies of technology emphasize that the process of acquiring and mastering technical know-how is evolutionary, gradual, laborious, uncertain and long term (Lall, 2000). Even seemingly simple cases of 'reverse engineering' are in reality complex and time-consuming. Economists' notions of infant industries may involve more protracted periods of endeavour than is commonly supposed. These features – uncertainty and long gestation periods in particular – underline the links between innovation and a country's political and institutional structures. That is, potential innovators need a nurturing environment, in which, at the very least, property rights, financial institutions and the political system are supportive. For this reason, the range of policies that impinge on technology and innovation is much broader than is commonly supposed. Indeed, a government's officially enunciated 'technology policy' may well be a relatively unimportant part of a country's broader technological effort. Countries may invest significant resources in technology, but if the economic policy environment is distorted, the pay-off may be minimal. The former USSR is an obvious example of this proposition. By contrast, small, nimble, highly open economies like Hong Kong may invest very little in formal R&D programmes, but still achieve rapid growth based on high-quality service delivery niches.

Is There a Case for Intervention?

The principal rationale for government intervention to develop a country's technological capability has to do with the latter's 'public good' characteristics. Left to the market, there is likely to be underinvestment in activities for which private agents are unable to appropriate adequately the returns from their investments. Such investments tend to enter the public domain quickly, a process facilitated in part by classic 'free-rider' problems. The returns are diluted, especially in developing countries, by a weak legal system that is unable to protect intellectual property rights. High worker mobility, especially for those whose skills are in limited supply, exacerbates the problem, since firms are reluctant to invest in R&D programmes if there is a high probability that key staff embodying these investments will quit before returns can be appropriated. The inability of firms to appropriate fully the returns on these investments results in social rates exceeding the private rates, and hence provides a justification for government intervention.

Other forms of market failure provide a further rationale for government intervention. Where markets are underdeveloped – incomplete or missing – price signals do not function adequately. This may arise because of poor information flows, limited entrepreneurial capacity, inadequate property rights or physical infrastructure bottlenecks. Institutions (such as producer cooperatives or industry associations) that foster the cooperative behaviour sometimes needed to achieve improved outcomes may not exist. Coordination functions, especially in dealing with international markets, may be poorly developed. The cooperative and coordination functions do not necessarily need to be supplied by government, but public policy can hasten their efficient evolution.

The particular issues associated with technological upgrading are examined in some detail in this volume. As argued above, different economies face different challenges, and therefore the policy challenges and approaches vary. In some cases, the major concern is with the provision of basic public goods, such as an adequate legal system and universal basic education and health. In others, it is developing scientific and research institutions to support a high-wage economy. In transition economies, establishing a sound commercial reputation in the global economy is a major challenge.

The International Environment

These issues need to be examined against the backdrop of a rapidly changing international environment.[4] Global integration is proceeding deeply and broadly. The international economy is larger and more competitive than ever before, providing both the challenge of intense competitive and the opportunity of ever-widening niches for newcomer firms and countries. More firms are linked to global production networks, including contract manufacturing, international buying groups, outsourcing and subcontracting. Depending on their competitive advantages, these may take the form of OEMs (original equipment manufactures), ODMs (original design manufactures) or OBMs (original brand manufactures), or a combination of these. Trade within both industries and firms is rising.

Several factors are driving these processes, and there is every likelihood that they will continue. Barriers to international trade and investment are falling, as are transport and communication costs. The pace of technological innovation is rising, and more countries are developing a broader capacity to supply goods and services to international markets. Productivity-enhancing managerial practices such as just-in-time delivery are proliferating. These trends are triggering the emergence of major cross-border geographic clusters, centred on cities with high-quality infrastructure and easy access to the international economy. Typically such clusters include both mass-production operations to exploit economies of scale, and a range of specialist ancillary services. Country reputation builds on these strengths and reinforces the pattern of geographic concentration. Within East Asia, the greater region of southern coastal China, including the high-income niches of Hong Kong and Taiwan, is the most important example of this regional concentration. Other examples include the Singapore-centred Sijori hub, with heavy electronics specialization, and greater Bangkok, which is the emerging Southeast Asian auto hub.[5]

Economies that are less connected to the international economy and are not able to attract these major concentrations of economic activity run the risk of falling behind. For low-income economies, labour-intensive export-oriented goods and services will be the major driver, and for which the prime requirements are reliable and competitive infrastructure, cheap and productive labour, and a simple and predictable investment climate. As wages rise, governments have to pay more attention to a range of science and technology issues. These include the quality of education, publicly funded research institutes, subsidies and incentives for R&D, and strategies to leverage the MNE (multinational enterprise) presence. Each of these areas involves a complex set of issues, many of which are addressed in the case studies that follow. For example, for both universities and research institutes, the challenge is not just to establish academic excellence but also to foster linkages with the private sector and thus 'commercialize' these strengths. To maximize the benefits from the MNE presence, a set of complementary fiscal and R&D initiatives needs to be in place. In the East Asian context, as Yusuf (2004, p. 20) notes, notwithstanding their diverse development strategies, a common element in the upgrading processes is that '[T]o a greater or lesser extent, [governments] are disengaging from old-style industrial policy and concentrating on augmenting the supply of skills, research, foreign capital flows, technology transfers, and trade'. It is not enough to set up R&D institutes and expect a local variant of Silicon Valley to flourish. These institutes need to connect to the business community, and they need to operate in an open and competitive environment that approaches international best practice in the provision of infrastructure, financial, educational, regulatory and legal services.

Against this backdrop, the link between trade and FDI (foreign direct investment) liberalization is central to such a transformation (Brooks and Hill, 2004). As trade barriers have fallen, the old 'tariff factory' model of FDI has given way to a new FDI-led, export-oriented paradigm. This is sometimes characterized as a switch from 'rent-seeking' to 'efficiency-seeking' FDI. This transformation has profound links for host countries' management of FDI. The regulatory apparatus which was constructed to manage (and frequently siphon off) the rents under the old regime is generally still present in most countries. Yet it has become largely irrelevant, and has usually been bypassed in the reform process, which typically has been driven by other parts of the bureaucracy. Rather, the contemporary challenge for developing countries is to develop a new approach to managing FDI. In a globalizing world, competition for FDI is no longer about rents but instead focuses on the establishment of an enabling, business-friendly commercial environment, consistent with national development objectives.

Concepts and Indicators

How are these abstract concepts translated into empirical observations that are both analytically informed and internationally comparable? A useful approach is to develop a set of measures around what one of us (Hill, 2004a) has termed the 'three *I*s': *incentives*, *institutions* and *infrastructure*. Incentives refer to domestic prices being more or less aligned with international price levels. Empirical proxies typically focus on openness to trade and foreign investment. Institutions revolve around the quality of governance and the legal system. Infrastructure embraces the pricing, regulation and quality of physical infrastructure, together with social infrastructure such as education and health that ensure that the benefits of growth are spread widely. In addition to these broad indicators, a number of specific measures related to technology and innovation are relevant to a study such as this one. The selection of these variables draws on the relevant literature, particularly as it relates to East Asia.[6]

Openness

Openness refers to all types of commercial engagement with the global economy: trade, investment, technology and labour flows. Three features dominate these flows. First, for most periods, global capital flows are growing faster than trade, which in turn is generally expanding faster than output. Second, as a corollary, for practically all countries, the relative importance of trade and foreign investment is rising. Third, while 'south–south' trade is expanding rapidly, and the sources of FDI proliferating, more than 90 per cent of global R&D is undertaken in high-income economies, and thus technology transfers either remain primarily within the North America–Japan–Europe triad or flow from the north to the south.[7]

We shall examine the issues and impacts associated with rising openness shortly, in the context of our East Asian case studies. But it will be useful to foreshadow some key elements:

1. Reform is inevitably uneven, giving rise to what may be termed 'dual policy regimes', across sectors and regions. The integration of 'enclaves of reform' and the rest of the economy is a major policy challenge in many countries.
2. These trends are central to an understanding of the rise of the global factory and cross-border, intra-firm and vertically integrated production networks. As noted, the more or less simultaneous

liberalization of both trade and FDI regimes has shifted MNEs' calculus from 'rent seeking' to 'efficiency seeking'. Lower transport and information costs, together with technological change in several major industries (most especially electronics), have been the other key drivers of these trends.

3. The presumption is that FDI is the most important channel of technology flows, but the empirical evidence suggests that these 'spillages' vary significantly across firms, industries, regions and countries.
4. There is no one model of openness that fits all countries' experiences. Analytically, there is a large literature on how selective and how rapid is the opening to the international economy. Empirically, the selection of indicators is inevitably somewhat arbitrary and subject to numerous caveats.

Human Capital and the Domestic R&D Base

The quality of a country's human capital base is arguably the most important determinant of its absorptive capacity, that is, an economy's ability to absorb, assimilate and diffuse imported technology, and to innovate on the basis of these imports. Over the course of economic development, priorities will differ. At early stages of development, mass literacy and universal primary education deserve priority, while upgrading strategies at middle-income levels requires that more attention be accorded to vocational and higher education, to quality, and to the composition of educational output among disciplines.

There are separate but inter-related issues relating to public support for R&D and the role of R&D institutions. These constitute the formal component of most governments' technology and innovation policies. Important questions here concern the scale of these institutions, their funding, and their interaction with enterprises and the education system. As real wages begin to rise, most governments recognize the case for public funding of R&D, to facilitate the upgrading process. For latecomer countries, the role of these institutions focuses almost entirely on the modification and diffusion of essentially 'off-the-shelf' technologies.

Most educational and all technology indicators have conceptual and empirical limitations, which explains why a basket of such measures is typically employed in comparative analysis. The usual approaches include both 'input' and 'output' measures. The former focus on resources devoted to human capital and the development of technological capacity, such as expenditure on education and R&D. The latter concentrate on outcomes of this capacity, and include such indicators as enrolments and years of schooling, patents and the proportion of output or exports originating from 'technology-intensive' activities.[8]

Infrastructure, Institutions and the Commercial Environment

These encompass a diverse set of factors. An efficient business operation able to connect to the global economy requires high-quality physical infrastructure. There needs to be a commercial climate that is conducive to long-term planning horizons and investments decisions. Property rights and an adequately functioning legal system are also important. Financial institutions, domestic and foreign, need to be present. Macroeconomic stability is crucial. Competition, domestic and foreign, is central to innovation, as a spur to raise productivity.

All three elements are important ingredients in the overall schema. Success in one or two may be insufficient, especially if other elements of the policy framework are not supportive. For example, a significant head-start in its educational standards was insufficient to maintain economic momentum in the Philippines. Countries may invest substantially in R&D, but these investments may be rendered ineffective by poor macroeconomic management or barriers to international commerce (e.g., China and India pre-reform). Governments may invest in expensive high-tech projects while starving the public education system of resources (e.g., India again, and Indonesia to some extent). Japan and Korea had little choice but to undertake major investments in public education and R&D given their restrictive policies towards FDI.

In other words, all three sets of variables are interdependent. For example, open economies require, and call forth, higher-quality institutions. With regard to industry policy, as economies open up, governments have to make the transition from protectionist/regulatory regimes to a new emphasis on promotion and efficiency. Thus, there needs to be more effective industrial extension, R&D and other support schemes; better physical infrastructure; legal reform; improved education; and administrative reform and simplifications. Hence, countries' performance according to a range of 'competitiveness' variables to be listed below in Tables 1.1 (for the middle–higher-income countries) and 1.2 (for the lower–middle-income countries) is the key to reaping the benefits of openness.

3. THE EIGHT ECONOMIES

Comparative Indicators

We present here a summary picture of industrial upgrading and technological capacity in the eight economies, drawing on the country studies and an illustrative set of indicators broadly consistent with the analytical framework presented above.

First, to set the scene, it will be useful to present some stylized facts on each economy according to a range of indicators that constitute the analytical building blocks of this chapter. These comprise three broad categories: openness to trade, FDI and human capital; supply-side capabilities relating to education, R&D, physical infrastructure and institutions; and other variables including the quality of macroeconomic management, ownership structures and the nature of the business regulatory environment.

Thus, to give a few examples, Korea emerges as having a historically restrictive trade, FDI and labour flows regime, except for the export sector. Its supply-side capabilities are generally very good. Its ownership structures are unusual in being dominated by the state-sponsored cabal conglomerates, its macroeconomic management has been moderately good, while its regulatory environment has been paternalistic but is being reformed. Malaysia has always been exceptionally open to the international economy. Its supply-side capabilities are generally good, though historically its educational base was much weaker than Korea's. Its macroeconomic management has been excellent, while since the early 1970s its ownership structures and regulatory environment have been dominated by the affirmative action objectives. As another example, the Philippines, has been a rather inward-looking economy, though with major liberalizations since the late 1980s. Labour inflows have been tightly regulated, but its outflows are, relative to the size of its population, the largest in East Asia. Its human capital base was historically one of the strongest in East Asia, though that edge is eroding, and its infrastructure and institutions are generally weak. Its macroeconomic management has been adventurous at times, while the business environment has been characterized by high levels of corruption and uncertainty. We elaborate further on these and other characteristics for all eight economies in this chapter.

Extending this analysis, Table 1.3 presents quantitative indicators for the eight economies, grouped according to the major variables listed above. We briefly consider each in turn.

Table 1.1 The Eight Economies – Some Stylized Facts for Middle–higher-income Economies

	Korea	Malaysia	Singapore	Taiwan
I. Openness:				
Trade	From 1960s, open for exports, but otherwise restrictive; major 1990s liberalizations	Consistently open, apart from some heavy industry & auto projects	Extremely open; world's highest trade/GDP ratio	Always open for exports; a fairly mild import-substitution period, with major liberalizations from 1970s
FDI	Very restrictive until 1990s; then gradual opening; major 1998 reforms; outward FDI generally exceeds inward	Consistently open, very large inflows	One of the world's highest FDI/GDP ratios; always very open; becoming a very large outward investor	Broadly open to FDI, especially since the 1970s; though not a large FDI presence; a significant outward investor, mainly to China
Human capital	Very restrictive, except for large, mainly US, diasporas; reverse brain drain in 1990s	Extremely open; very large inflows, mainly unskilled, much quasi-legal	Very open labour market, across all skill groups; though closely monitored	Increasingly open labour market; large, mainly US, diasporas, reverse brain drain in 1990s
II. Supply-side capacities:				
Education	Extremely high education achievement at all levels; though limited international connections	Generally quite good; major affirmative action programme; very large tertiary numbers abroad; continuing loss of 'non-bumiputeras' skills	Comprehensive public education system; very large tertiary numbers abroad; recent emphasis on high quality	Consistently high education performer; though concern that universities not at international standards

R&D / extension	Major R&D effort from 1980s; very strong capabilities, though limited international connections	Historic strengths in tropical agriculture, but limited industrial capabilities	Aggressive promotion of R&D-centred activities from 1990s, both government and MNE based	Effective industrial extension networks in public and private sectors; ITRI widely seen as a model for others
Physical infrastructure	Generally excellent	Generally excellent	Consistently ranks among world leaders	Generally excellent
Institutional quality	Generally good, aspiring tc OECD standards	Generally high	Extremely high; though some concerns about limited accountability	Generally good
III. Other:				
Ownership structures	Predominantly private; large cabal presence and high concentration; limited SME sector	Always large foreign presence, major and controversial bumiputeras business promotion	Extremely large foreign presence; major SOE presence; concern about weak domestic private sector	Substantial foreign presence, alongside historically dynamic SMEs; early SOE role diminished
Macro management	Moderately good, though occasionally adventurous	Consistently very low inflation	Consistently excellent; huge public sector surpluses	Mostly low inflation and prudent fiscal policy
Regulatory complexity	Major 1990s reforms, but still quite politicized business environment	Relatively business-friendly environment; moderate corruption; affirmative action priorities	Simple, minimal regulatory environment; though on major policy directions an activist government	Relatively business-friendly environment; still quite politicized; moderate corruption

Source: Revised and extended by authors from Hill (2004a).

Table 1.2 *The Eight Economies – Some Stylized Facts for Lower-middle-income Economies*

	China	Indonesia	Philippines	Thailand
I. Openness:				
Trade	Closed to 1978; then progressive opening, especially for exports; landmark 2002 WTO accession	Broadly open since 1967, and especially since mid-1980s reforms	Early import substitution, but reform gathered momentum during 1980s, and now broadly open	Consistently fairly open, after mild import-substitution phase
FDI	Closed to 1978, rapid increase from 1980s, especially in southern coastal regions; now world's largest FDI recipient	Sudden opening in 1967; restrictive in 1970s, then major 1980s reforms; negative flows since 1997 owing to uncertainty	US-dominated until mid-1970s, but limited inflows from other sources; major 1990s liberalizations	Consistently fairly open; rapidly increasing flows from late 1980s, sustained during the crisis
Human capital	Restrictive, except for huge Chinese diasporas	Officially restrictive, though porous borders	East Asia's major labour exporter; very large diasporas	Porous borders, and significant labour inflows, mainly unskilled
II. Supply-side capacities:				
Education	Mass education; uneven quality, with pockets of excellence; rapid catch-up	Mass education, but generally low–medium quality	An early East Asian leader in universal public education; still high relative to income, though early advantage is eroding	Mass primary education, but low continuation rates; now catching up
R&D / extension	Pockets of excellence, including from past, mostly uneconomic, SOE heavy	Mostly weak, except for some agricultural activities; huge concentration on	Generally weak, except for IRRI-related agricultural strengths	Limited, though increased government support from 1990s

	industry	uneconomic aerospace project		
Physical infrastructure	Spatially uneven, though rapidly improving	Major Soeharto-era investments, but deteriorating post-crisis	Continuing underinvestment in all sectors; 1990s reforms eased some bottlenecks	Generally very good; although international connections heavily Bangkok-centred and congested
Institutional quality	Uneven though improving; still in transition to market economy	Generally weak, and much less predictable post-Soeharto	Generally weak; though highly vocal civil society	Improving and functionally predictable, though uneven bureaucratic quality
III. Other:				
Ownership structures	Dominant but rapidly declining SOEs; large foreign presence; private firms now growing fast	Once large SOE sector shrinking; modest foreign presence; continuing concern with ethnic business divide	Significant foreign presence in export industries; high levels of corporate concentration	Predominantly private, Sino-Thai; rising foreign presence
Macro stability	Low inflation; generally prudent fiscal policy, but large potential contingent liability from SOE and banking sector	Moderate inflation, and generally prudent fiscal policy, but large public debt from past crises	Moderate inflation since mid-1980s; but recurring precarious fiscal deficits	Consistently low inflation and prudent fiscal policy
Regulatory complexity	Continuing reform, though considerable vestiges from command economy; high levels of corruption; rapidly decentralizing	High levels of corruption; unlike Soeharto era, now also unpredictable; major 2001 decentralization	Politicized, corrupt and (for major investment projects) unpredictable business environment; considerable decentralization	Generally predictable and business-friendly environment, but significant corruption

Source: See Table 1.1.

China is by far the largest economy in the sample, while Singapore and Taiwan have the highest incomes, around seven times that of the poorest, Indonesia. China has also grown the fastest since 1980, by a significant margin. Over that period, all economies have at least doubled their per capita income, except for the Philippines which has hardly progressed. The two poorest countries, Indonesia and the Philippines, are also the most indebted and have averaged the highest inflation rates.

According to the various 'indicators of openness', Singapore stands out as easily the most open to trade and investment. Malaysia also registers very high levels of trade and investment ratios. This partly reflects their size – small economies by definition trade more than do larger ones – but it also indicates open FDI regimes, and in the case of Singapore very low formal trade barriers. Singapore is also the largest outward investor in the sample, relative to the size of its economy. Among the other economies, trade barriers are generally low, certainly in comparative developing country context. Barriers are somewhat higher in China, Korea and Thailand, though the difference between them and the others is not great. There are numerous caveats to be attached to these data. In most cases, exporters are placed on something close to a free trade footing. The tariff data do not indicate the dispersions, or the incidence of non-tariff barriers. Where electronics dominate exports, as in the Philippines, the trade to GDP shares are inflated by thin domestic value added in the components trade and high levels of inter-industry trade.

Even the two poorer Southeast Asian economies and the notionally socialist China have received significantly more FDI than Korea and Taiwan. The last two, and Singapore, are also major outward investors. In an earlier period, Korea and Taiwan received more capital in the form of non-equity investment, as compared to Southeast Asia. However, this pattern is changing following the liberalization of FDI regimes, and the recent Southeast Asian picture is complicated by capital flight during the 1997–98 economic crisis.

'Human capital and innovation indicators' present a mixed picture. Most East Asian economies score well on basic education achievements, certainly in comparative developing country context. The three NIEs (i.e., Korea, Singapore, Taiwan) are already at OECD levels, with Korea the most striking. Very high levels of secondary enrolments are evident among the NIEs and the Philippines. At tertiary levels, the NIEs also generally stand out, while China and Indonesia lag. Indonesia's public expenditures are well behind the others. The tertiary enrolment figures for Malaysia and Singapore are greatly understated, because about one-third of their tertiary enrolments are abroad. These quantitative indicators reveal little about quality. International rankings (admittedly confined to maths and science examinations) do confirm general impressions that quality is much higher among the NIEs. The Philippines slips markedly according to this criterion.

As would be expected, the R&D indicators show the NIEs to be mostly well ahead of the ASEAN (Association of South East Asian Nations) Four. The farmer's formal R&D effort is generally at median OECD levels or more, with Korea among the highest in the world. Taiwan (and now Singapore) invests proportionately more in R&D than several middle-income OECD economies. Indonesia, the Philippines and Thailand invest practically nothing in R&D. Malaysia is now investing substantially, more than double the figure reported in Table 1.3. China's R&D investments are unusually large for a country of its per capita income. Taiwan leads the field in a measure of US patents data, followed by Singapore and Korea. The others record very little or nothing. Korea leads in terms of information technology usage, where there is a large gap between the NIEs and Malaysia as compared to the other four countries.

It will be obvious from the data that inferences on technological capacity based on export composition and 'revealed comparative advantage' can be very misleading. This is so owing to the rapidly expanding 'slicing' of international production activities, and the dominance of electronics in East Asian exports. Even finely disaggregated export statistics cannot now capture the fact that the electronics industry encompasses the full gamut of factor intensities, ranging from the highly R&D intensive to the technologically simple packaging and assembly operations. Thus, in Table 1.3, the country that exhibits the highest proportion of 'hi-tech exports' is the Philippines, which has one of the weakest technological capacities in East Asia.[9]

It is also evident that the growth of total factor productivity (TFP) and technological progress are at best loosely connected, and that the former is a weak indicator of the latter. The limitations of TFP growth are well known (see Chen (1997) for an East Asian survey). The TFP numbers mask the microeconomics of technical progress, for example the mastery of products and processes which in many cases did not exist in these countries a decade or more ago.[10] As an illustration, the proposition that there was little technological progress or innovation at the enterprise level in Singapore during the 1960s and 1970s because measured TFP growth was negligible contradicts all available evidence pointing to that country's evolution towards technologically advanced, export-oriented industrialization.

A range of 'institutional, infrastructure, finance and commercial policy indicators' are also presented in Table 1.3. This is a very large area, data are in some cases perceptions based, and thus the variables presented here

*Table 1.3 The Eight Economies – Quantitative Indicators**

	PRC	Indonesia	Korea	Malaysia	Philippines	Singapore	Thailand	Taiwan
General economic indicators								
GDP, 2003 ($ billion)	1,410	208	605	103	79	91	143	286
GDP per capita PPP, 2003 ($)	4,995	3,364	17,908	9,696	4,321	24,480	7,580	24,560
GDP per capita growth, 1990–2003 (%)	8.2	3.1	5.1	3.9	1.0	3.5	4.2	4.4
Annual average inflation, 1990–2003(%)	5.6	12.8	5.0	3.0	8.0	1.6	4.0	2.1
Total external debt/GDP, 2002 (%)	13	76	23	51	76	N.A.	47	12
GDP per capita 2003/1980	6.1	2.2	3.7	2.2	1.1	2.5	2.9	3.3
Openness								
Trade								
(Exports + imports)/GDP, 1990 (%)	29.9	54.5	58.4	154.8	62.9	373.8	79.7	88.5
(Exports + imports)/GDP, 2003 (%)[1]	65.7	64.0	74.3	213.9	113.9	352.0	128.7	109.0
Export growth, 1990–2003 (%)[2]	18.0	8.1	9.3	11.3	10.7	9.1	10.3	8.2
Average tariff rate, 2001[3]	14.3	5.4	9.2	5.8	4.0	0.0	9.7	3.5
Index of economic freedom, 2004[4]	3.6	3.8	2.7	3.2	3.1	1.6	2.9	2.4
Investment								
FDI as % of total capital inflows, 1990–2003[5]	93.0	−48.2	16.6	129.0	52.2	33.4	−19.2	19.1
Total FDI inward stock, $ billion, 2003[6]	501.5	57.2	47.5	59.0	11.5	147.3	36.9	33.9
Total FDI outward stock, $ billion, 2003[7]	37.0	2.7	34.5	29.7	1.0	90.9	3.3	65.2
Total FDI inward stock as % of GDP, 2003	35.6	27.5	7.8	57.2	14.5	161.3	25.8	11.9
Total FDI outward stock as % of GDP, 2003	2.6	1.3	5.7	28.8	1.2	99.5	2.3	22.8
Human capital and innovation								
Years of education, 2000[8]	5.7	4.7	10.5	7.9	7.6	8.1	6.1	8.5
Gross tertiary enrolment rate (%), 2001	12.6	14.6	77.6	28.2	31.2	46.0	35.3	83.4
R&D expenditure as % of GDP, 2002[9]	0.6	–	2.6	0.2	–	1.4	0.1	2.3
Number of internet users as % of total	6.3	3.8	61.3	34.5	4.4	54.8	9.7	39.0

inhabitants, 2003								
Public spending on education as % of GDP, 2001[10]	2.1	1.3	3.6	7.9	3.2	3.1	5.0	6.4
Average mathematics scale scores of eighth-grade students[11]	–	411	589	508	378	605	–	585
Average science scale scores of eighth-grade students[12]	–	420	558	510	377	578	–	571
Total R&D employment, thousand, 2001[13]	956.5	–	165.7	10.1	–	19.5	14.0	138.4
R&D employment per 1 million population, 2001[14]	752	–	3,500	432	–	4,709	236	6,177
Utility patents data, 2003[15]	0.2	0.0	82.7	2.0	0.3	99.3	0.4	234.4
% of manufactured exports which are 'high tech', 2002[16]	23.3	16.4	31.5	58.3	65.3	60.3	31.4	43.2
Physical infrastructure[17]	60	45	23	22	87	6	39	21
Institutional quality and risk								
Corruption								
Corruption perceptions index, 2003[18] (country ranking)	3.4 (66)	1.9 (122)	4.3 (50)	5.2 (37)	2.5 (92)	9.4 (5)	3.3 (70)	5.7 (30)
Country risk								
Composite risk ranking, 2002	75.0	58.3	79.8	77.5	71.0	90.0	76.3	82.0
Property rights								
Index of economic freedom[19]	4	4	2	3	4	1	3	2
Bureaucratic quality								
Public institutions index, 2004[20] (country ranking)	4.39 (55)	4.12 (68)	4.81 (41)	5.06 (38)	3.21 (99)	6.21 (10)	4.71 (45)	5.56 (27)
Fiscal/finance								
Stock market capitalization as % of GDP, 2003[21]	48.3	26.2	54.5	163.2	29.3	115.4	84.4	130.6
Highest corporate tax rate, 2003 (%)	33	30	27	28	32	22	30	25

Notes:
* All data covering a period of time, e.g., 'GDP per capita growth, 1990–2003,' refer to the yearly average among all years during the period; 'N.A.' means 'not applicable'.

1. Data for Singapore are for 2002.
2. Data for Singapore are for 1990–2002.
3. Data for Indonesia, Korea, Malaysia and Thailand are for 2000, 2002, 1997 and 2000, respectively.
4. Index of economic freedom ranges from 0 (mostly free) to 5 (highly restricted).
5. FDI : Foreign Direct Investment.
   Total capital inflows = Direct investment + (Increases in) portfolio investment liabilities + (Increases in) other investment liabilities.
   FDI as per cent of total capital inflows = FDI inflows / total capital inflows.
6. FDI inward stock:
   Data for Korea are accumulated since 1962.
   China, Korea, the Philippines and Thailand: Stock data after 2002 are estimated by adding flows.
   Indonesia: Stock data after 1999 are estimated by adding flows.
   Malaysia: Stock data after 1994 are estimated by adding flows.
   Taiwan: Stock data after 1988 are estimated by adding flows.
7. FDI outward stock:
   Data for Korea are accumulated since 1968.
   China: Stock data after 1989 are estimated by adding flows.
   Indonesia: Stock data after 1999 are estimated by adding flows.
   Korea, Malaysia: Stock data after 2002 are estimated by adding flows.
   Taiwan: Stock data after 1988 are estimated by adding flows.
8. Years of education are average years of school over age 25.
9. Data for Malaysia are for the year 2001. Data for Indonesia and Philippines are unavailable.
10. Data for the PRC, Singapore and Taiwan are for 1999, 1995 and 2003, respectively.
11. International average mathematics score is 466. Data for China and Thailand are unavailable.
12. International average science score is 473. Data for China and Thailand are unavailable.
13. Data for Malaysia and Thailand are for 2000 and 1997, respectively. Data for Indonesia and the Philippines are unavailable.
14. Data for Malaysia and Thailand are for 2000 and 1997, respectively. Data for Indonesia and the Philippines are unavailable.
15. US utility patents granted per million of population.
16. Data for Thailand and Taiwan are both for 2001.
17. Based on the 1–104 country ranking in the Global Competitiveness Report (1 = best).
18. The index ranges from 0 (highly corrupt) to 10 (highly clean). The world average for the 133 countries covered is 4.2, with a maximum of 9.7 and a minimum of 1.3.

19. The property rights index is a composite from the index of economic freedom developed by the Heritage Foundation. The range is from 0 (very good) to 5 (very poor).
20. The public institutions index is based on survey data and ranges from 2.47 to 6.59 across 104 countries. The higher the index, the higher the quality.
21. Data for Singapore are for 2002.

Sources:

General economic indicators	
GDP, 2003 ($ billion)	*The Global Competitiveness Report 2004–2005.*
GDP per capita PPP, 2003 ($)	*The Global Competitiveness Report 2004–2005.*
GDP per capita growth, 1990–2003 (%)	World Bank, *World Development Indicators* (Online).
Annual average inflation, 1990–2003(%)	World Bank, *World Development Indicators* (Online).
Total external debt/GDP, 2002 (%)	Total External Debt: World Bank, *World Development Indicators* (Online). Inter-Agency Task Force on Finance Statistics, 'Joint BIS-IMF-OECD-World Bank Statistics on External Debt', http://www.oecd.org/dac/debt/
GDP per capita 2003/1980	World Bank, *World Development Indicators* (Online).
Openness	
Trade	
(Exports + imports)/GDP, 1990 (%)	World Bank, *World Development Indicators* (Online). National Statistics, ROC, http://www.dgbas.gov.tw/
(Exports + imports)/GDP, 2003 (%)	
Export growth, 1990–2003 (%)	
Average tariff rate, 2001	Heritage Foundation, http://www.heritage.org/research/features/index/
Index of economic freedom, 2004	Heritage Foundation, http://www.heritage.org/research/features/index/

Investment	
FDI as % of total capital inflows, 1990–2003	FDI: UNCTAD, *World Investment Report*, http://stats.unctad.org/fdi/ Total capital inflows: IMF, *International Financial Statistics*, http://www.imfstatistics.org/ Central Bank of China, ROC (Taiwan), http://www.cbc.gov.tw/
Total FDI stock, $ billion, 2003	
Total FDI stock as % of GDP, 2003	
Human capital and innovation	
Years of education, 2000	Barro-Lee Education Data, http://www.nber.org/pub/barro.lee/
Gross tertiary enrolment rate (%), 2001	*The Global Competitiveness Report 2004–2005.*
R&D expenditure as % of GDP, 2002	World Bank, *World Development Indicators* (Online). National Statistics, ROC (Taiwan), http://www.dgbas.gov.tw/
Number of internet users as % of total inhabitants, 2003	*The Global Competitiveness Report 2004–2005.*
Public spending on education as % of GDP, 2001	World Bank, *World Development Indicators* (Online). UNESCO Institute for Statistics, http://www.uis.unesco.org 'Indicators of Educational Statistics of the Republic of China', Ministry of Education, ROC
Average mathematics scale scores of eighth-grade students	National Centre for Education Statistics, http://www.nces.ed.gov/timss/
Average science scale scores of eighth-grade students	
Total R&D employment, thousand, 2001	UNESCO Institute for Statistics, http://www.uis.unesco.org National Statistics, ROC, http://www.dgbas.gov.tw/
R&D employment per 1 million population, 2001	
Utility patents data, 2003	*The Global Competitiveness Report 2004–2005.*

% of manufactured exports which are 'high tech', 2002	World Bank, *World Development Indicators* (Online). 'Monthly Statistics of Exports and Imports Taiwan Area, ROC', Ministry of Finance, ROC
Physical infrastructure	*The Global Competitiveness Report 2004–2005.*
Institutional quality and risk	
Corruption	
Corruption perceptions index, 2003 (country ranking)	Transparency International, http://www.transparency.org/
Country risk	
Composite risk ranking, 2002	UNCTAD, *World Investment Report*, http://stats.unctad.org/fdi/
Property rights	
Index of economic freedom	Heritage Foundation, http://www.heritage.org/research/features/index/
Bureaucratic quality	
Public institutions index, 2004 (country ranking)	*The Global Competitiveness Report 2004–2005.*
Fiscal/finance	
Stock market capitalization as % of GDP, 2003	World Bank, *World Development Indicators* (Online). Taiwan Stock Exchange Corporation.
Highest corporate tax rate, 2003 (%)	Heritage Foundation, http://www.heritage.org/research/features/index/

can provide only a glimpse of the bigger picture (see also Hill, 2004b). In general, the NIEs perform quite well, and are well ahead of the lower-income ASEAN countries, although Malaysia in some respects belongs more to the former than to the latter. Thus, for example, Singapore tops international ranking, while Korea, Malaysia and Taiwan all score well. The Philippines is very low, reflecting in large part its successive fiscal crises. Similar rankings apply for the incidence of corruption, country risk and bureaucratic quality, except that Indonesia scores the lowest for the first two. The rankings according to economic freedom correlate closely with those of per capita income. Malaysia, Singapore and Taiwan appear to have more attractive corporate taxation rates and more developed stock markets.

Finally, and focusing particularly on the theme of this volume, the high-tech drive in East Asia, it will be useful to examine in more detail the skill intensity of the region's exports since the late 1980s, when the issue first became prominent.

How can the technology or skill intensity of exports be measured?

According to a two-digit definition of 'high-tech' given by the OECD, the sectors that belong to the 'high-technology industry' are (i) aircraft and spacecraft, (ii) pharmaceuticals, (iii) office, accounting and computing machinery, (iv) radio, television and communications equipment, and (v) medical, precision and optical instruments. 'Medium–high-technology industries', on the other hand, include (i) electrical machinery and apparatus other than those listed above, (ii) motor vehicles, trailers and semi-trailers, (iii) chemicals excluding pharmaceuticals, (iv) railroad equipment and transport equipment other than motor vehicles, trailers and semi-trailers, and (v) machinery and equipment other than those listed above (see Directorate for Science, Technology and Industry, OECD, 2004). However, as noted above, researchers using this or similar definitions sometimes run into problems. The entire electronics sector belongs to the category of 'high- or medium–high-technology', so countries that specialize in such exports would have high shares of high-technology products in exports regardless of whether it is an exporter of predominantly lower-end assembly or higher-end niche products.[11]

To overcome such a difficulty, here we follow Chu's (2004) efforts by compiling a 6-digit data set for exports, using Taiwan as a benchmark. Specifically, we compute the percentage of workers with college or higher degrees in total employment for each 6-digit HS commodity category (manufactures excluding oil or petroleum products). The export sector is considered high-skilled labour intensive (HS) if this percentage is high, and vice versa. Taiwan's exports were then divided into three groups of equal size, with HS, MS (medium-skilled labour intensive), and LS (low-skilled labour intensive) describing these three groups in descending order of technological intensity. The same ranking from Taiwan's exports is then applied to all other economies.[12] Table 1.4 reports the results.[13] According to this table, the rank of the eight economies in descending order of the share of high-skilled labour-intensive goods in total exports in 2002 would be Singapore, Korea, Taiwan, Malaysia, the Philippines, Thailand, China and Indonesia. Although these results are far from conclusive, we shall present our eight economies in this order.

In addition, it will be useful to compare changes in both per capita income and the skill intensity of exports. We do this in Table 1.4 and Figure 1.1 for the 1989–2002 period. Note that we define the 'net' skill intensity of exports as the HS share less the LS share of exports. The arrows in Figure 1.1 show the movement over the period. The results indicate that Korea has been quite successful in its high-tech drive. The (net) skill intensity of its exports, measured by the difference between the HS and LS shares, was lower than Taiwan's in 1989, but surpassed it by 2002, although its per capita GDP still lagged behind Taiwan. Singapore was ahead in both its net skill intensity of exports and per capita GDP for both years, in spite of the slowdown during the Asian financial crisis that occurred in this period. According to the World Bank classification, these three economies belong to the 'high-income' group,[14] for which the average growth rate (among all such economies in the world) was 2.7 per cent per annum. All three economies outperformed this benchmark, Korea and Taiwan by more than 4 percentage points and Singapore by 2.4 percentage points.

Malaysia, the only upper–middle-income economy in the group of our eight economies, performed well in both per capita GDP growth and increases in skill intensities of its exports. In spite of the Asian financial crisis, its growth rate outperformed the average growth rate among all upper–middle-income economies of 2.9 per cent per annum between 1989 and 2002 by 2.1 percentage points. Among the four lower–middle-income economies, China's income growth outperformed the world average for this group of economies of 1.3 per cent per annum by 12 percentage points. Its net skill intensity also rose substantially. Thailand's exports also displayed a substantial change in net skill intensity and its growth rate exceeded the world average by 3.7 percentage points in spite of the financial crisis. The two laggards in the group of eight are the Philippines and Indonesia. The former saw a substantial rise in net skill intensity of its exports but very little growth in its per capita GDP between 1989 and 2002. The latter enjoyed some growth during the period but its exports

Table 1.4 Skill Intensity of Exports: Eight Economies, 1989–2002 (%)

		1989	1990	1991	1992	1993	1994	1995	1996	1997	1998	1999	2000	2001	2002
Singapore	HS	49.2	51.5	53.9	57.7	62.4	64.0	65.6	69.4	72.4	75.8	76.5	70.5	73.8	70.2
	MS	36.6	36.5	34.1	32.2	29.6	29.8	29.1	26.2	23.5	20.7	19.8	25.9	22.4	26.7
	LS	14.2	12.0	12.0	10.1	8.0	6.2	5.2	4.4	4.1	3.6	3.7	3.6	3.8	3.1
Korea	HS	22.6	22.6	25.0	29.0	32.6	34.6	38.9	44.5	49.2	53.8	54.7	55.6	58.9	62.3
	MS	31.5	32.2	31.6	32.8	36.5	41.5	43.9	39.4	35.4	31.8	31.9	33.2	29.9	27.2
	LS	46.0	45.2	43.4	38.3	30.9	23.8	17.3	16.2	15.4	14.5	13.4	11.2	11.3	10.6
Taiwan(ROC)	HS	27.3	31.2	32.3	35.8	37.8	39.8	43.2	46.6	47.4	51.4	52.7	52.7	54.5	53.2
	MS	35.5	36.4	35.3	34.7	35.5	36.9	37.9	36.1	36.2	33.8	32.6	34.4	32.3	34.8
	LS	37.2	32.4	32.4	29.5	26.7	23.4	18.9	17.4	16.5	14.8	14.7	12.9	13.2	12.1
Malaysia	HS	17.0	16.2	18.2	22.7	23.6	26.4	28.8	33.9	39.2	47.2	48.7	45.6	47.4	42.6
	MS	33.9	38.1	41.8	40.0	44.7	47.0	48.4	43.7	39.5	34.6	34.5	38.2	36.7	42.2
	LS	49.1	45.7	39.9	37.4	31.6	26.6	22.8	22.5	21.4	18.2	16.7	16.2	15.9	15.3
Philippines	HS	17.3	19.3	16.8	25.0	27.7	25.8	29.0	34.7	45.3	53.4	49.6	46.0	46.7	36.9
	MS	18.5	19.0	20.4	20.2	24.7	31.0	33.7	31.6	27.8	27.1	30.9	35.6	33.7	46.3
	LS	64.3	61.7	62.8	54.9	47.6	43.2	37.3	33.7	26.9	19.6	19.5	18.4	19.7	16.9
Thailand	HS	10.1	9.6	11.1	12.3	15.8	18.2	22.5	26.9	30.5	31.7	31.1	31.5	33.3	31.3
	MS	23.0	23.4	24.0	23.6	26.3	27.8	28.2	27.4	27.5	27.2	28.8	30.4	28.3	31.4
	LS	66.9	66.9	64.9	64.1	57.9	54.0	49.3	45.7	42.0	41.1	40.2	38.1	38.4	37.3
PRC	HS	11.9	11.3	10.8	11.2	12.0	14.2	17.8	17.4	18.4	20.4	21.6	23.7	25.1	25.9
	MS	17.5	18.8	19.1	20.4	21.4	23.0	25.3	26.3	27.3	28.5	28.8	30.3	29.9	32.1
	LS	70.6	69.9	70.1	68.4	66.7	62.8	56.9	56.3	54.4	51.1	49.6	45.9	45.0	42.0
Indonesia	HS	14.0	12.1	10.9	12.4	11.5	13.3	15.0	16.5	16.5	21.9	19.5	19.5	19.5	20.8
	MS	5.3	6.9	7.6	9.1	12.5	14.9	15.8	19.1	19.6	19.4	21.5	22.6	23.2	23.6
	LS	80.7	81.1	81.4	78.5	76.0	71.9	69.2	64.4	63.9	58.7	59.0	57.9	57.3	55.5

Source: Computed from trade data by the authors.

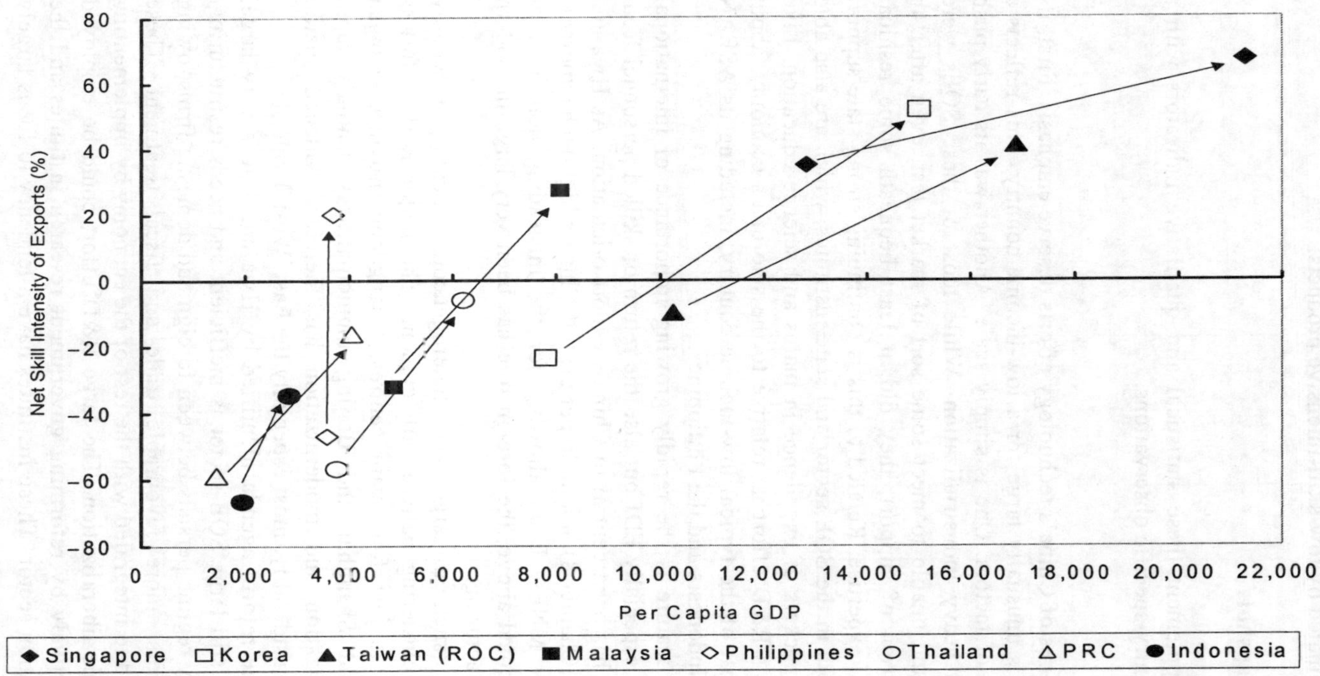

Note: 'Net Skill Intensity of Exports' (%) is measured by the HS share minus LS share in exports; as such it could be either positive or negative.

Figure 1.1 *Changes in Per Capita GDP and Net Skill Intensity of Exports: 1989–2002*

are still dominated by low-skill-intensive products.

Country Snapshots

We now supplement these statistical and qualitative indicators with a range of country-specific observations.

China

Several features of China's technology efforts deserve emphasis. First, its R&D effort is unusually large for a low-income country. This reflects a combination of factors. One is simply scale. Another was its early push for state-led heavy industrialization. While most of these SOEs were inefficient, and unable to meet some sort of market test, even after an extended period of support, they did at least bequeath some residual technological expertise. Related to this is continuing strong state support for R&D, both in the SOE sector and state institutes, which are also able to build on pockets of excellence in maths and science education. Like India, China's R&D effort is, relative to the size of its economy, about five times that of the typical low-income country, including its ASEAN neighbours, Indonesia and the Philippines.

A second feature is the rapidly growing importance of international connections, especially FDI but also the return of skilled personnel from abroad. FDI has been central to China's export orientation. As Hwa and Liu observe (Chapter 8), about 86 per cent of China's high-tech exports is generated by MNEs. Even allowing for the misleading definition of 'high-tech' noted above, the foreign role has been very large in modern sector manufacturing.

Third, in such a large and heterogeneous country, inevitably pronounced clustering patterns are evident. The major R&D effort is concentrated in just four main centres, Guangdong province, Jiangsu province, and Shanghai and Beijing municipalities). China's rapid internationalization and modernization has therefore widened spatial disparities considerably, most especially the East–West divide.

Finally, the reform agenda identified by Hwa and Liu is very large. Much of the still large SOE sector is inefficient and needs restructuring. A dual policy regime persists between foreign and domestic firms, owing to the farmers' more favourable trade and fiscal treatment. These enclaves need to integrated with the rest of the economy by implementing economy-wide liberalization. The large R&D effort could be employed more productively by reforming government research institutes and the higher education sector. These institutes have a tendency to focus largely on basic research, and they generally have weak ties to the private sector (see also Yao, 2003 on this issue). There has also been a rapid proliferation of policy initiatives – high-tech parks, government venture capital funding, various commercial incentives – but it is too early to evaluate their effectiveness.

Indonesia

Several features of Indonesia's R&D policies stand out.[15] First it was the most severely affected by the economic crisis of 1997–98, and the slowest to recover. For this reason, and because the transition to a more advanced industrial structure is still decades away, technology and innovation rank very low in government priorities. The crisis also resulted in a profound shift in both the development trajectory and the political economy of economic policy. Bird and Hill argue (Chapter 9) that these changes render more difficult the challenge of returning to high growth.

Second, Indonesia is a 'latecomer' in practically all respects. Its educational base was hampered by colonial neglect and the absence of economic progress until the mid-1960s. It hardly possessed a modern industrial sector prior to the late 1960s, by which time almost all the private foreign capital had been driven out of the country. According to almost all indicators of science, education and technology, it is the laggard among the group of economies covered in this volume. The country also has a stronger natural resource base than most of its neighbours, hence dictating a somewhat different industrialization trajectory.

Third, technology policy has been sporadic and lacking in coherence. Prior to 1997, the government had begun to pay attention to technology issues, but in a most unusual manner. For 20 years, its research and technology minister, Dr B.J. Habibie (who was also briefly president, 1998–99), completely dominated Indonesian technology policy. His plans centred on a number of highly ambitious heavy industry projects, most especially a showcase aircraft factory, IPTN. State investment in the latter totalled at least \$3 billion, and involved a 'back-to-front' approach to high-tech industrialization involving aircraft assembly without the base of supplier industries to support it. Not surprisingly, in the wake of the crisis, state support for the venture evaporated, and the project has collapsed. Like China's earlier push for heavy industry, there are residual, albeit high-cost, technological benefits.

Finally, in other respects, the picture is mixed. Indonesia achieved 30 years of rapid economic development through to 1997, mainly owing to prudent macroeconomic management, reasonably open trade and investment policies, and major investments in physical infrastructure and to a lesser extent mass education. However, institutions remain weak, and most micro-level interventions are highly vulnerable to capture by vested interests. Industry policy in particular has been driven by rent-seeking rather than efficiency considerations.

Korea

Korea's development record has been both fantastically impressive and controversial. From a very low base in the 1950s, it grew as fast as any country in recorded history until the early 1990s. Shortly thereafter, it experienced a serious economic crisis in 1997–98, by far the worst among the Asian NIEs. It achieved universal and high-quality education by the 1970s, at a comparatively low level of income. In the 1990s, its R&D effort was one of the strongest among the OECD countries. Its industrial and technology policies were 'Japan-like' and quite unlike its developing East Asian neighbours: a restrictive trade and FDI regime, except for export-oriented and technology-intensive sectors, the nurturing of privately owned but state-sanctioned industrial conglomerates, and an unparalleled commitment to growth targets. During the authoritarian era at least, this highly interventionist regime appeared to be predominantly productivity oriented.[16]

In assessing Korea's development experience, it is useful to identify those elements that are either relatively uncontroversial or amenable to replication, and those that are either country specific or problematic.

Among the former, as Shin emphasizes (Chapter 3; see also Kim, 1997), were at least six elements. Firms had to focus on exports. Export orientation was employed as a means of ensuring that assisted firms would quickly meet some sort of market test. There was an early and strong commitment to education. Macroeconomic management was reasonably sound, if sometimes adventurous. FDI was discouraged wherever possible, in favour of arm's-length licensing agreements, capital goods imports, and other channels of technology transfer. Trade policy pursued the twin-track objectives of vigorous import substitution and active export orientation, depending on an industry's (or even a firm's) circumstances. Finally, there was a very strong government commitment to the development of an indigenous R&D base, both within the cabals and through government research institutes.

The principal controversy centres on whether the highly interventionist, firm-specific guidance was the key to the country's extraordinarily rapid growth. Authors attracted to this view of Korean history (e.g., Amsden, 1989) rightly point to instances of very rapid learning by doing. However, there were clear costs associated with the strategy, too. The major cabal also made costly mistakes. Their links to universities and small and medium-sized enterprises (SMEs) remained rather weak. There were many opportunities for bureaucratic corruption, especially as the Korean state became more democratic and less 'insulated'. The restrictive FDI regime hampered Korean firms' access to international know-how, and was part of the reason why the country had to invest so heavily (and perhaps overinvest) in scientific education and know-how. Moreover, the crisis of 1997–98 highlighted significant vulnerabilities in Korea's commercial and financial systems. As Kim (2000, p. 336) observe, in the 1990s the government shifted 'from an effective orchestrate to a rigid regulator', and hence the reforms commensurate with a modern OECD economy were slow to be implemented.[17]

The Korean experience suggests that the earlier authoritarian governments were remarkably successful in lifting the country from very low levels of income to middle-income status, but that this policy framework has been less useful as the country approaches the technological frontiers. More generally, Korea's political and historical development was unique and therefore difficult to replicate. The harsh Japanese colonial rule, the Korean civil war and the ever-present threat from the North bequeathed a regime with an unparalleled development commitment, an intrusive, authoritarian political system and a government bureaucracy that actively cajoled firms and individuals to meet highly ambitious targets. In any case, it is unlikely that, in the current international commercial policy environment, such a restrictive regime would even be possible in the latecomer industrializers.

Malaysia

Malaysia's strengths include its relatively good institutions, its openness to both international trade and foreign investment (apart from, recently, the special case of short-term capital), its good infrastructure, and its political stability and policy consistency. Its human capital base is good, but not outstanding, and has been complicated by its ongoing loss of high-level 'non-bumiputeras' human capital. Its business environment is open, with the one major qualification that the government's ethnic restructuring objectives do impose some restrictions on firms' commercial freedom. The country had notable early strengths in agricultural extension services (particularly for cash crops), which for many years underpinned its status as the world's largest and most efficient producer of natural rubber and palm oil.

Malaysian industry comprises several distinct segments. One is electronics, which has been extremely successful in terms of growth and exports, but now must make the transition upstream, integrate more with the rest of the economy, and develop its human resource base. Another segment is the automotive industry, whose record is the subject of considerable controversy. Here Malaysia has attempted a prolonged period of costly import protection, resulting in Thailand overtaking it as Southeast Asia's leading automotive producer. A third segment is the agro-processing industries, where Malaysia has been able to develop on the basis of its traditional production strengths and a high-quality, publicly funded (or at least mandated) R&D base.

Malaysia is in the midst of a major upgrading effort, only briefly interrupted by the crisis of 1997–98, outlined in detail by Tham and Ragayah (Chapter 5).[18] This current 20-year plan seeks to lift the country to OECD status, particularly in knowledge-based activities, including computer infrastructure, 'info structure', education and training. R&D and technology strategies include a number of schemes, ranging from R&D grants to double deduction of R&D expenditures and commercialization incentives. Just prior to the Asian financial crisis, the government announced its plans to develop a 'Multimedia Super Corridor' (MSC), which included the aspiration to develop a 'Silicon Valley of Southeast Asia'. The government is taking the initiative in this project, through the provision of subsidized telecommunications infrastructure, fiscal incentives, university-based incubators and related government investments. Its electronics industries have particular strengths in assembly, testing and packaging, but it lacks the technological sophistication of the NIEs, particularly outside the dynamic Penang region. Universities are being promoted, but are still constrained by ethnic quotas in staff and student selection. There is a significant skills shortage.

The Philippines

Industrial growth has been the slowest among the sample, and the share of manufacturing has been shrinking. Tecson (Chapter 6) draws attention to several of the key features in this puzzling record (see also Medalla et al., 1995/96; Hill, 2003). It was the first country in developing East Asia to embark on a comprehensive strategy of import substitution. But protection was always more about rents than about industrial planning and, despite attempts at partial export-oriented industrialization from the early 1970s, it failed to make the transition to an export-oriented regime. Attempts at liberalization began in earnest in the early 1980s, and again in the latter part of the Aquino regime. The Ramos administration then implemented quite a bold reform programme, to the point where the Philippines has a reasonably open industrial sector. The historically restrictive policies towards FDI were also relaxed. Export growth in the

1990s was impressive, and was a factor contributing to economic resilience during the 1997–98 crisis. Quite comprehensive services reform was also undertaken, in banking, telecommunications and transport.

However, paradoxically, this extended period of protection resulted in more, not less, export concentration. Electronics now dominate exports, constituting about 75 per cent of total merchandise exports, and accounting for virtually all the growth in merchandise exports since 1990. In spite of the relatively good educational indicators, most electronics exports are at the low end of the industry, and backward linkages from MNEs and beyond the zones are weak. The latter arises in part owing to the country's dual trade regime. SMEs are generally located outside the zones and where regulatory procedures are more complex and infrastructure poorer. The country's earlier educational advantage is fading, owing to slower growth and misdirected public expenditure priorities. Infrastructure remains underfunded. Tecson also argues persuasively that fiscal incentives lack analytical justification.

Singapore

Singapore as a city state has notable strengths in the area of technology and innovation policy. It is an extremely open economy to both trade and FDI, with practically no barriers to entry. The open trade regime confers a discipline on sometimes quite interventionist firm-level promotion, while the FDI regime connects to global R&D centres. Its bureaucracy is one of the best (and best paid) in the world, and apparently immune from political pressures, thus enabling it to contemplate a range of 'first-best' interventionist strategies. Its physical infrastructure is world class, with its port and airport regularly being ranked at the top of global comparisons. Its education system is good and being rapidly upgraded. Owing to its exceptional fiscal prudence, it has the resources to implement a range of ambitious upgrading programmes. Its high salaries and open labour market enable research institutions and firms to recruit on the international labour market, often quite selectively.

As the comparative chapter by Shin and Chu (Chapter 2) points out, Singapore's technology and innovation policy has evolved rapidly, as it has shifted quickly from using to creating technologies.[19] It began to lose comparative advantage in labour-intensive activities in the late 1970s, while the government embarked on its first serious R&D effort in the late 1980s. A big push for higher education, high-tech industry and basic R&D got underway in the late 1990s. R&D expenditure has increased rapidly, from minimal levels through to the mid-1980s to almost 2 per cent of GDP currently. Most of it occurs in the private sector, but with aggressive state inducement. During the 1990s, funding for public research institutes and universities has increased rapidly.

MNEs have traditionally dominated Singapore's manufacturing, generating about three-quarters of value added, as pointed out by Shin and Chu. The government has actively sought to induce these firms to upgrade, to use the country as a regional headquarters, to undertake product and process innovation, and to build and fund joint-venture technology institutes to diffuse their technology. Electronics has been the key to its industrialization. A particular success story has been the hard disk drive industry, where Singapore spearheaded its relocation to Southeast Asia (see also McKendrick et al., 2000, for a detailed case study).

Apart from the major emphasis on R&D programmes, schemes were introduced for the training of employees and the development of supplier industries. The government has employed fiscal incentives to induce MNEs to use the country as a regional headquarters and to undertake R&D. A Skills Development Fund levied on firms provided funds for training. Government officials have a well-deserved reputation for being attentive to the needs of foreign companies, and to thinking creatively about means of leveraging their presence. Start-up times for new ventures are among the quickest in the world.

Among future challenges, Wong (2001 and 2003) identifies developing links between enterprises and both the universities and the public research institutes. Links among the last two are already strengthening. Venture-capital funds are emerging, but remain rather limited. International R&D and educational links need to be developed. There is the question of what R&D to fund given that Singapore's small size dictates specialization. The government laments the lack of domestic entrepreneurship, while vigorously recruiting the best and the brightest from tertiary institutions. With a large and well-paying MNE presence, it is perhaps not surprising that local privately owned firms are the country's weak link. Firm-level selectivity appears to continue to be important, principally under the auspices of the Economic Development Board and the huge government-linked corporations. It is very difficult to judge the efficacy of these interventions since little information on their operations is publicly available. The presumption is that openness and low levels of corruption are the major checks on policy mistakes and abuse.

Taiwan

Taiwan's rapid industrialization and economic development have been underpinned by several key factors. Chu (Chapter 4) identifies at least eight. Its economy has always been broadly open to trade and investment. It has had an exceptionally strong and resilient SME base. Its physical infrastructure is mostly excellent. It has achieved very high saving rates, which have been intermediated reasonably effectively by its financial sector, including since the 1980s venture capital and the stock market. There has been a good general education system, with strong emphasis on science and education (Lin, 1998). Increasing numbers of students went abroad from the 1960s, principally to the US. Many stayed abroad for decades, returning not only with formal education but also with extensive international commercial experience. The government adopted a range of effective innovation strategies as the economy began to lose comparative advantage in labour-intensive activities.

Taiwan is frequently grouped with Korea in development comparisons, but in many respects the two are quite different (Smith, 2000). Its economic policy was consistently less interventionist than Korea's, and the government embarked on fewer ambitious heavy industry projects. It lacked the large conglomerates within which major R&D initiatives could be funded. Historically, government-supported R&D was not large. Its FDI regime was neither as restrictive as that of Korea, nor as open as the city states. Taiwan also weathered the Asian economic crisis more successfully than Korea.

The political economy of industry policy has been dominated by two key parameters. First, as in Korea, but perhaps even more so, there has been a shift from a developmental and insulated state to one that is increasingly subjected to populist pressures. Hence, the notion of an overbearing and effective state guiding firms' strategic planning decisions is not an accurate description of reality, if it ever was. Second, the process of 'hollowing out' and loss of comparative advantage in labour-intensive industries has occurred extremely rapidly, owing to the high-growth trajectory, and it has more or less coincided with China's export-led hyper-growth. Commercial reality, driven by extremely high complementarity, together with proximity and cultural familiarity, has dictated that Taiwanese firms going abroad locate overwhelmingly in the PRC. Policy makers may fret over the geo-strategic implications of this trend, but there is little in practice that they can do about it.

What they can do is to strengthen the technology base of Taiwan's industry and so maintain its technological superiority over China. This will enable Taiwanese firms to supply the technologically more advanced heavy-chemical materials and key components to China, just as Japan has done in the process of Taiwan's development. So far the evidence shows that such a strategy is working for Taiwan: as its exports become more technologically advanced, at the same time it is becoming more dependent on the PRC market.

Whether the strategy will work in the future depends very much on how successful Taiwan will be in enhancing its science and technology infrastructure. It is not short of high-calibre researchers. What is worrisome, as Chu points out, is that most of them reside in the academic community and they have not been interacting actively with industry. The science-technology branch of the government, which is relatively free from the populist politics that have been weakening the development-oriented characteristics of the state, has adopted various policies to strengthen university–industry collaborations. Time will tell whether these policies will be effective in augmenting the technological capacity of Taiwan's industries at a rate that is high enough to keep its exports competitive in a market increasingly dominated by Mainland Chinese imports.

Thailand

Thailand's rapid economic growth was interrupted by the crisis of 1997–98, but the factors explaining its stellar performance through the mid-1990s are reasonably well understood, as Medhi (Chapter 7) points out. Its fiscal and monetary policies have always been prudent, with the result that it has never experienced serious inflation since 1950. Its commercial policies have been 'always open' in the Sachs–Warner (1995) sense, although it experienced a relatively mild import-substitution period.[20] Physical infrastructure is generally good, especially its rural road networks. The commercial policy regime is largely predictable, notwithstanding frequent coups and changes of government through to the 1990s. The spread of universal primary education ensured that economic growth was reasonably inclusive, although inequality has risen since the 1970s.

This formula for growth worked very well through to the 1990s. However, the crisis exposed flaws in the financial sector, and the exchange rate regime. Moreover, the rapid increase in real wages from the late 1980s began to expose the country's historic underinvestments in post-primary education and industrial extension services. All indicators of higher education, innovation and technological capability underline this conclusion. Moreover, the approach to FDI and fiscal incentives has been largely passive and lacking in purpose. Of course, there have been some notable successes, most especially the auto industry. Thailand has become the leading Southeast Asian automotive nation, on the basis of its size, and the fact that it liberalized more quickly than its neighbours.

Medhi outlines the wide array of programmes adopted by the

government to promote industrial upgrading (see also Brimble, 2003). The government has issued an ICT Master Plan. There is a software development programme. A wafer fabrication plant is being developed. In biotechnology, there are plans to build on traditional agricultural strengths, in food crops, horticulture and marine products. A science park has been established, and Board of Investment incentives are encouraging firms to innovate. These initiatives will be crucial in determining how successfully Thailand manages the upgrading process. Since they are all very recent, it is too early to evaluate their effectiveness.

4. KEY ISSUES AND CHALLENGES

Having briefly surveyed the country indicators and profiles, in this section we examine a number of cross-cutting thematic issues with special reference to the East Asian experience. We consider in turn FDI and technology, the role of publicly funded R&D institutes, and the issue of industry policy.

FDI and Technology

This is arguably the central issue associated with industrial upgrading. For most of the twentieth century, global capital flows grew more quickly than global trade, which in turn grew faster than world output. The East Asian economies differ enormously in their international commercial orientation. However, a common feature is that all of them, in varying degree, have become more open to both international trade and investment during at least the past decade and sometimes longer. However, their approach to FDI continues to differ significantly. As Lall (2003) observes, in spite of converging policy approaches, at least four postures towards FDI have been evident in East Asia. One, illustrated historically by Japan and Korea, may be termed 'autonomous' in the sense of actively eschewing FDI and building strong domestic capabilities. Another is 'strategic FDI dependent', entailing an open-door approach accompanied by deliberate strategies to leverage the MNE presence. Singapore since the mid-1980s is the best East Asian illustration of this approach. A third group is 'passive FDI dependent', which has characterized the Malaysian and Thai approaches until recently. A final example is 'ISI (import substitution industrialisation) restructuring', and particularly refers to transitional economies, attempting re-engage with the global economy, often after a period of extended commercial isolation. China after 1978, Indonesia after 1966 and Vietnam after 1986 are all broadly illustrations of this phenomenon.

The indicators in Table 1.3 underline these differences. Thus, for example, Korea has a modest FDI presence but very strong R&D and human capital indicators. Thailand is largely the reverse.

The key issue is the interaction between MNEs and domestic technological capabilities. Foreign investors introduce a package of highly productive resources into the host economy, including production and process technology, managerial expertise, accounting and auditing standards, and knowledge of international markets. The challenge for the host economy is to benefit from the MNE presence, through processes of imitation, assimilation, adaptation and absorption, and to thereby appropriate some of the increased income accruing from the resultant productivity growth. The large literature on FDI impacts (see, for example, Blomstrom and Kokko, 1998) concludes that host-economy benefits are quite uneven, both across and within countries and over time. This suggests that host-country policies are an important factor in the distribution of these benefits. We therefore briefly survey some of the East Asian evidence on this issue, including both firm-level material and econometric evidence.

The literature on FDI in China generally concludes that it has been beneficial for growth (Fan, 2002). This has occurred through the usual channels: augmenting investment, connecting firms in China to global markets (very important for a country which hitherto had experienced several decades of international commercial isolation), and facilitating a rapid transition from the uneconomic SOE-dominated heavy industry investments. Yao (2003) and others argue that the most important spillages from foreign firms have been in worker training and stimulating domestic competition. In the first decade of reform, much of the FDI went into joint-venture projects with uneconomic SOEs, the benefits from which were minimal. As the liberalizing reforms took root, MNEs shifted to labour-intensive, export-oriented investments with more beneficial spillover effects. The benefits of FDI were also limited initially owing to institutional (particularly regulatory) weaknesses, the enclave-like FDI regime, and the weak domestic base of SMEs. In all respects, continuing economic reforms are reducing the barriers between MNEs and the domestic economy.

The quantitative literature on FDI impacts in Indonesia and the Philippines is rather limited. Both countries initially attracted MNEs to their import-substituting industrial strategies, in addition to very large mining investments in Indonesia. In such a distorted environment, while there were many instances of micro-level learning and technology transfers,[21] the national benefits were not maximized. Econometric research on Indonesia through to the early 1990s suggested mixed results, with benefits depending on the intensity of domestic competition (Sjoholm, 1999). In the Philippines, much of the FDI since the late 1980s was directed into the export-oriented electronics industry and to the liberalized services sectors (banking, telecommunications). The former connected a dormant industry to international market channels, while the latter introduced competitive pressures and new technologies into hitherto regulated monopolies. As noted, the principal debate in the case of the electronics success story has been the pace and depth of backward linkages, and the policy barriers that reinforce the emerging dual economy.

A puzzling feature of the literature on FDI impacts in Korea is that the econometric analysis suggests that the spillages are minimal, and possibly even negative, while case studies of the electronics, retail trade and banking sectors point to largely positive conclusions (Lee et al., 2004). Part of the explanation appears to be to do with timing. Most of the econometric analysis was conducted for earlier periods when the FDI regime was significantly more restrictive, and hence the benefits presumably smaller. By contrast, the case studies have focused on the liberalizations of the late 1990s. While inevitably less extensive in their coverage, the case studies have intuitive appeal, since they illuminate the channels through which the spillages occur (e.g., more competition in retailing). By contrast, spillover studies in distorted and restrictive environments leave much that is unexplained.

The Malaysian experience aptly illustrates the challenges associated with maximizing technology transfers in the context of an MNE and export-led development strategy which is initially based on export-zone enclaves. Linkage development and spillages proceeded slowly, too slowly in the view of some domestic analysts. However, such a pattern was not surprising. In the 1970s, Malaysia's industrial history was very recent; its human capital base was quite limited, especially so for technical skills; firms had great difficulty complying with government requirements to hire *Bumiputeras* employees, or source from their companies; and the incentives regime anyway discouraged firms in the zones from sourcing outside the enclave. But the evidence suggests that, over time, MNEs began to put down local roots, particularly as supply-side capacities improved. By the 1990s, the country was quickly losing its comparative advantage in labour-intensive activities, although the government sought to prolong this advantage by opening up the labour market. From a low base, expenditure on R&D is increasing rapidly, and is now approximately 0.5 per cent of GDP. MNEs are shedding low-skill activities, and shifting to higher-value segments. Although the government will arguably have to be more activist in developing supply-side capacities, Malaysia's experience with FDI and electronics has clearly been more successful than the case of its auto industry, where a highly protected and politicized quasi-monopoly has achieved few of the original technology and export objectives.

Malaysia also illustrates another aspect of FDI impacts. Namely, that MNEs' local linkages are quite selective by firm and region, and develop much more rapidly in conducive environments. The state of Penang, for example, has been the best regional performer, as the home of the earliest export-oriented investments, and where the strongest SME subcontracting base has developed. The state's commercial and business history, together with effective political leadership and public institutions, are generally regarded as the keys to its success (see Rasiah, 2003 and Ernst, 2004). Similar patterns are evident in China, in the sense that the spillages from MNEs have been more pronounced in the non-SOE sector, and that linkages are far more developed in the early reforming coastal zones.

Taiwan has been cited by Lall (2003) as a case of adopting an 'autonomous' industrial development strategy like Korea because its domestic R&D expenditure has been high while FDI has occupied a small share of its gross fixed capital formation. During 1994–97, for example, FDI as a percentage of domestic gross investment was 3.1 per cent for Taiwan, while it was 1.1 per cent for Korea, 27.8 per cent for Singapore, 13.2 per cent for China, 12.5 per cent for Malaysia, 8.2 per cent for the Philippines, 6.6 per cent for Indonesia, and 4.2 per cent for Thailand (see Lall, 2003, Table 2.9). However, these statistics, taken at their face value, could lead to an underestimation of the importance of FDI, at least in the case of Taiwan. In the sectional studies of textile and electronics, the two leading industries in Taiwan during most of the export-led growth era, Chen et al. (2001a and 2001b) clearly demonstrate the importance of acquiring technologies through joint ventures with foreign companies (in the case of textiles), and that of the technological (and even managerial) spillover of the production and marketing expertise of foreign-invested electronic firms that typically sell their products in the international market. An earlier work by Schive (1990) also emphasizes the positive contribution of FDI to Taiwan's industrial development, as he finds these enterprises typically imported new technologies, and were gradually increasing their local linkages by using a higher ratio of domestic materials and components. Econometric work by Tsou and Liu (1997) investigating the spillover effects from FDI among Taiwan's four-digit industries (123 and 152 industries in 1986 and 1991, respectively) finds significantly positive effects for the low technology gap sector (meaning the gap of value added per employee between foreign and domestically owned firms was relatively large) in 1986. However, in the high-gap sector, such spillover effects disappear statistically for both 1986 and 1991.

The Thai experience also illustrates that the econometric and case-study approaches do not necessarily point in the same direction. There is a general perception that Thailand has reaped only limited benefits from the substantial MNE presence owing to its weak R&D efforts and 'passive' approach to FDI. However, Archanun (2003) found that the spillages from foreign to domestically owned firms in the manufacturing industry were significant. He also concluded that they were more likely to occur in less-protected sectors, both because these sectors are more attractive to MNEs and because of the presumed competitive spur of lower protection. Furthermore, there is clear evidence that MNEs have been instrumental in establishing the country as the region's major automotive nation. As noted, its large and rapidly growing domestic market and earlier liberalization were the key elements in inducing the major auto MNEs to establish a presence, and to create clusters among both assemblers and suppliers. By contrast, although Thailand has attracted much export-oriented FDI in electronics, it has not emerged as a major player in this industry, on the scale of either Malaysia or Singapore. This is because it has been slower to provide the major prerequisites required in this MNE-dominated, internationally integrated industry: export zones with high-quality infrastructure, linked in a seamless manner to international markets, and allowing 100 per cent foreign ownership.[22]

Three additional points deserve emphasis in assessing FDI impacts. First, 'technology' needs to be defined broadly, to include access to international production networks and markets. MNEs are playing an ever-more prominent role in the global economy, with crucial implications for countries pursuing an export-oriented development strategy. Compared to a decade or more ago, FDI is now typically (i) more export oriented, (ii) less likely to be attracted by the old 'tariff factory' model of production for a protected domestic market, and (iii) accounting for an increasing share of host economy exports, and (iv) more likely to be part of sometimes very large transborder production and marketing networks.[23]

It is therefore much more difficult for latecomer industrializers to achieve high rates of export growth without MNE participation. The earlier literature on this subject, in which Nayyar (1978) was the major study, argued that MNE involvement in export expansion from the NIEs was mostly low by international standards. (As a corollary, therefore, developing countries could achieve rapid economic development while maintaining relatively restrictive FDI policies.) While this generally remains the case for Korea and Taiwan, it is important to note that in both economies the MNE share in exports has increased significantly since the mid-1970s. In any case, however, there is clear evidence that the strong export performance of developing countries since the 1970s has been closely associated with MNE involvement. Nayyar estimated the share of MNEs in total manufactured exports from developing countries to be not more than 15 per cent *circa* 1974. By contrast, a similar calculation, based on unpublished estimates prepared by Professor Chandra Athukorala of the Australian National University, suggests that MNEs accounted for 24 per cent of total manufactured exports from developing countries *circa* 1980. This figure had increased to 36 per cent *circa* 1990, and to over 50 per cent by the turn of the century.

A second concluding point is that countries will assess the range of costs and benefits from FDI differently. As argued above, the stages of development and commercial histories do matter. For countries with prolonged commercial isolation, attracting MNEs is an indication that they are being recognized in the global economy. This especially applies to communist states such as China and Vietnam, but it is a broader phenomenon. It also explains why these late reformers frequently 'overdo' their FDI liberalizations, and offer excessively generous fiscal incentives, at the expense of domestic private firms. As a corollary, the expectations of host countries towards FDI differ over the course of economic development. For low-income countries, the principal attraction is employment generation. As noted, when governments adopt export-oriented strategies, MNEs may play a very important role in exploiting a country's (hitherto latent) comparative advantage in labour-intensive activities. For countries shifting out of this phase of development, or with a stronger domestic R&D base, interest focuses on the potential contribution of FDI to enhance the process of structural adjustment towards higher-value-added activities. These shifts are amply illustrated among our sample of economies. In the late 1960s, for example, a strategically vulnerable Singapore sought FDI as a means of coping with a serious unemployment problem. By the late 1980s it was regarded as an instrument of technological upgrading.

Finally, it needs to be noted that 'spillages' are not just about production technology or commercial expertise. There may also be broader implications for institutional development and the quality of corporate governance. MNEs are presumed to employ higher standards of accounting, auditing and reporting. In a liberalized FDI regime, international providers of these business services may also establish local operations. The entry of foreign banks in several of the economies under study, for example, appears to have raised standards in the industry, in addition to increasing competitive pressures and providing access to a broader range of risk-sharing portfolio instruments. There are, of course, caveats that need to be attached to this argument. The causality between MNE entry and institutional quality is arguably bi-directional, that is, better-quality institutions are both a cause and a consequence of an increased MNE presence. Moreover, there is considerable evidence to suggest that MNEs adapt, often quite quickly, to local corporate cultures: if the prevailing rules of the game are characterized by extensive corruption and poor corporate transparency, MNEs do tend to adjust their behaviour accordingly.

Government Research Institutes

Government research institutes are critical actors in all stages of upgrading and innovation. Typically, most commercial entities are too small to sustain a major R&D effort. A culture of R&D has generally not yet taken root in developing East Asian economies, and thus these institutes can act as catalysts. Moreover, the institutional supports necessary to maintain a modern R&D effort – from protection of intellectual property rights to a venture-capital market – are generally underdeveloped.

East Asian governments have experimented with a wide variety of institutional arrangements in support of these institutes. The general record has been one of limited success, since the institutes are typically poorly funded, their links to universities and firms are weak, and they tend to be supply rather than demand driven. Indonesia is a case in point. Two interrelated weaknesses are generally present (Thee, 1998). First, their funding base is generally inadequate. This results in salaries being too low to attract high-calibre staff, facilities that do not meet best-practice requirements, and (especially in those agencies working in remote rural communities) a limited capacity to engage in meaningful outreach activities for the targeted client groups. The second problem relates to the objectives and functioning of these institutes. Their mission statements and philosophies are typically supply rather than demand driven. Their ties with the private sector are generally weak: there is little staff mobility, and even their physical location may not be based on the needs of their client groups. Not infrequently, the institutes have a welfare rather than an efficiency orientation.

There are, however, some notable exceptions to this generalization. One, as noted, was Malaysia's tropical crops research institutes, in both rubber and palm oil. Both were well staffed, had clear mission statements, and underpinned the country's achievements as the leading global producer for extended periods (Barlow, 2000). Both, it should also be noted, had strong political support, since they were regarded as key instruments to advance the socio-economic status of the predominantly small-holder *Bumiputeras* producers.

Another success story of clever and nimble intervention, widely cited, is Taiwan's ITRI, the Industrial Technology Research Institute (Lin, 1998). Founded in 1973, the institute expanded rapidly in the 1980s. ITRI is the largest industry-oriented research institution in Taiwan, and has working relations with some 20,000 companies. It has received significant public sector funding, absorbing about 25 per cent of the government's non-defence technology projects between 1983 and 1994. ITRI entirely funded Taiwan's first DRAM (dynamic random access memory), which in turn underpinned much of the subsequent success in electronics (Mathews and Cho, 2000). The Hsinchu Science-based Industry Park, established in 1980, has been an effective means of disseminating new technologies. In both cases, Taiwan's industrial structure was such that government funding of these initiatives was a necessary prerequisite for upgrading.[24] Possessing few very large conglomerates, the Korean strategy of promoting R&D within large firms was not an option.

Central to the success, according to Lin, were six factors: a strong national human resource base on which to draw; access to international technology markets; a competent domestic R&D base; an emphasis on diffusion and commercialization of domestic and international know-how; strong ties to the private sector; and supportive intellectual property rights. The technology diffused very quickly to SMEs, through the movement of personnel between firms and research institutes, and direct observation and learning. Joint ventures with foreign firms adjacent to these scientific facilities hastened the process.

The Role of Industry Policy

Two intersecting literatures related to this topic sit uncomfortably alongside each other. One focuses on technological development and innovation, and emphasizes the importance of establishing a commercial environment and institutions that are conducive to technological progress at the firm level. It sees governments as important in this process, in creating a national research infrastructure, particularly in countries about to lose comparative advantage in labour-intensive activities. It stresses the importance of getting the 'fundamentals' right: macroeconomic stability, predictable and stable policy regimes, improved physical infrastructure and education, a reasonably adequate system of property rights and legal infrastructure, and increasing openness to international commerce.

An alternative literature accepts some or all of the above prescriptions, but argues that it is an insufficient recipe for industrial success. This school rejects an emphasis based primarily on economic liberalism and static comparative advantage. It argues that Korea, Taiwan and Japan owe their success to selective industrial policies – targeting industries, 'picking winners', and deliberately 'getting prices wrong' through fiscal incentives, subsidized credit, import protection and direct investment. It argues for the desirability of governments deliberately introducing a non-neutral incentives regime between industries, and sometimes between firms.

This is therefore a hotly contested field of research, with significant differences between those who advocate the importance of 'getting the fundamentals right' and those who favour a more interventionist approach. Stiglitz (2001, p. 521) comments on this literature,[25] and also offers a potential way out of the conundrum:

> In the light of *market* and *government* failures, there are two alternative strategies: to focus on one and ignore the other or to try to address the weaknesses in each, viewing the public and private sectors as *complementary*. Singapore illustrates nicely the advantages of the latter approach.

Most of developing East Asia has industrialized rapidly since 1960, and most governments have intervened extensively, at the industry and even firm level. That is, there have been pronounced variations in inter-industry protection in several countries. Subsidized finance and fiscal incentives have been directed to favoured projects. SOEs, mostly in heavy industry, have sometimes been regarded as instruments on industrial policy. Superficially at least, therefore, the evidence might be interpreted as supporting the notion that selective industrial policies have contributed to their success. However, on closer examination such an argument receives limited empirical support. For the arguments of the interventionist school to be convincing, it is obviously insufficient merely to point to the coexistence of selectivity and rapid growth. There has to be a demonstration of causality, and most of the research on this topic finds little evidence of this kind.

In reality, however, and especially in non-Singapore Southeast Asia, promotional measures have been prone to abuse, implementation has been sporadic and often short-lived, and there has been little systematic attempt to prescribe conditionality, in the sense of linking incentives to tightly defined performance criteria. It is therefore hardly surprising that studies which have examined the relationship between inter-industry variations in government assistance (for example, through protection and credit subsidies) and subsequent (lagged) performance, according to a variety of measures, have found little evidence of causality. In the case of Indonesia, a survey of selective policy instruments introduced or extended in the 1970s and 1980s – protection, credit subsidies, state enterprises – detected very little evidence of such a strategy having 'worked' according to a range of subsequent performance criteria (Hill, 1996).[26] Similarly, Warr's (1995) study of several policy correlates of export performance in Thailand over the 1970–89 period revealed a remarkably consistent picture of negative coefficients for all variables and time periods. In the case of Malaysia, it is very difficult to find any evidence in support of the proposition that the government's promotion of heavy industry over the 1978–86 period promoted efficient industrial growth.

The Northeast Asian experience is more complex and the empirical evidence somewhat ambiguous. In their authoritarian phases, the governments of Korea and Taiwan were significantly insulated from business pressures; there was a fierce desire to succeed; the human capital base was stronger; conditionality was more likely to be attached to incentives; and firms receiving support were expected to meet some sort of market test, generally related to export performance. Both Chu and Shin in this volume document cases where major public investments in science and technology underpinned industrial success.

Yet even here the counterfactual needs to be confronted: was it the general factors adumbrated above that explains their success, in addition to the especially strong support for R&D? The record at the industry level was mixed, in the sense that bureaucrats supported both winners and losers. Smith's (2000) detailed work on the subject finds plenty of examples of both, including across industries an inverse correlation between incentives and a range of lagged performance indicators. The economic crisis has weakened still further the case for an interventionist industrial policy. That is, in the wake of Korea's severe economic crisis and Japan's protracted recession, the 'Northeast Asian model' is looking a lot less attractive than it was a few years ago. Attempts to finance uneconomic, not to mention politicized, industrial projects through a shaky financial system, or through foreign borrowings with implicit government guarantees, have contributed to Korea's current difficulties.

Of course, as noted above, markets do fail. Economic agents will tend to underinvest in training and innovation, for well-known reasons associated with free-rider problems and an inability to appropriate firm-specific investments in technology. The issue for policy makers is how to intervene effectively, and how to overcome 'market failure' while not introducing 'government failure'. The following would appear to be some useful guiding principles based on the East Asian experience.

First, the 'fundamentals' need to be in place, and not just taken for granted (as is frequently evident in the literature). That is, there needs to be good macroeconomic management, a broadly open economy, a good and broad-based system of universal education, good quality infrastructure, effective financial supervision, protection of property rights and so on. Without these building blocks, industry policy is bound to fail.

Second, some broad industry-level selectivity may be justified. It will arguably make sense for a country strongly dependent on marine products, or garments, or electronics to invest in industry-wide R&D, especially if it is losing comparative advantage in labour-intensive segments.

Third, if there is to be selectivity, it is important to establish just what, precisely, is the infant industry case for intervention. That is, is there an externality, a market failure, or the inadequate provision of a public good? And what is the source of the externality and/or problem? Is it to do with a commodity (final product) market imperfection, a capital market imperfection, labour skills and training, the acquisition and diffusion of knowledge, penetrating international markets, establishing a country reputation internationally, or something else?

To the maximum extent possible, R&D efforts should be demand, not supply, driven. Research institutes should include major private sector participation, and perhaps be run by industry bodies, providing they are able to operate effectively. Tax incentives can be employed to reward private R&D effort, assuming that the tax administration is able to implement such a policy.

Fifth, the selection of policy modalities and instruments is crucial. For example, as much as possible, assistance should:

1. be industry rather than firm based (and as a corollary it should be 'contestable');
2. contain clear provisions for a sunset clause;
3. be provided in the form of subsidies rather than tariffs and (especially) quantitative restrictions;
4. be completely transparent and fully costed; and
5. be insulated from political processes, once broad priorities and budgets are specified.

Finally, purely for political economy reasons, there may be a stronger case for supporting export industries, simply because a market test (i.e., sales in internationally competitive markets) is more readily available. This is especially the case for countries with a long history of protection.

5. CONCLUSION

We conclude with the following two broad generalizations. First, there is no single 'East Asian model' of industrialization and technology development. The economies examined in this volume vary significantly in their levels of development, resource endowments, historically embedded openness, and institutional quality. For lower-income countries like Indonesia and the Philippines, the principal task of government is to provide an enabling commercial environment and to supply crucial, growth-enhancing public goods. For the much richer economies of Korea and Taiwan, the policy challenges centre around the development of high-quality institutions and capabilities which will enable its firms to support OECD-level wages. For middle-level upgrading economies like Malaysia and Thailand, a major issue concerns how to upgrade and move beyond the labour-intensive industrialization phase. For China and other transitional economies, a major preoccupation is the transition from plan to market, and to develop the institutions which support a modern capitalist economy.

Second, although each country's circumstances are different, and therefore 'one-size-fits-all' prescriptions should be approached with caution, the case studies in this volume do elaborate on, and affirm, much of the emerging consensus on the broad gamut of industry and technology policy. This literature takes as given the basic building blocks of sound macroeconomic policy, an internationally integrated economy and effectively functioning institutions, without which all interventions will fail.

The quality of the education system is central to technology and innovation. This implies universal basic education at early stages of development, which was a particular Northeast Asian strength. As income levels rise, attention inevitably shifts towards tertiary education, including its creativity, research orientation, and the science and technical content. The internationalization of the education system is also a key issue.

Related to this last point, openness is essential and is about much more than reducing barriers to international trade and investment, essential though these are. For example, the most successful and dynamic urban and industrial clusters are those which are thoroughly international, and connect in a seamless manner to the global economy. This implies regions which are culturally open and diverse, and which offer the highest-quality physical infrastructure. (Hence, within East Asia, the continuing success of Hong Kong and Singapore.) Such structures also enable countries with large international diasporas to tap into a talent pool abroad. In the words of Yusuf (2004, p. 33), 'an open urban milieu is vital for creativity' in these dynamic clusters.

Finally, the specifics of government innovation programmes need to be carefully formulated. An 'input-based' approach is insufficient. As several contributors to this volume note, even well-funded government research institutes are unlikely to be effective without some key structural reforms. For example, these institutes need to be well resourced, closely linked to business, and with clear mission statements. They need to be connected to major international networks of relevance, and be free to recruit staff accordingly. Tax incentives for R&D can be employed as a complementary initiative. Where there is a substantial MNE presence, a range of 'leveraging' arrangements can be employed.

Moreover, there needs to be great caution in politicians and bureaucrats attempting to pick winners. Even the best intentioned are just as likely to back losers. Introducing political economy constructs immediately highlights the reality that 'losers are better at picking governments than are governments at picking winners'. This is particularly the case as poor institutional quality and high regulatory complexity often go hand in hand, as illustrated for example by the two laggards in our sample, Indonesia and the Philippines. Some selectivity is inevitably involved in industry and technology policy, and in the previous section we offered a framework which might facilitate successful outcomes.

NOTES

1. See, e.g., Amsden (2001) and Lall (2000).
2. The role of institutions in economic development is a much discussed and contested topic. See Rodrik (2003) for a diverse and interesting set of case studies. See also Chu et al. (2001) who analyse the problem from the viewpoint of the demand for and the supply of effective government intervention; with the demand depending on the extent of market failures and the supply depending on the capability and reliability of the government.
3. Here we mean competition either in the consumer market (as owners of different brand names) or in the producer goods market (as suppliers of raw materials and components). Alongside the latter pattern of competition there exists, of course, a simultaneous cooperation relationship, in the sense that many exporters in Korea and Taiwan are OEM suppliers to the OECD brand name-owning firms.
4. See Yusuf (2004) for an excellent overview of the international environment as it impinges on cross-border flows of trade, investment and technology.
5. These trends are particularly pronounced in the most internationally integrated industries, such as the automotive and electronics industries. See the industry case studies in Yusuf et al. (2004).
6. In addition to the literature cited elsewhere in this chapter, see in particular Kim and Nelson (2000).

7. Of these three generalizations, the last will change the most quickly if the two developing giants, China and India, both with strong if handicapped R&D bases, maintain their current strong growth rates for the next few decades.
8. There is much discussion of these indicators and their meaning in the literature. The work of Lall (2000) is particularly important. See UNIDO (2002) for a recent and comprehensive set of comparative indicators.
9. We shall present some alternative estimates of the technology intensity of exports below.
10. See Nelson and Pack (1999) for a detailed discussion of this issue.
11. In Lall (2003), whose definition of high-technology manufactures is similar to the OECD's, the Philippines has the highest share of high-technology exports of manufactures among the 13 largest developing economy exporters (Lall, 2003, Figure 2.2).
12. This assumes implicitly that the 6-digit HS sector-specific skill intensity is similar across all eight economies.
13. For all eight economies, only exports to the US, Japan and the EU are included.
14. For the definitions of 'high-income' and 'middle-income' countries, see World Bank's website, https://publications.worldbank.org/subscriptions/WDI/Help.htm.
15. In addition to the Bird–Hill chapter, see the contributions to Hill and Thee (1998) for detailed analyses of the pre-crisis situation. Okamoto and Sjoholm (2003) provide a recent survey of the issues.
16. The literature on Korea's industrialization record and strategies is extensive. In addition to Shin's chapter in this volume, see for example Amsden (1989), Chang (1994), Kim (1997, 2003) and Dahlman and Andersson (2000).
17. The post-crisis Korean literature therefore focuses heavily on elements of such a reform agenda rather than its past industrial successes. A good example is Dahlman and Andersson (2000), who include the following recommendations: a more modern legal system; more competition in product markets; better supervision and transparency in financial markets; more flexible labour markets; removal of the still significant informal barriers to FDI; greater autonomy for educational institutions; stronger linkages between educational institutions and research agencies; and liberalization and improved regulation of telecommunications.
18. See also *The Third Outline Perspective Plan, 2001–2020* (Economic Planning Unit, Kuala Lumpur, 2000), chapter 5, and the contributions to Jomo and Felker (1999), and Jomo et al. (1999).
19. See also Wong (2001, 2003) for a useful analytical history.
20. This openness has extended to social assimilation, as evidenced by the domination of much of the domestic business sector by the Sino-Thai community.
21. See Hill (1988, chapter x) for a survey of the Indonesian evidence through to the mid-1980s.
22. In passing, it is worth noting that MNEs have not been central to another of Thailand's agro-processing export successes, where several large domestic firms, most notably the CP group, have been major players. This illustrates a more general proposition, that MNEs are particularly important in industries which are characterized by high levels of cross-border vertical integration, and where knowledge of international market chains may be firm specific.
23. See Yusuf and Associates (2003, chapter 7) for a recent analysis of this phenomenon in an East Asian context.
24. The role of ITRI in Taiwan's industrial upgrading may be changing, however. For one thing, some private firms have become large enough to establish their own R&D departments, or they could acquire technology directly from MNEs. Moreover, large-scale spin-offs from the ITRI seen in the case of the establishments of UMC (United Microelectronics Corp) and TSMC (Taiwan Semiconductor Manufacturing Company Ltd.) have been rarer in recent years, and there have been complaints that ITRI itself has become 'aged' and supply rather than demand oriented. See, e.g., Tsai and Chu (2004).

25. The literature on this subject is now very large. Several of the chapters in this volume refer to the issue, and offer contrasting perspectives. Yusuf (2001, pp. 20–25) provides a succinct recent summary. The most detailed interventionist case is developed by Amsden (2001). Smith (2000) provides a comprehensive survey of the East Asian literature through to the mid-1990s from a largely neoclassical perspective.
26. In fact, what little systematic work to have been done on the political economy of trade policy in Indonesia has found a 'crony' variable to be consistently the most significant explanatory variable of inter-industry variations in protection (Basri and Hill, 1996).

REFERENCES

Amsden, A. (1989), *Asia's Next Giant: South Korea and Late Industrialization*, New York: Oxford University Press.

Amsden, A. (2001), *The Rise of the Rest: Challenges to the West from the Late-Industrializing Economies*, New York: Oxford University Press.

Archanun, Kohpaiboon (2003), 'Foreign Direct Investment and Technology Spillover: A Cross-Industry Analysis of Thai Manufacturing', unpublished paper, Australian National University, Canberra.

Balisacan, A.M. and H. Hill (eds) (2003), *The Philippine Economy: Development, Policies, and Challenges*, New York: Oxford University Press, and Manila: Ateneo de Manila University Press.

Barlow, C. (ed.) (2000), *Institutions and Economic Change in Southeast Asia*, Cheltenham: Edward Elgar.

Basri, M.C. and H. Hill (1996), 'The Political Economy of Manufacturing Protection in LDCs: An Indonesian Case Study', *Oxford Development Studies*, 24 (3), pp. 241–59.

Blomstrom, M. and A. Kokko (1998), 'Multinational Corporations and Spillages', *Journal of Economic Surveys*, 12 (3), pp. 247–77.

Brimble, P. (2003), 'Foreign Direct Investment, Technology and Competitiveness in Thailand', in Lall and Urata (eds), pp. 334–74.

Brooks, D.H. and H. Hill (eds) (2004), *Managing FDI in a Globalizing Economy: Asian Experiences*, London: Palgrave Macmillan.

Chang, H.J. (1994), *The Political Economy of Industrial Policy*, London: Macmillan.

Chen, Been-Lon, Tain-Jy Chen and Yun-Peng Chu (2001a), 'The Role of Textiles and Man-made Fibre in the Process of Industrialization: The Case of Taiwan', in Wong and Ng (eds), pp. 283–321.

Chen, E.K.Y. (1997), 'The Total Factor Productivity Debate: Determinants of Economic Growth in East Asia', *Asian-Pacific Economic Literature*, 11 (1), pp. 18–38.

Chen, Tain-Jy, Been-Lon Chen and Yun-Peng Chu (2001b), 'The Development of Taiwan's Electronics Industry,' in Wong and Ng (eds), pp. 245–82.

Chen, Tain-Jy and Joseph Lee (eds) (2004), *The New Economy in Taiwan*, Cheltenham: Edward Elgar.

Chu, Yun-Peng (2001), 'Equalization Effects of the Expansion of Labour-Intensive Exports: The Case of Taiwan', *The Developing Economies*, 39 (3), pp. 235–66.

Chu, Yun-Peng (2004), 'Transformation of Traditional Industries in Taiwan', in Chen and Lee (eds), pp. 103–17.

Chu, Yun-Peng, Been-Lon Chen and Tain-Jy Chen (2001), 'Rethinking the Development Paradigm: Lessons from Taiwan – The Optimal Degree of Government Intervention', in Wong and Ng (eds), pp. 197–244.

Dahlman, C.J. and T. Andersson (2000), *Korea and the Knowledge-Based Economy: Making the Transition*, Washington DC: World Bank/OECD.

Directorate for Science, Technology and Industry, OECD, (2004), *STAN Indicators: 1980–2002*, http://www.oecd.org/dataoecd/60/28/21576665.pdf.

Ernst, D. (2004), 'Global Production Networks in East Asia's Electronics Industry and Upgrading Perspectives in Malaysia', in Yusuf et al. (eds), pp. 89–158.

Fan, E.X. (2002), 'Technological Spillages from Foreign Direct Investment – A Survey', ERD Working Paper Series, No. 33, Manila: Asian Development Bank.

Hill, H. (1988), *Foreign Investment and Industrialization in Indonesia*, Singapore: Oxford University Press.

Hill, H. (1996), 'Indonesia's Industrial Policy and Performance: "Orthodoxy" Vindicated', *Economic Development and Cultural Change*, 45 (1), pp. 147–74.

Hill, H. (2003), 'Manufacturing Industry', in Balisacan and Hill (eds), pp. 219–53.

Hill, H. (2004a), 'Six Asian Economies: Issues and Lessons', in Brooks and Hill (eds), pp. 29–77.

Hill, H. (2004b), 'Technology and Innovation in Developing East Asia', in Yusuf et al. (eds), pp. 353–93.

Hill, H. and Thee Kian Wie (eds) (1998), *Indonesia's Technological Challenge*, Singapore: Institute of Southeast Asian Studies.

Jomo, K.S. and G. Felker (eds) (1999), *Technology, Competitiveness, and the State: Malaysia's Industrial Technology Policies*, London: Routledge.

Jomo, K.S., G. Felker and R. Rasiah (eds) (1999), *Industrial Technology Development in Malaysia: Industry and Firm Studies*, London: Routledge.

Kim, Linsu (1997), *Imitation to Innovation: The Dynamics of Korea's Technological Learning*, Boston: Harvard Business School Press.

Kim, Linsu (2003), 'The Dynamics of Technology Development: Lessons from Korea', in Lall and Urata (eds), pp. 143–67.

Kim, Linsu and R.R. Nelson (eds) (2000), *Technology, Learning, and Innovation: Experiences of Newly Industrializing Countries*, Cambridge: Cambridge University Press.

Lall, Sanjaya (2000), 'The Technological Structure and Performance of Developing Country Manufactured Exports, 1985–98', *Oxford Development Studies*, 28 (3), pp. 337–63.

Lall, Sanjaya (2003), 'Foreign Direct Investment, Technology Development and Competitiveness: Issues and Evidence', in Lall and Urata (eds), pp. 12–56.

Lall, S. and S. Urata (eds) (2003), *Competitiveness, FDI and Technological Activity in East Asia*, Cheltenham: Edward Elgar.

Lee, Seong-Bong et al. (2004), 'Republic of Korea', in Brooks and Hill (eds), pp. 153–88.

Lin, O.C.C. (1998), 'Science and Technology Policy and its Influence on Economic Development in Taiwan', in Rowen (ed.), pp. 185–206.

Mathews, J.A. and D.-S. Cho (2000), *Tiger Technology: The Creation of a Semiconductor Industry in East Asia*, Cambridge: Cambridge University Press.

McKendrick, D.G., R.F. Doner and S. Haggard (2000), *From Silicon Valley to Singapore: Location and Competitive Advantage in the Hard Disk Drive Industry*, Stanford: Stanford University Press.

Medalla, E.M. et al. (1995/96), *Philippine Trade and Industrial Policies: Catching Up With Asia's Tigers*, volumes I and II, Makati: Philippine Institute for Development Studies.

Nayyar, D. (1978), 'Transnational Corporations and Manufactured Exports from Poor Countries', *Economic Journal*, 88, pp. 59–84.

Nelson, R.R. and H. Pack (1999), 'The Asian Miracle and Modern Growth Theory', *Economic Journal*, 109, pp. 416–36.

Okamoto, Y. and F. Sjoholm (2003), 'Technology Development in Indonesia', in Lall and Urata (eds), pp. 375–96.

Oshima, H.T. (1987), *Economic Growth in Monsoon Asia: A Comparative Survey*, Tokyo: University of Tokyo Press.

Rasiah, R. (2003), 'Industrial Technology Transition in Malaysia', in Lall and Urata (eds), pp. 305–33.

Rodrik, D. (ed.) (2003), *In Search of Prosperity: Analytic Narratives on Economic Growth*, Princeton: Princeton University Press.

Rosenberg, N. (1994), *Exploring the Black Box: Technology, Economics, and History*, Cambridge: Cambridge University Press.

Rowen, H. (ed.) (1998), *Behind East Asian Growth: The Political and Social Foundations of Prosperity*, London: Routledge.

Sachs, J. and A. Warner (1995), 'Economic Reform and the Process of Global Integration', *Brookings Papers on Economic Activity*, 1 (25), pp. 1–95.

Schive, Chi (1990), *The Foreign Factor: The Multinational Corporation's Contribution to the Economic Modernization of the Republic of China*, Stanford: Stanford University Press.

Sjoholm, F. (1999), 'Technology Gap, Competition and Spillages from Direct Foreign Investment: Evidence from Establishment Data', *Journal of Development Studies*, 36, pp. 53–73.

Smith, H. (2000), *Winning with the Market*, Cheltenham: Edward Elgar.

Stiglitz, J.E. (2001), 'From Miracle to Crisis to Recovery: Lessons from Four Decades of East Asian Economic Experience', in Stiglitz and Yusuf (eds), pp. 509–26.

Stiglitz, J.E. and S. Yusuf (eds) (2001), *Rethinking the East Asian Miracle*, New York: Oxford University Press for the World Bank.

Thee, Kian Wie (1998), 'The Determinants of Indonesia's Industrial Technology', in Hill and Thee (eds).

Tsai, C.-Y. and Yun-Peng Chu (2004), 'Building Bridge among Universities, Research Institute, and Industry in Taiwan', presented at the Stanford-ITRI Seminar on Industrial Technology Development in Stanford, California, September 2004.

Tsou, Meng-Wen and Jin-Tan Liu (1997), 'The Spillover Effect from Foreign Direct Investment: Empirical Evidence from Taiwan's Manufacturing Industries', *Taiwan Economic Review*, 35, pp. 155–81.

UNIDO (United Nations Industrial Development Organization) (2002), *Industrial Development Report 2002/2003: Competing Through Innovation and Learning*, Vienna.

Warr, P. (1995), 'Myths about Dragons', *Agenda*, 1 (2), pp. 215–28.

Wong, Poh-Kam (2001), 'Leveraging Multinational Corporations, Fostering Technopreneurship: The Changing Role of S&T Policy in Singapore', *International Journal of Technology Management*, 22/56, pp. 539–67.

Wong, Poh-Kam (2003), 'From Using to Creating Technology: The Evolution of Singapore's National Innovation System and the Changing Role of the Public Policy', in Lall and Urata (eds), pp. 191–238.

Wong, Poh-Kam and C.-Y. Ng (eds) (2001), *Industrial Policy, Innovation and Economic Growth: The Experience of Japan and the Asian NIEs*, Singapore: Singapore University Press.

World Bank (1993), *The East Asian Miracle: Economic Growth and Public Policy*, Washington DC: World Bank.

Yao, Yang (2003), 'In Search of Balance: Technological Development in China', in Lall and Urata (eds), pp. 239–67.

Yusuf, Shahid (2001), 'The East Asian Miracle at the Millennium', in Stiglitz and Yusuf (eds), pp. 1–45.

Yusuf, Shahid (2004), 'Competitiveness Through Technological Advances Under Global Production Networking', in Yusuf et al. (eds), pp. 1–34.

Yusuf, Shahid, M.A. Altaf and K. Nabeshima (eds) (2004), *Global Production Networking and Technological Change in East Asia*, New York: Oxford University Press.

Yusuf, Shahid and Associates (2003), *Innovative East Asia*, New York: Oxford University Press.

2. Three Paths for High-Technology Catch-Up: Singapore, Korea and Taiwan

Jang-Sup Shin and Yun-Peng Chu

1. INTRODUCTION

This chapter aims to delineate some model features of successful high-tech industrialising economies in East Asia, namely, Singapore, South Korea (henceforth Korea), and Taiwan (Republic of China; henceforth Taiwan) by focusing on their strategies and institutions. It does not aspire to provide universal prescriptions for high-tech industrialisation, which are impossible to attain and often misleading. Instead, by comparing both similar and dissimilar aspects of their innovation systems and their different configurations, the discussion attempts to provide other later emulators with some policy implications when they shift the emphasis of industrialisation from lower-tech areas to higher-tech ones.

In a sense, embarking on low-tech industrialisation is a relatively easy task, especially in the current era of accelerated globalisation. It is because latecomers can take part in global production networks (GPNs) of multinational corporations (MNCs) simply by providing low-cost labour, which is their natural competitive asset compared with that of more advanced economies. The semiconductor industry is a case in point. Global production in the industry began in the 1960s as Silicon Valley firms shifted their low-skilled assembly lines to developing countries in their attempts to reduce production costs. By the middle of the 1970s, the American semiconductor firms set up their assembly lines not simply in the three East Asian economies of our concern, but also in other East Asian countries such as the Philippines, Malaysia, Indonesia and Thailand, as well as in Latin American countries like Mexico and Brazil (Henderson 1989; Shin 1996). Despite the fact that the semiconductor industry is one of the most advanced high-tech industries of the day, it was possible for latecomers to participate in it at their earliest stages of industrialisation with little technological capability and capital investment of their own. This was because low-tech segments of the industry required mainly unskilled or semi-skilled labour.

However, shifting towards higher-end production processes within the semiconductor industry later took place only in the three economies while it did not progress significantly in other developing countries. Similarly, the three economies were also more successful in establishing themselves as major players in other high-tech areas such as computers and computer peripherals, chemicals, automobiles and so on, while other countries were slower or failed in upgrading their economies.

This belies the oriental belief that 'the beginning is the half'. Sustaining and upgrading industrialisation may be a more daunting task than starting industrialisation. This difficulty has mainly to do with risks involved in high-tech industrialisation. Knowledge involved in high technologies is more difficult to imitate or create. MNCs are also less willing to transfer it to latecomers because it constitutes basic sources of their competitive edge, on the one hand, and competition in high technologies is less determined by labour costs, on the other. Moreover, commercial application of this knowledge often requires concurrent heavy investments in skill formation, facilities, organisation building and so on. The necessity of acquiring this complex knowledge and related investments drives the latecomers to build a system different from what they had at the beginning of industrialisation.

It seems to us that the success or failure in high-tech industrialisation hinges on devising an appropriate strategy and establishing required institutions to take risks involved in assimilation and deployment of high technologies. Higher returns associated with high-tech industries are no more than the flip side of high risks of the same coin. By focusing on the risk-taking strategy and institutions, this chapter investigates factors behind successful high-tech industrialisation in the three economies.

The three economies share some similarities as they embarked on export-oriented industrialisation. Unlike Latin American countries or India, which adopted import-substitution strategies relying on bigger domestic markets and export earnings from primary products, they had smaller domestic markets and few primary products to export, and had to export manufactured products to finance their industrialisation.[1] Attaining and maintaining export competitiveness of manufactured products therefore became a major concern in building and managing economic systems of these economies.

With the backdrop of these similarities, they adopted quite different systems, reflecting their differences in history, size, geography, internal politics and so on.

1. Singapore chose the most internationalist path for high-tech industrialisation by making MNC investments the main thrust of upgrading its economy.
2. Korea went for the most nationalistic path for high-tech industrialisation by focusing on establishing internationally competitive local industries that are in direct competition with the forerunners.
3. Taiwan's system was semi-internationalist in the sense that it promoted export competitiveness of local firms while seeking alliances actively with MNCs in the process of industrial upgrading.

Although Hong Kong is one of the newly industrialising economies (NIEs) along with the three economies, we exclude it in this chapter because it has experienced a rapid de-industrialisation from the 1980s and '[i]n terms of capabilities, the Hong Kong industry is grouped with the new low-cost manufacturing bases in Thailand, Malaysia and the People's Republic of China (henceforth China) as a cost-based competitor with relatively few capabilities in product innovation and development' (Boston Consulting Group 1995, quoted in Chiu and Lui 2001: 467). It seems to us that the case of Hong Kong provides few positive lessons for high-tech industrialisation, though it may offer some negative lessons to caution against.

In what follows, the case of Korea will be presented first, that of Singapore second, followed by the third case of Taiwan, although the case of Singapore will be discussed in somewhat more detail. In this way the Singaporean path of development can be viewed more clearly in the mirror of the other two cases.

2. THE KOREAN PATH: A NATIONALISTIC MODEL

Among the three economies, Korea developed through the most nationalistic route for industrialisation. It imposed heavy regulations on foreign firms' activities within its territory and tried to keep foreign direct investment (FDI) at a minimum, maintaining the lowest ratio of FDI to fixed capital formation among East Asian NIEs (Table 2.1). Korea instead relied heavily on foreign debts to finance its industrialisation and had the highest foreign debt to GNP ratio among East Asian NIEs. At the same time, the country employed systematic import-substitution policies under an overarching framework of industrial policy. Constructing an 'independent economy' has been explicitly stated as a primary aim of the

Table 2.1 Ratio of FDI Inflows to Gross Fixed Capital Formation in Selected East Asian Countries, 1971–97 (per cent)

Country	1971–1980	1981–1990	1991–1997
Japan	0.1	0.1	0.2
Hong Kong	5.1	9.9	8.7
Republic of Korea	1.2	0.9	1.0
Singapore	15.8	26.2	25.9
Taiwan	1.3	1.3	2.7
Indonesia	3.5	1.5	5.5
Malaysia	13.6	11.3	17.2
Philippines	1.0	3.8	7.3
Thailand	2.3	4.8	4.1
China	0.0	1.5	12.0

Source: Shin and Chang (2003, Table 2.1)

state's economic policy agenda, and consequent import-substituting policies were combined with export-promoting policies. Establishing internationally competitive *local companies* was the main thrust of this 'export-promotion *cum* import-substitution' strategy of Korea.

Basic institutions for Korea's industrialisation were created in the early 1960s after the authoritarian regime led by President Park Chung Hee came to power. The Korean state thereafter began drafting 'The Five-Year Economic Development Plans'. It also nationalised commercial banks and subordinated their lending decisions to its industrial policy. The state drew foreign loans directly or often indirectly by guaranteeing credibility of domestic banks and firms. Through the banking system over which the state maintained controlling power, financial resources were channelled mainly to the *chaebols*, the family-owned diversified business groups (Figure 2.1). This state bank–*chaebols* nexus certainly helped propel export-oriented growth mainly of light industrial goods in the 1960s. However, its role was most pronounced in the 1970s when the country began to develop high-tech industries through heavy and chemical industrialisation (HCI thereafter).

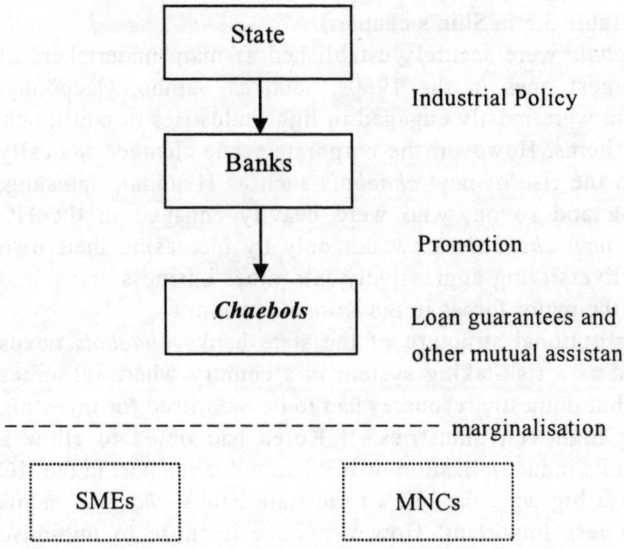

Source: Shin and Chang (2003, figure 2.2).

Figure 2.1 Korea's Nationalistic Model

First, the state's role in steering the economy was significantly strengthened. HCI was, from the beginning, on the state's initiative. Many private firms, which were earning comfortable profits from their exports of light goods, were not really interested in moving into the HC industries. However, the state drafted the HCI plan and induced or sometimes even forced selected firms to enter some designated HC industries by providing them with an array of incentives in the form of tax benefits, subsidised loans, preferential access to limited raw materials and so on. In the case of the petro-chemical and steel industries, which would serve as basic supply industries for other HC industries, the state established them directly by public enterprises.

Second, related to the above, the banking sector was more extensively utilised for financing industrialisation. Most foreign debts, which jumped from 10.7 per cent of gross national product (GNP) in 1965 to 44.4 per cent of GNP in 1980 and again to 52.1 per cent in 1985 at its peak, was guaranteed by the state and distributed to the *chaebols* through the banking system (Table 3.1 in Shin's chapter). Commercial banks substantially increased loans to the *chaebols* at subsidised rates. As a result, the corporate debt–equity ratio in the manufacturing sector shot up from 119 per cent on average during 1963–69 to 342 per cent on average during 1970–79 (Table 3.5 in Shin's chapter).

Third, the *chaebols* were securely established as main undertakers of the HCI. The largest ones in the 1960s, such as Samho, Gaepoong, Donglip and so on, were mostly engaged in light industries or consumer-related importing items. However, the corporate scene changed radically in the 1970s with the rise of new *chaebols* such as Hyundai, Samsung, Lucky, Ssangyong and so on, who were heavily engaged in the HC industries. These new *chaebols* grew not only by increasing their own size but also by diversifying aggressively into other business areas, and have remained as the major forces in the Korean economy.

The overall institutional structure of the state banks–*chaebols* nexus can be understood as a risk-taking system in a country where resources were constrained but domestic resources had to be mobilised for investing in and sustaining high-tech industries. If Korea had opted to allow a bigger FDI role in its industrialisation or if it had not taken part in the HC industries in such a big way, the role of the state banks–*chaebols* nexus would have been less important. However, once it chose to minimise foreign ownership of local industries and precipitated itself into the HC industries, the nexus became an effective institutional arrangement for high-tech industrialisation.

To elaborate on this point, it seems necessary to examine requirements when the latecomers compete with the forerunners in high-tech industries by adopting a Korean style 'substituting strategy'. One requirement is large-scale financing. As Gerschenkron (1962; 1970) points out from the experience of late industrializers in Europe in the nineteenth century such as Germany and Russia, the utilisation of 'economies of scale' to a fuller extent is a key to attaining competitiveness in the HC industries, and this entails high initial set-up costs. In the case of new high-tech industries in the twentieth century, notably the electronics industry, the technology frontier moves rapidly and it is not sufficient for the latecomers to embark on large-scale capital investment in the beginning. They should also take part in a prolonged race of investments both in capital facilities and R&D to catch up with the pace of changes in technologies (Shin 1996).

Another requirement is a relatively long gestation period in assimilating knowledge involved in these high-tech industries and translating it into material gains. In the case of Korea's HCI, it was estimated that the gestation period was often over a decade (Jacobsson 1993). It is also necessary to invest heavily in building marketing networks, especially when the latecomers are in direct competition with the forerunners. Companies and banks are forced to go through a long tunnel of deficit accumulation while keeping investments. Financing capability is again crucial during this period. It is also important to step up efforts to assimilate, deploy and improve on imported technologies to reduce time and costs spent in surviving this difficult stage.

In a country with limited financial and technological resources, the state banks–*chaebols* nexus functioned as the central institutions to meet these imperatives of financing and technological learning while bearing the risks involved here. First, as in most late-industrialising countries, the state was the only entity with 'deep pockets' and coercive power, which are related to taxation, money issuance, guaranteeing or transferring property rights and so on, in mobilising resources for this high-tech industrialisation. Second, the banking system provided the *chaebols* with 'patient money' under the auspices of industrial policy and state ownership. Third, the *chaebol* structure itself functioned as a credit-creation and risk-taking mechanism. Through internal transactions and loan guarantees, the *chaebols* further created credits. Limited technological, financial and managerial resources were also more intensively utilised through internal transfer within business groups (refer to Shin's chapter for details).

What should be noted is that the Korean system thus constructed was inherently a high-risk one. Introducing foreign debts was to expose the economy to foreign exchange risks as well as repayment obligations, which were less problematic in countries that grew by introducing equity participation of foreign companies. Heavier reliance on bank financing, which resulted in low (ordinary) profit rates in the corporate sector, also made Korean firms more vulnerable to financial shocks. Therefore, the Korean system had to be equipped with mechanisms to bear effectively these higher financial risks. As for external financial risks, the Korean government managed them by controlling tightly cross-border capital flows, exchange rates, and even the size of foreign debts. Higher financial risks in the corporate sector were borne by long-term commitment from commercial banks as well as by the intra-group support structure within the *chaebols*.

Another system requirement was a disciplinary mechanism that would induce technological learning at the firm level and force those favoured *chaebols* to gain and maintain competitiveness. As a developmental state which set its national objective at maximising economic growth and was relatively free from domestic interest groups, the Korean state was able to discipline the *chaebols*, rather than letting them survive on continued government favours. The history of Korean high-tech industrialisation was thus associated with a continual reshuffle of leading *chaebols*. Competition among the *chaebols* was also maintained, in the form of 'oligopolistic competition'. Export-orientation also provided the state with a better yardstick to gauge the performance of those favoured *chaebols*, than in the case of simple import-substitution strategies (See Shin's chapter for details).

As we shall see later in detail (Section 5), higher financial risks and a stronger necessity for maintaining a disciplinary mechanism over domestic actors are relative weaknesses of the Korean system, compared with the Taiwanese or Singaporean system. However, its relative strength lay in enabling a rapid growth of local technological and managerial capabilities. The *chaebols*, though they were children of the state-led HCI in the 1970s, rapidly began taking their own initiatives for new large-scale projects from the 1980s. For instance, the *chaebols*' foray into the semiconductor industry in the 1980s can be better understood as a result of oligopolistic competition among themselves in spite of the initial reluctance of the government to support it (Yoon 1990; Shin 1996). The pace of the *chaebols*' expansion was partly reflected in the phenomenal growth of research and development (R&D) expenditure by the private sector, which increased 128-fold from 21.7 billion won ($24.6 million, $1 = 850 won) in 1976 to 2,698.8 billion won ($3,175.0 million) in 1990. The public sector share of R&D accordingly dropped from 64 per cent in 1976 to 19 per cent in 1990, similar to the level in Japan in the 1990s (Shin 1996). In the 1980s, the *chaebols* securely established themselves as the major promoters of high-risk projects in Korea although their financing was still heavily dependent on banks which were under the state's control.

3. THE SINGAPOREAN PATH: AN INTERNATIONALIST MODEL

Singapore entered high-tech industries in a polar opposite way to that of Korea. Instead of nurturing local enterprises as major undertakers of high-tech industries, Singapore developed the industries by attracting and upgrading MNC investments into its territory. Contrary to their counterparts in Korea who wanted to compete directly with the forerunners and establish internationally competitive local companies, Singaporean policy makers were not interested in competing with their forerunners, and attempted instead to develop their economy by establishing complementary relations with MNCs.

Singapore chose this 'complementing strategy' of economic development mainly due to the country's special economic and political conditions. It had developed as an entrepôt open to foreign trade, investment and labour, and, as a former British colony, it had a large English-speaking population, which formed an environment conducive for MNCs to operate in. On the other hand, its relations with its neighbours were not easy from the start of its nationhood and it could expect a greater possibility of success by connecting its economy directly to the 'First World', rather than relying on these unstable and often hostile neighbours for markets and resources.

In the 1960s when Singapore started its industrialisation, this kind of complementing strategy was an exception among developing countries. However, due to the rapid progress in globalisation thereafter, the strategy has been increasingly popular and was promoted vehemently by the multinational organisations (Dunning and Hamdani 1997; Lipsey 1997; World Bank 1999). Even Britain, its former colonial ruler, adopted this strategy of attracting FDIs when it attempted to revitalise its economy in the 1980s. In this respect, the complementing strategy was a Singaporean innovation that has later spread broadly to other parts of the world.

In the beginning, the focus of Singapore's complementing strategy lay in solving unemployment problems, which was likely to deteriorate by the imminent withdrawal of the British armed forces in the late 1960s, and the Singaporean government attempted to attract any kind of FDI as far as this would create jobs in the country. However, once unemployment problems were eased and its economy began picking up, Singapore became more selective in attracting FDIs and shifted its strategic focus towards high-tech industries in the 1970s.

Specifically, Wong (2004) divides the development of Singapore's industry from the technological development perspective into four distinctive stages. The first stage is 'industrial take-off' from the early 1960s to the mid-1970s, when the MNCs were invited into that country to provide employment opportunities. The second stage is 'local technological deepening' from the mid-1970s to the late 1980s, which is characterised by the rapid growth of local process technological development within MNCs and the development of local supporting industries. In the third stage of 'applied R&D expansion' from the late 1980s to the late 1990s, there was rapid expansion of applied R&D by MNCs, public R&D institutions and later local firms. In the latest fourth stage from the late 1990s onwards, there has emerged emphasis on high-tech start-ups and the shift towards technology-creation capabilities. In all three stages since the mid-1970s, the government has taken an active role in the development of technological capabilities with a view to high-tech catch-up.

Wong (2003) highlights such a change as a shift in policies from an emphasis on using technology to creating it. In the process, there was a phased building-up of MNCs, local manufacturing enterprises particularly

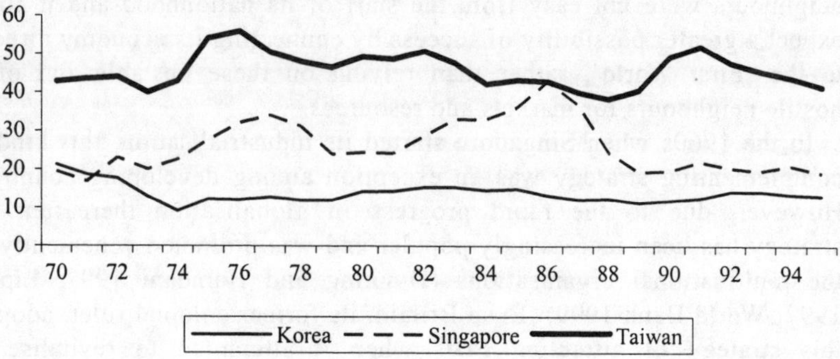

Source: Shin and Chang (2003, figure 2.9).

Figure 2.2 Public Share to Gross Fixed Capital Formation in Korea, Taiwan and Singapore (per cent)

in an electronics supporting capacity, public research institutes and centres, and university R&D. In terms of technology capability development, there was a shift from 'learning to use' relying primarily on internal transfer by MNCs to 'learning to adapt and improve' mainly through 'learning by doing' within MNCs as well as 'learning by transacting' in local firms acquiring external technology; and then to 'learning to innovate' through applied R&D in product or process, and finally to 'learning to pioneer' by creating indigenous intellectual property and commercialising it in the marketplace.

In understanding system elements of the Singaporean economy, the role of public enterprises, which are known as government-linked companies (GLCs) in Singapore, should not be ignored. Although MNCs were a major engine of growth for Singapore, there were business areas where MNCs showed little interest in but which the Singaporean government regarded as important for the country's development. These industries, such as shipbuilding, steel-making, some chemicals and so on were developed through the GLCs. There were also other business areas where it was more convenient to assign GLCs due to their monopolistic positions in the domestic economy, such as telecommunication (SingTel), port management and development (PSA), airlines (SIA) and so on. Therefore, the public share of gross fixed capital formation in Singapore was 35.6 per cent in the 1960s, 26.7 per cent in the 1970s, and 30.3 per cent in the 1980s, much higher than that of Korea though it was lower than that of Taiwan (Figure 2.2).

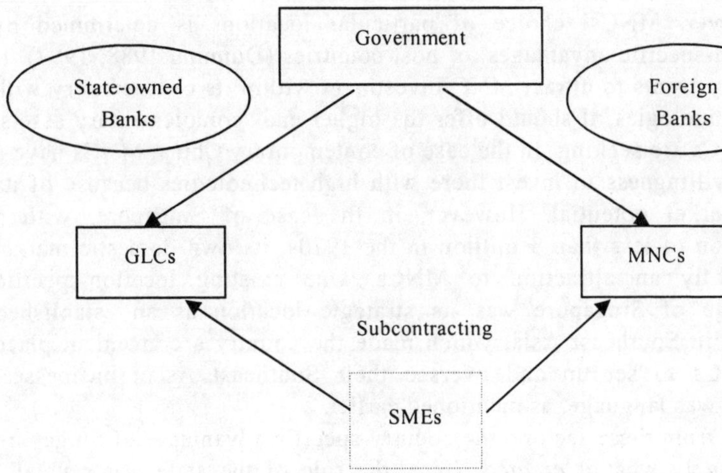

Source: Shin and Chang (2003, figure 2.4)

Figure 2.3 Singapore's Internationalist Model

Thus seen, Singapore's economic system consisted of two pillars, that is, MNCs and GLCs, with local private small and medium-sized enterprises (SMEs) serving as subcontractors to MNCs or GLCs. Despite the dominance of MNCs, the role of the state was crucial as the prime agent in designing and leading this system, responsible for attracting MNC investments as the major provider of 'complementary assets'. The running of GLCs was also, by nature, dictated by the state (Figure 2.3).

Similar to Korea, the basic institutions for economic growth were established in the 1960s when it began growing through participating in labour-intensive industries, and they were strengthened as the country moved to high-tech industries. However, Singapore's way of shifting to high-tech industries in the 1970s and onwards was quite different from that of Korea. As we pointed out above, common requirements in entering high-tech industries include assimilating complex knowledge in the industries and undertaking concurrent investments in capital equipment, skill formation and so on. If a latecomer attempts to do this by relying on domestic resources, it becomes a high-risk game, as in Korea. However, Singapore entered the high-tech industries but took fewer financial risks by sharing them with MNCs. When MNCs come in, they bring with them technologies, capital, marketing networks, managerial capabilities and so on, and this definitely reduces the investments and risks for the latecomers.

However, MNCs' choice of particular locations is determined by 'location-specific advantages' of host countries (Dunning 1988; 1997). If a country wants to attract MNC investment within its own territory with high technologies, it should offer the higher-end 'complementary assets' that MNCs are seeking. In the case of contemporary China, MNCs have a higher willingness to invest there with high technologies because of its huge market potential. However, in the case of Singapore, with a population of less than 3 million in the 1970s, its own domestic market was hardly an attraction to MNCs. One existing location-specific advantage of Singapore was its strategic location as an established entrepôt in Southeast Asia, which made the country a convenient place for MNCs to set up and oversee their Southeast Asian businesses. Another was language, as mentioned earlier.

Apart from these factors, the country-specific advantages of Singapore were largely what it *created*. Here, the role of the state was crucially important. It formulated and implemented policies for high-tech industrialisation. The existence of pro-business government and their policies conducive to FDIs were important complementary assets. And it was responsible for creating various other higher-end complementary assets.

One of them was skill formation. To meet the demand of skilled labour from MNCs, the Singaporean government set up the Skill Development Fund (SDF) and subsidised MNCs in training their workers. The overall training system was also geared to supplying the necessary labour force that will be readily available for MNCs. Wong (2001: 548) argues:

> A distinguishing feature of many of these technical training programs is that they were established and run as collaborative venture between EDB and reputed overseas partners, either in form of well-known MNCs (e.g. Philips, ABB and Seiko), or highly-regarded industrial training institutes (e.g. the German Agency for Technical Cooperation). ... [T]he government did not hesitate to recruit foreign expatriates with significant MNC experiences to head new industrial training institutes.

Flexibility in the application of immigration policy also enabled MNCs to recruit engineers and managers easily from all over the world. For this purpose, the Singaporean government changed the name of the Ministry of Labour to the Ministry of Manpower and endowed it with the role of attracting foreign talent. This is quite different from other countries where the Ministry of Labour is normally against importing foreign workers to maintain domestic employment. Mainly thanks to this aggressive immigration policy, Singapore's population doubled from just over 2 million in 1970 to over 4 million in 2000. Singapore's R&D manpower also increased substantially, with its RSE (R&D Scientists and Engineers, Researchers) per 10,000 of labour force rising from less than 10 in the 1970s to over 50 by the mid-1990s and further to 73.5 by 2002 (see Table 2A.1).

The Singaporean government was particularly active in building industrial and science parks to attract investment from MNCs. Here, the government's ownership of more than 80 per cent of total land of the country played an important role. With this ownership, the Singaporean government was able to act quickly and aggressively in supplying the necessary factory and office space for MNCs to operate. During the period of high-tech industrialisation, the government employed the cluster approach and ensured related industries would congregate in the vicinity.

Singapore's strategy towards high-tech industries was to create an environment in which MNCs take risks in investing in high-tech industries within the country, rather than to build its own system to finance and take risks in high-tech industrialisation. The system it developed was thus a low-risk one. Therefore, it has been less concerned with problems of controlling financial risks and building Korean-style risk-management mechanisms. However, it had relative weaknesses in that domestic technological capability was relatively underdeveloped. This was because the country did not have strong incentives to invest in R&D, as MNCs were major sources of technologies and investments. Reflecting this, Singapore's expenditure on R&D investment had been very low even until the early 1990s when it reached the advanced country status. The gross expenditure on R&D (GERD) of Singapore was 0.21 per cent in 1981 and increased only to 0.84 per cent in 1990. During the same period, Korea's figure jumped from 0.65 per cent to 1.95 per cent while Taiwan's rose from 0.84 per cent to 1.66 per cent (Hou and Gee 1993; Wong 2001).[2] In addition, the weakness in the private sector R&D capability is more pronounced in comparison with other East Asian countries, as we shall discuss further in Section 5.

Singapore was also in need of a functional equivalent to Korea's mechanism of disciplining domestic actors. Similar to Korea, it had to build a disciplinary mechanism over its GLCs. By exposing them to international competition as well as instituting a good governance system, Singapore has kept its GLCs relatively efficient. It also needed mechanisms to encourage MNCs to upgrade their investments within the country, rather than to become 'footloose', as production costs in Singapore were rising with the progress in economic development.

It seems to us that the relative weakness of domestic technological capability and the stronger need to provide MNCs with continuously upgrading complementary assets have made it necessary for the state to work harder in Singapore than in Korea, especially after the country reached a certain level of high-tech industrialisation. As we shall discuss later (Section 5), the state had to invest heavily in R&D to compensate for the relative lack of domestic capability. The state also had to maintain its role as the active promoter of FDIs into Singapore through a close understanding of global trends and interaction with MNCs.

4. THE TAIWANESE PATH: A SEMI-INTERNATIONALIST MODEL

Like Korea and Singapore, the state was the prime initiator of industrialisation in Taiwan (Wade 1990; Cheng 1993). However, compared with those of Korea and Singapore, the Taiwanese path for high-tech industrialisation can be understood as a semi-internationalist one. It promoted development of local enterprises, but was not as focused as Korea on big business groups. It formed an international alliance of equity partnership, especially in high-tech industries, but not as much as Singapore did. A balanced mixture between local and foreign companies has been a characteristic feature of the Taiwanese system.[3]

With the state at the top, the economic system in Taiwan consisted of four main actors: public enterprises, MNCs the *guanxiqiye* (family-owned business groups), and SMEs. The first three sometimes cooperated with one another by forming joint ventures (JVs). As in Korea, commercial banks were nationalised and tightly controlled by the government. However, in contrast to Korea, the banking system was not so heavily utilised for industrial financing. The state's support for industries was mainly demonstrated through fiscal measures, technical assistance from public research institutes, coordinating (sometimes initiating) reorganisation of industries and so on (Figure 2.4).

The role of public enterprises in Taiwan was most pronounced among the three economies of our comparison, reflecting the country's peculiar history and social composition. From the beginning of the Kuomintang government in 1947, Taiwan was a heterogeneous society in which there was a schism between the minority mainlanders and the majority local Taiwanese, with the former maintaining political domination over the latter by repressive measures.[4] Fearing that the local Taiwanese might politically challenge the mainlanders' domination, the Kuomintang government restrained the growth of big businesses in the private sector and developed heavy and chemical industries related to national security

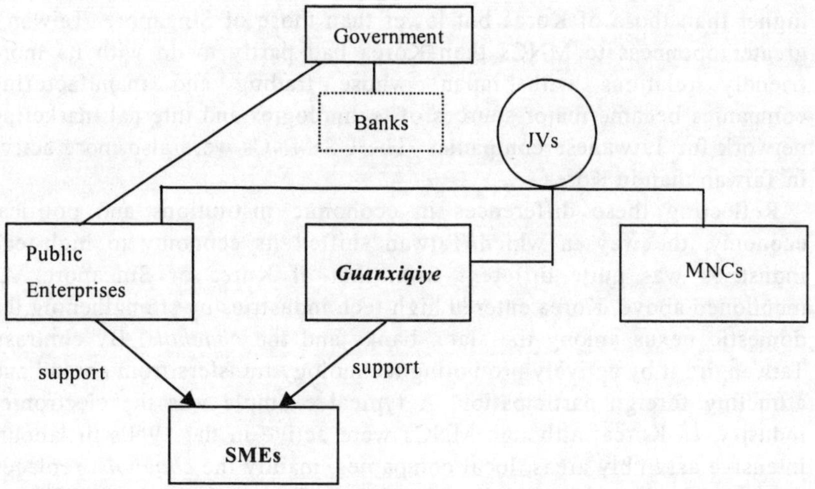

Source: Shin and Chang (2003, Figure 2.6).

Figure 2.4 Taiwan's Semi-internationalist Model

through public enterprises whose top managerial positions were monopolised by the mainlanders. Public enterprises contributed significantly to the capital formation of the country, with their share of total fixed capital formation being the highest among East Asian NIEs (See Figure 2.2). According to Whitley (1992, p. 58), 'from the 1950s to the 1970s Taiwan had one of the largest public enterprise sectors outside the Communist bloc and sub-Saharan Africa'.

On the other hand, the private sector economy was mostly run by the local Taiwanese in the form of SMEs or the *guanxiqiye*. SMEs were heavily engaged in exports, for instance, accounting for 65 per cent of total exports in 1985 (Wade 1990, p. 70). The *guanxiqiye*, though much smaller in terms of size and scope of businesses as compared with the Korean *chaebols*, also established themselves as a pillar of the Taiwanese economy, despite the fact that they were 'handicapped initially by their lack of access to the Nationalist regime and persistent Nationalist suspicions of their political aspirations' (Fields 1995, p. 69).[5]

Taiwan did allow a bigger role of MNCs than Korea, although it maintained several regulations over FDIs, including limits to foreign shareholding and local content regulations, unlike Singapore. Thus, MNCs accounted for 20 per cent of exports during 1974–82 and 16 per cent of employment in manufacturing in 1975 in Taiwan (Wade 1990), higher than those of Korea but lower than those of Singapore. Taiwan's greater openness to MNCs than Korea had partly to do with its more friendly relations with Japan, whose trading and manufacturing companies became major sources of technologies and internal marketing network for Taiwanese companies. The US MNCs were also more active in Taiwan than in Korea.

Reflecting these differences in economic institutions and political economy, the way in which Taiwan shifted its economy to high-tech industries was quite different from that of Korea or Singapore. As mentioned above, Korea entered high-tech industries by strengthening the domestic nexus among the state, banks and the *chaebols*. By contrast, Taiwan did it by actively promoting technology transfers from abroad and attracting foreign participation. A typical example was the electronics industry. In Korea, although MNCs were active in the 1960s in labour-intensive assembly areas, local companies, mainly the *chaebols*, replaced them as they moved into high-tech areas. In Taiwan, however, local firms were upgraded by strengthening their ties with MNCs. For instance, Chung-Hua Picture Tube, the largest monitor manufacturer in the world, was formed in 1970 as a joint venture between major local TV manufacturers and RCA of the US (Lim and Pan. 1991).

In contrast with Singapore, however, Taiwan shifted to high-tech industries by simultaneously promoting the development of local companies. For instance, the Taiwanese government used local content regulations as a tool for MNCs to transfer their technologies to and increase purchase from local subcontractors. The local content target 'was set at a level which was attainable through reasonable effect ... [w]as gradually raised in line with the progress of the local parts industry', and 'was phased out ... by the time it had apparently become redundant' (Chen et al. 2001b).

What is notable is that the Taiwanese state stepped up its coordinating efforts when the need arose. For instance, the formation of Chung-Hua Picture Tube in 1970 was initiated by the state and it was a response to emerging bottlenecks in achieving a higher level of local content in TV assembly.

In the case of the semiconductor industry, the state initially wanted existing local electronics companies such as Tatung and Sampo to enter the higher-end chip manufacturing business. However, they 'opted not to enter the semiconductor market' in the 1970s because of high risks in the industry (Mody 1990, p. 299). Therefore, the state, through a public institution, the Electronics Research and Service Organization (ERSO), has kept the leading role in the Taiwanese semiconductor industry.

Taiwan's major semiconductor manufacturers emerging in the 1980s were all 'spin-offs' of ERSO (Mody 1990; Hou and Gee 1993; Mathews and Cho 2000). ERSO was also a main conduit of introducing sophisticated technologies from abroad.

The famous example was the birth of UMC (United Microelectronics Corporation) and TSMC (Taiwan Semiconductor Manufacturing Company Limited). Originally the ERSO sent a team to RCA in the US to learn integrated circuit (IC) manufacturing technology. After the team returned to Taiwan, the members spun off from ERSO to form UMC, which began chip manufacturing. Later, the government invited Morris Chang to head the Industrial Technology Research Institute (ITRI), of which ERSO is a part, and then accepted Chang's idea of establishing a foundry industry in Taiwan. Chang led a team spun off from ITRI to form TSMC in 1987. The new business model proved effective, and TSMC became the largest semiconductor foundry in the world with $5.3 billion of sales in 2000. TSMC was therefore mainly a Taiwanese creation with state participation in ownership (48 per cent in the beginning). The government invited many local private businesses to invest, but many of the traditional *guanxiqiye* were sceptical and did not become involved. In the end, Chang had to resort to MNCs for ownership participation, and Phillips became a large shareholder (27 per cent).

In this regard, it is worthwhile to compare the changing role of the state in Taiwan and Korea (see Chu 2001 and Chu et al. 2001, for a general discussion of the case of Taiwan). In the 1970s, both countries set up public research institutes to support technological development in the electronics industry: ERSO in Taiwan and the Korea Institute of Electronics Technology (KIET) in Korea. Both institutes played a leading role in providing their incipient local electronics manufacturers with technologies and information.

In Korea, KIET transferred its major R&D facilities to the *chaebols* in the 1980s. In Taiwan, R&D on most chip manufacturing technologies is now being developed by the previously spun-off foundries such as TSMC and UMC. ERSO still does R&D in electronics, but chooses not to concentrate on areas where the private companies are doing well. Over time, the private companies in both economies took over the responsibilities of R&D and the pace was faster in Korea than in Taiwan.

Taiwan became the world's fourth-largest non-utility US patent holders in 2000, and the second largest on a per capita basis, next only to the US. The groups that were granted the largest number of US patents were TSMC, UMC, Honhai (a local, completely private original equipment manufactures (OEM) system assembler), and ITRI. This highlights the

Table 2.2 Taiwan's R&D Expenditure: Allocation among Implementing Organisations (per cent)

	R&D Expenditure	Enterprises			Research Institutions				Colleges & Universities		
		Total	SOEs	Private	Total	Public	Private	Semi-public Non-profit Organisations	Total	Public	Private
1983	100.00	49.76	11.74	38.02	30.23	16.24	0.00	13.98	20.01	15.04	4.97
1984	100.00	48.88	12.25	36.63	32.64	14.71	0.03	17.89	18.49	16.38	2.10
1985	100.00	50.06	15.14	34.92	33.07	12.53	0.00	20.55	16.86	14.62	2.24
1986	100.00	55.47	16.30	39.17	29.12	11.27	0.00	17.84	15.41	13.71	1.70
1987	100.00	59.17	11.52	47.65	26.76	10.19	0.00	16.57	14.06	12.47	1.60
1988	100.00	47.78	9.47	38.32	35.64	17.18	0.13	18.33	16.57	13.00	1.29
1989	100.00	56.02	7.76	48.27	28.89	8.19	0.19	20.51	15.09	11.82	1.45
1990	100.00	59.04	6.61	52.43	27.72	10.37	0.38	16.96	13.25	10.61	1.24
1991	100.00	53.61	6.41	47.20	30.92	9.02	0.08	21.82	15.47	12.87	2.60
1992	100.00	52.54	6.06	46.48	33.02	11.55		21.48	14.43	11.33	3.10
1993	100.00	56.91	5.72	51.18	28.66	12.09		16.57	14.43	11.98	2.45
1994	100.00	57.48	5.64	51.84	27.42	11.80		15.62	15.10	12.34	2.76
1995	100.00	57.56	4.85	52.71	28.44	11.27		17.17	14.00	11.35	2.64
1996	100.00	57.85	4.60	53.25	29.42	11.28		18.15	12.73	10.24	2.49
1997	100.00	61.39	5.45	55.94	26.88	11.50		15.38	11.72	9.27	2.45
1998	100.00	63.14	5.29	57.85	25.72	12.04		13.68	11.14	8.94	2.20
1999	100.00	63.30	4.58	58.72	25.05	11.59		13.46	11.65	9.46	2.20
2000	100.00	62.66	3.94	58.72	25.33	11.26		14.07	12.01	9.54	2.47

Source: National Science Council, Taiwan, ROC, *Science & Technology Statistics*, 2002.

importance of the new private (although with some government participation in ownership) HT actor in R&D activities (Table 2.2).

Reflecting the relative strength of its local firms compared with those in Singapore, Taiwan established its own domestic risk-taking system earlier than Singapore. While ERSO assumed a leading role in high-tech industrialisation of Taiwan in the 1980s, this kind of R&D investment push from the public sector began earnestly in Singapore only in the 1990s.

5. COMPARISONS AND DISCUSSIONS

We have discussed strategies and institutional underpinnings of the three East Asian economies in their entry into high-tech industries. Korea adopted the most nationalistic strategy and achieved its objective by strengthening the state banks–*chaebols* nexus. Singapore chose the most internationalist strategy and became successful by attracting and upgrading MNC investments with the provision of continuously upgraded complementary assets. Taiwan went for a semi-internationalist strategy and actively formed an alliance with MNCs while upgrading local firms' technological capabilities.

These differences in strategy and institutions resulted in different strengths and weaknesses of these economies. As pointed out above, the Korean system is a higher-risk one and it should be equipped with mechanisms to shoulder the risks. On the other hand, Singapore and Taiwan built less risky systems as they avoided direct competition with their forerunners and spread the financial risks among partners of the equity ownership.

Therefore, Korean firms' reliance on debts was much higher than their counterparts in Singapore or Taiwan. For instance, during 1974–95, the average corporate debt–equity ratio for the Korean manufacturing sector was 342.2 per cent while it was 95.1 per cent for the Taiwanese counterpart during the same period. In Demigruc-Kunt and Maksimovic's (1996) study covering the 1980-91 period, the debt–equity ratio of Singaporean firms was 123.3 per cent while that of Korean firms was 366.2 per cent. Singapore's figure should be discounted because most of the bank loans were drawn by MNCs with their own credit from international financial institutions. This seems to indicate that the government perceived less financial risk of corporate debts in Singapore than in Taiwan, though the actual debt–equity ratio was higher in Singapore than in Taiwan (Figure 2.5).

Similarly, Korea relied much more heavily on foreign debts than Taiwan or Singapore. Korea's foreign debt to GDP ratio shot up to over 50 per cent at the height of the HCI while Singapore's figures hovered around the 10–20 per cent level and Taiwan's figures were even lower

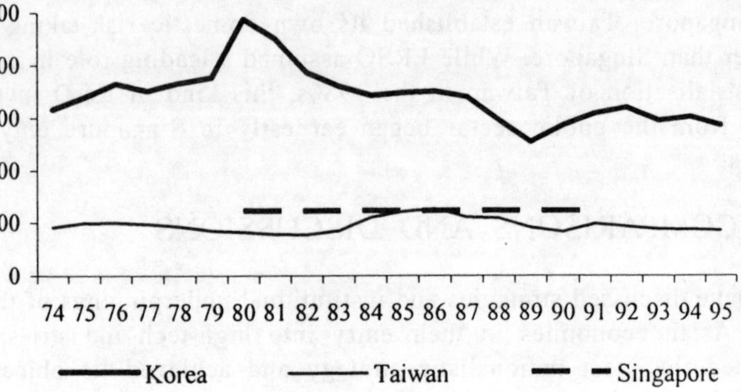

Note: Singapore's figure is 123.3 per cent on average during 1980–91 (Demigruc-Kunt and Maksimovic 1996).

Sources: BOK (Bank of Korea) website, Fields (1995, p. 108, Table 4-5), Bank of China in Taiwan, quoted in BOK, *Financial Statement Analysis* (1999)

Figure 2.5 Debt–Equity Ratio of Manufacturing Firms in East Asia (per cent)

Table 2.3 External Debt to GDP Ratios of Korea, Taiwan and Singapore (per cent, selected years)

	1976	1982	1985	1993	1996	1997
Korea	36.7	52.0	52.1	12.7	20.2	25.5
Taiwan	13.6	12.8	14.5	7.6	8.0	9.3
Singapore		22.0	22.8	9.5	10.7	16.5

Source: Shin and Chang (2003, Table 2.3).

(Table 2.3). Here again, Singapore's foreign debts were mostly shouldered by MNCs and the Singaporean government did not regard them as its own obligations. In view of this fact, it cannot be said that Singapore has been exposed to higher external financial risks than Taiwan.

On the other hand, Korea was ahead of Taiwan or Singapore in building its own system to take risks in R&D investments and large-scale projects, although Taiwan is quickly catching up. This is partly reflected in the trend of R&D expenditure. From the 1980s, Korea constantly recorded a higher expenditure of R&D over GDP than Taiwan or

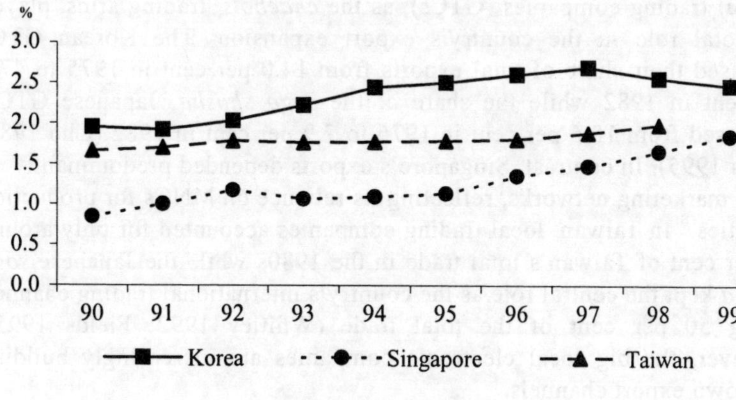

Source: Shin and Chang (2003, Figure 2.7).

Figure 2.6 Trend of GERD/GDP: Korea, Taiwan and Singapore

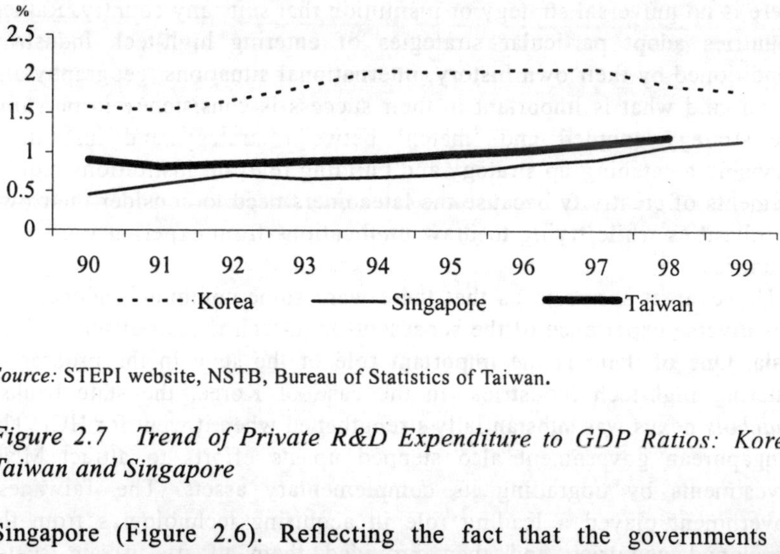

Source: STEPI website, NSTB, Bureau of Statistics of Taiwan.

Figure 2.7 Trend of Private R&D Expenditure to GDP Ratios: Korea, Taiwan and Singapore

Singapore (Figure 2.6). Reflecting the fact that the governments of Taiwan and Singapore originally played a leading role in R&D investment to compensate for the relative lack of risk-taking capability of their private sectors, a gap in private R&D expenditure existed between Korea *vis-à-vis* Taiwan and Singapore as shown in Figure 2.7.

As for the differences in local technological capability, the Korean system was stronger in building its own marketing network. In Korea, the general trading companies (GTCs), as the *chaebols*' trading arms, played a pivotal role in the country's export expansion. The Korean GTCs increased their share of total exports from 14.0 per cent in 1975 to 47.9 per cent in 1982 while the share of the *sogo shosha*, Japanese GTCs, decreased from 15.6 per cent in 1976 to 7.9 per cent in 1982 (Cho 1986; Fields 1995). In contrast, Singapore's exports depended predominantly on MNC marketing networks, reflecting its reliance on MNCs for production activities.[6] In Taiwan, local trading companies accounted for only around 20 per cent of Taiwan's total trade in the 1980s while the Japanese *sogo shosha* kept the central role as the country's international trading channel, taking 50 per cent of the total trade (Whitley 1992; Fields 1995). However, the big local electronic companies are increasingly building their own export channels.

What lessons then can we draw from the experience of the three economies? What we have emphasised above is that there were diverse paths for high-tech industrialisation and the sample economies were successful in their endeavour mainly because they were able to build and manage institutions appropriate to their own strategies. In this analysis, there is no universal strategy or institution that suits any country. Rather, countries adopt particular strategies of entering high-tech industries conditioned by their own history, international situations, geography and so on, and what is important in their success is consistency in pursuing the strategy adopted and 'match' between strategy and institution. Devising a catching-up strategy and building relevant institutions require elements of creativity because the latecomers need to consider their own peculiarities while trying to draw implications from experience of other countries.

However, it seems to us that there were some common tendencies in this diverse experience of the successful high-tech industrialisers in East Asia. One of them is the important role of the state in the process of entering high-tech industries. In the case of Korea, the state banks–*chaebols* nexus was substantially strengthened when it went for HCI. The Singaporean government also stepped up its efforts to attract MNC investments by upgrading its complementary assets. The Taiwanese government played a leading role in acquiring technologies from the developed countries and then provided them to its private sector (sometimes after improving upon them) on the one hand, and in forming joint ventures between local firms and MNCs on the other. This was because there should be a system to take risks involved in assimilating high technologies and deploying them in production processes. Whether an economy built the system on its own or shared the risks with MNCs, the private sectors in those latecomers were not strong enough to initiate building the risk-taking system, though they had been successful in engaging in export competition of lower-skilled labour-intensive items.

Related to the above, the other tendency to be noted is the formation of local technological capability. Even in the case of Singapore, which developed mainly through attracting MNC investments, it was able to do so because it upgraded local capability and complemented the MNC need for increasingly higher-end complementary assets. In the beginning of its high-tech industrialisation, it was enough for Singapore to focus on skill formation and training, along with providing MNCs with fiscal incentives, infrastructure and so on. Its government later began investing heavily in R&D mainly through public institutes and promoting local entrepreneurs in order to match the MNC need for higher-end complementary assets.[7] In the case of Taiwan, the pool of competent local firms and engineers was a main attraction for MNCs to form alliance relationships (through OEM or ODM (Original design manufactures) arrangements) or joint ventures with the local companies. In Korea, which pursued a nationalistic strategy, the acquisition and upgrading of local technological capability was a matter of survival.

We can explain the de-industrialisation and shrinking of high-tech industries in Hong Kong from the 1980s in the same vein. As production costs rise in a location and other locations offer lower-cost labour, MNCs or domestic firms have incentives to move their production sites away from their original locations. De-industrialisation is in this sense a more natural phenomenon than upgrading the level of industrialisation. Especially when an economy is at the threshold of shifting from low to high-tech industrialisation, counter-incentives for MNCs and domestic firms to remain and upgrade their investments within the location become more important. Owing to the entrenched *laissez faire* tradition and the integration into China, Hong Kong did not act fast enough to strengthen the role of its administration in providing and creating higher-end complementary assets for its manufacturing sector. It has tried to do this recently but the results have yet to be seen.

Apart from common tendencies pointed out above, what else can we suggest for current developing countries that aspire to emulate the successes of the three economies? One important difference between the three economies and the later industrialisers lies in the international environment, that is, the fact that the pace of globalisation has been radically accelerated since the 1980s. When the three economies began industrialisation in the 1950s and 1960s, and also moved to high-tech industries in the 1970s, export promotion versus import substitution was a matter of choice among developing countries. However, with the wider opening of the global market, an exclusive import-substitution strategy of domestic markets is no longer available. Even if one may argue that the success of East Asian NIEs is attributed to export promotion in comparison with Latin America or India during their developmental periods, export promotion itself cannot be a prescription for current developing countries as everybody is taking part in the export game. They need to find their own ways of entering high-tech industries with the open international market as a given condition.

Moreover, with the widespread adoption of the World Trade Organisation (WTO) regime, it has become more difficult to implement a Korean-style nationalistic development model. It is increasingly necessary for current developing countries to find ways of working closely with MNCs in their development processes, rather than promoting their own local firms under protections and subsidies, though they are not completely unavailable under the WTO regime.[8] The overall balance of growth strategy seems to have skewed towards the direction of the Taiwanese or Singaporean model.

However, there are two caveats in pursuing a more internationalist path for high-tech industrialisation. First, there are more developing countries competing for MNC investments. In the 1960s and 1970s, the Singaporean way of growing through inviting MNC investments was an exception, rather than a norm, for developing countries. Taiwan also adopted a strategy of forming alliances with MNCs and utilising complementary relationships with them relatively earlier than other developing countries. Therefore, they rarely faced serious competition in attracting FDI or otherwise acquiring advanced technologies from other developing countries.[9] As far as they were able to provide what MNCs were seeking, such alliances could occur. In a sense, they benefited greatly from the rapid expansion of global production networks by MNCs thereafter because they were pioneers of complementing strategies. However, current developing countries need to survive competition among them. It is of course a matter of empirical investigation whether MNCs' supply of FDIs has grown enough to ease competitive pressure from the increasing demand for FDIs by current developing countries. However, it seems that the overall supply of FDI has not grown as much as the demand from the developing world, and its distribution of FDI has remained very uneven.

Second, there are business areas in which MNCs are not interested in relocating or outsourcing to the latecomers while the latter may find great potential for their own growth. In the case of the electronics or textile industries where spatial division of labour did occur more easily, there is an incentive on the part of MNCs to shift part of their production process to developing countries. However, in the industries whose competitiveness is crucially determined by close integration of the production process and resulting economies of scale, such as chemicals, steel, a large part of microprocessor and DRAM manufacturing and so on, there may be no complementary assets available for the latecomer to persuade MNCs to part with their production processes or technologies.

Therefore, a crucial question for developing countries becomes how sufficiently a specialisation in the complementary areas will meet their needs of economic growth. The answer will ultimately be determined by the size of the market created by the international subcontracting network at any given time. The size of the countries concerned will also be a factor. For bigger countries, engaging only in the complementary areas may not suffice to attain a desirable rate of economic growth. For smaller countries, it should be relatively easier to achieve a desirable rate of economic growth through specialising in limited areas. As the Taiwanese case shows, this scope has been greatly expanded with the acceleration of globalisation. Nevertheless, it remains to be seen whether it will work for bigger countries like Korea, China and so on, although it should be noted that the lure of the domestic market would be an element in the decision making of MNCs. Even Singapore, a city state with just over 4 million people, needed a great role of GLCs to play a significant role in developing non-complementary business areas.

APPENDIX 2A

Table 2A.1 R&D Expenditure and Manpower in Singapore

Year	GERD (S$m)	GERD/GDP (%)	RSEs	RSE/10,000 labour force
1978	37.80	0.21	818	8.4
1981	81.00	0.26	1,193	10.6
1984	214.30	0.54	2,401	18.4
1987	374.70	0.86	3,361	25.3
1990	571.70	0.84	4,329	27.7
1991	756.80	1.00	5,218	33.6
1992	949.50	1.17	6,454	39.8
1993	998.20	1.06	6,629	40.5
1994	1,174.98	1.10	7,086	41.9
1995	1,366.55	1.13	8,340	47.7
1996	1,792.14	1.39	10,153	56.3
1997	2,104.56	1.49	11,302	60.2
1998	2,492.26	1.76	12,655	65.5
1999	2,656.29	1.84	13,817	69.9
2000	3,009.5	1.89	14,483	66.1
2001	3,232.7	2.11	15,366	72.5
2002	3,404.7	2.19	15,654	73.5

Source: Wong (forthcoming, Table 2a), with data compiled from *National Survey of R&D in Singapore*, various years, Agency for Science, Technology and Research (previously National Science and Technology Board).

Table 2A.2 Patent Indicators: Singapore vs. Selected Countries

	All Patents[1] per 1000 population	Utility Patents[6] Growth Rate	Utility Patents[7] / GERD (no. per US$ million)	
	2001	1996–2001	1990	2000
Singapore	**0.90**	**27.50**	**0.07**	**0.13**
G7 Nations				
G7 Average	1.61	8.70	0.19	0.21
Canada	1.41	10.10	0.24	0.27
UK	0.85	10.10	0.15	0.14
USA	3.46	7.50	0.31	0.32
Japan	2.79	7.60	0.23	0.21
Germany	1.56	10.60	0.18	0.22
Italy	0.38	7.30	0.09	0.14
France	0.83	7.70	0.10	0.14
Asian NIEs				
Hong Kong	0.82	21.90	na	0.24
South Korea	0.81	18.80	0.07	0.27
Taiwan	3.00	23.10	0.35	0.74
Selected European OECD				
Finland	1.65	10.50	0.12	0.15
Denmark	1.21	14.70	0.10	0.12
Netherlands	1.09	10.80	0.20	0.15
Norway	0.72	13.90	0.06	0.10
Sweden	2.36	15.30	0.15	0.17
Switzerland	2.63	5.00	0.25	0.16

Notes:

1 All patents include utility, plant and design patents and SIRs. Nationality of patent is determined using the 'at least one inventor resides in specified country' convention.

2 Nationality of patent is determined by residence of first-named inventor at the time of grant.

3 Nationality of patent is determined using the 'at least one inventor resides in specified country' convention.

4 Locally owned patents are patents that are assigned to locally based establishments in each country.

5 Patents invented abroad are patents where none of the inventors is a resident in the specified country.

6 Nationality of patent is determined by residence of first-named inventor at time of grant.

7 Nationality of patent is determined by residence of first-named inventor.

Source: Wong (forthcoming), with data compiled from US Patent and Trademark Office.

Table 2A.3 The Foreign Share of Equity Investment in Singapore-based Companies: 1984–2000 (%)

	Manufacturing	Total Economy	Electronics Industry
1984	65.0	na	na
1987	70.8	39.6	na
1990	69.4	37.2	na
1993	60.2	33.2	85.3
1994	62.6	32.8	84.8
1995	66.9	31.5	87.2
1996	67.9	32.1	86.4
1997	68.8	33.2	88.1
1998	69.8	35.1	86.9
1999	75.5	36.6	na
2000	75.7	36.8	na

Sources: Wong (forthcoming, Table 11), with data compiled from Department of Statistics, *The Extent and Pattern of Foreign Investment Activities in Singapore*, 1995; Department of Statistics, *Foreign Equity Investment in Singapore*, 1997 and 2000, 2001; Department of Statistics, *Singapore's Corporate Sector*, 2002.

NOTES

1. Manufacturing products for instance have comprised more than 90 per cent of total export in Korea during its developmental period (Jones and Sakong 1980, p. 86).
2. However, due mainly to deliberate efforts by the government later, Singapore's GERD/GDP ratios rose substantially in the late 1990s (over 1.5 per cent during 1998–99) and the early 2000s (2.19 per cent by 2002); see Table 2A.1. Similar improvements occurred in the area of patenting, as Table 2A.2 reveals: Singapore's growth rate of utility patents granted by the US patent office grew by 27.5 per cent over 1996–2001, and its total number of patents granted by that office became 0.90 per 1000 of population by 2001, a figure bigger than those of Korea and Hong Kong, although lagging behind Taiwan.
3. Here we are mainly dealing with transition to high-tech industrialisation in the 1970s and the 1980s; see Chapter 4 for a discussion on the post-1990 situation.
4. The country was ruled by martial law until 1986.
5. This said, it must be noted that the situation changed greatly after the mid-1980s when the ruling party KMT became indigenised. After Chiang Ching-Kuo's death in 1988 and the rise to power of Lee Teng-huei, a Taiwan-born successor, the transition accelerated. Moreover, the private enterprises have had a good relationship not only with the then ruling KMT party, but also with the opposition Democratic Progressive Party (DPP). When the DPP won the presidential election in 2000, the problem of mainlanders hanging on to power was almost completely eliminated. The new situation was that public enterprises, while still controlling a substantial amount of economic activities in the communication, steel, petro-chemical and banking sectors, are less and less important as vehicles of industrial development. They retained their political functions – acting as organisations

through which political support can be mobilised. Private enterprises, which occupy the other sectors, became much more powerful economically and politically. Moreover, after the 1990s, one new actor should be added to the scene. This is the electronics/computer-related sector. Led by the Semi-conductor Association, this sector has become the engine of Taiwan's economic growth since the 1990s, when almost all other sectors began either declining or growing less rapidly than before. This sector, often called the 'high-tech' (HT) sector, is mixed both in terms of ethnicity and in terms of state/private ownership. Given its size and its popularity, this sector became both economically and politically powerful, overshadowing the importance of traditional *guangxiqiye,* whose main sphere of influence is the domestic, not global market.

6. As can be seen in Table 2A.3, the foreign share of equity investment in Singapore-based companies was over two-thirds in manufacturing most of the time from 1984 to 2000. In the electronics industry which has been the leading export sector, the share has been as high as over 85 per cent. For the economy as a whole, the share has been about one-third, reflecting the importance of GLCs in the domestically oriented service sectors. Similar patterns emerge in Wong (forthcoming) discussion of the top 30 electronics manufacturing companies and the top ICT companies. Among the former, all were foreign in 1991, while 25 were foreign in 2001; with two of the five Singaporean companies being GLCs (Chartered Semiconductor Manufacturing and Tech Semi) and three being genuinely independent entrepreneurial firms (Creative, Venture Manufacturing, and GES). Among the last, which has been more domestic market oriented, 19 out of the 30 top companies were foreign in 1991. Foreign share shrank to 12 out of 30 in 2001, but the majority of the emerging local big ICT companies were GLCs, only three being genuinely private firms (Informatics, ECS Holdings and Frontline Technologies).
7. According to Wong (forthcoming), in the area of strengthening the innovation system, the government of Singapore substantially increased public funding of R&D over 1991–2002 and through 2005 (planned), re-organised the public funding system along the lines of the US National Science Foundation in 2001, and made major public investments in life science R&D and Biopolis since 2001, in addition to the existing and ongoing programmes of assistances in the forms of tax and other incentives to R&D and manpower training for innovation by the MNCs and by the local companies including SMEs. In the area of promoting entrepreneurial activities, in 1999 it established co-investment funds to attract local and foreign venture capitalists to build up operations in Singapore, and to attract angel investing into new start-ups; it also provided incentives to attract FDI in various KIBS (Knowledge Intensive Business Services) industries including media content, intellectual property services and so on. It established in 2003 a public–private consultation forum on entrepreneurship (Action Community For Entrepreneurship or ACE), the activities of which included promoting interests in entrepreneurship in schools and reviewing government red tape and unnecessary regulations by receiving public feedback. Various liberalisations were undertaken as part of T21 (Technopreneurship 21, initiated in 1999) including allowing business operation from home and relaxation of bankruptcy law and so on; other barrier-removing measures were adopted as part of the Free Trade Agreement with the US, including liberalisation of financial services and public procurement. The government has also been reviewing the role of GLCs in the domestic economy with a view to enhancing competition and private participation.
8. For a discussion on the possibilities of employing protections and subsidies, see Chang (2001).
9. See Chen et al (2001), for a discussion of the choice between simple technology purchases and joint ventures in the case of the textile industry in Taiwan.

REFERENCES

Bank of China, Taiwan (various years), *Financial Report of the Manufacturing Sector in Taiwan, Republic of China* (in Chinese).

Bank of Korea (BOK) (various years), *Financial Statement Analysis.*

Bank of Korea (BOK) (various years), *Economic Statistics Yearbook*, website, http://www.bok.or.kr.

Bureau of Statistics of Taiwan (various years), *Taiwan Statistical Data Book*, website, http://www.stat.gov.tw/.

Chang, Ha-Joon (2001), 'Rethinking East Asian Industrial Policy – Past Records and Future Prospects', in P.-K. Wong and C.-Y. Ng (eds.), *Industrial Policy, Innovation and Economic Growth: The Experience of Japan and the Asian NIEs*, Singapore: Singapore University Press

Chen, Been-Lon, Tain-Jy Chen and Yun-Peng Chu (2001a), 'The Role of Textiles and Man-made Fibre in the Process of Industrialization: The Case of Taiwan', in P.-K. Wong and C.-Y. Ng (eds.), *Industrial Policy, Innovation and Economic Growth: The Experience of Japan and the Asian NIEs*, Singapore: National Singapore University Press

Chen, Tain-Jy, Been-Lon Chen and Yun-Peng Chu (2001b), 'The Development of Taiwan's Electronics Industry', in P.-K. Wong and C.-Y. Ng (eds.), *Industrial Policy, Innovation and Economic Growth: The Experience of Japan and the Asian NIEs*, Singapore: National Singapore University Press, pp. 245–82.

Cheng, Tun-Jen (1993), 'Guarding the Commanding Heights: The State as a Banker in Taiwan', in S. Haggard, et al. (eds.), *The Politics of Finance in Developing Countries*, Ithaca, NY: Cornell University Press

Chiu, Stephen and Tai-Lok Lui (2001), 'Hong Kong Industry Case Studies', in P.-K. Wong and C.-Y. Ng (eds.), *Industrial Policy, Innovation and Economic Growth: The Experience of Japan and the Asian NIEs*, Singapore: Singapore University Press

Cho, Dong-Sung (1986), *The General Trading Company*, Lexington, MA: Lexington Books.

Chu, Yun-Peng (2001), 'Market Grew and Matured as a Result of Government Actions', *International Journal of Development Planning Literature*, vol.16, No. 3 & 4, September–December, pp. 209–34.

Chu, Yun-Peng, Been-Lon Chen and Tain-Jy Chen (2001), 'Rethinking the Development Paradigm: Lessons from Taiwan – the Optimal Degree of Government Intervention', in P.-K. Wong and C.-Y. Ng (eds.), *Industrial Policy, Innovation and Economic Growth: The Experience of Japan and the Asian NIEs*, Singapore: National Singapore University Press, pp. 197–244.

Demigruc-Kunt, A. and V. Maksimovic (1996), 'Stock Market Development and Firm Financing Choices', *World Bank Economic Review*, vol. 10, no. 2.

Dunning, John H. (1988), *Explaining International Production*, London: Unwin Hyman.

Dunning, John H. (1997), *Alliance Capitalism and Global Business*, New York: Routledge.

Dunning, J.H. and Khalil A. Hamdani (eds.) (1997), *The New Globalism and Developing Countries*, Tokyo: United Nations University Press.

Fields, Karl J. (1995), *Enterprise and the State in Korea and Taiwan*, Ithaca, NY and London: Cornell University Press.

Gerschenkron, A. (1962), *Economic Backwardness in Historical Perspective*, Cambridge Mass: Harvard University Press.

Gerschenkron, A. (1970), *Europe in the Russian Mirror, Four Lectures in Economic History*, Cambridge: Cambridge University Press.

Henderson, J. (1989), *The Globalisation of High Technology Production*, London: Routledge.

Hou, C.M. and S. Gee (1993), 'National Systems Supporting Technical Advance in Industry: The Case of Taiwan', in R.R. Nelson (ed.), National Innovation Systems: A Comparative Analysis, Oxford: Oxford University Press, pp. 384–413.

Jacobsson, S. (1993), 'The Length of the Learning Period: Evidence from the Korean Engineering Industry', *World Development*, vol. 21, no. 3, pp. 407–20.

Jones, L.P. and I. Sakong (1980), *Government, Business, and Entrepreneurship in Economic Development: The Korean Case*, Cambridge, MA: Harvard University Press.

Lim, L. and E.F. Pan (1991), *Foreign Direct Investment and Industrialisation in Malaysia, Singapore, Taiwan, and Thailand*, Paris: OECD Development Centre.

Lipsey, R.G. (1997), 'Globalization and National Government Policies: An Economist's View', in J.H. Dunning (ed.), *Governments, Globalization, and International Business*, Oxford: Oxford University Press

Mathews, John A. and Dong-Sung Cho (2000), *Tiger Technology: The Creation of a Semiconductor Industry in East Asia*, Cambridge: Cambridge University Press.

Mody, A. (1990), 'Institutions and Dynamic Comparative Advantage: The Electronics Industry in South Korea and Taiwan', *Cambridge Journal of Economics*, vol. 14, pp. 291–314.

National Science and Technology Board (NSTB) various issues, *National Survey of R&D in Singapore.*

National Statistical Office (Korea), website, http://www.nso.go.kr .

Science and Technology Policy Institute (STEPI), website, http://www.stepi.re.kr .

Shin, Jang-Sup (1996), *The Economics of the Latecomers: Catching-Up, Technology Transfer and Institutions in Germany, Japan and South Korea*, London and New York: Routledge.

Shin, Jang-Sup and Ha-Joon Chang (2003), *Restructuring Korea Inc.: Financial Crisis, Corporate Reform, and Institutional Transition*, London and New York: Routledge.

Singapore Department of Statistics website, http://www.singstat.gov.sg .

Wade, R. (1990), *Governing the Market: Economic Theory and the Role of Government in East Asian Industrialization*, Princeton, NJ: Princeton University Press.

Whitley, Richard D. (1992), *Business Systems in East Asia*, London: Macmillan.

Wong, Poh-Kam (2001), 'Leveraging Multinational Corporations, Fostering Technopreneurship: The Changing Role of S&T Policy in Singapore', *International Journal of Technology Management*, vol. 22, no. 56, pp. 539–67.

Wong, Poh-Kam (2003), 'From Using to Creating Technology: The Evolution of Singapore's National Innovation System and the Changing Role of the Public Policy', in S. Lall and S. Urata (eds.), *Competitiveness, FDI and Technological Activity in East Asia*, Cheltenham: Edward Elgar

Wong, Poh-Kam (forthcoming), 'The Re-Making of Singapore's High Tech Enterprise Ecosystem', in H. Rowen, W. Miller and M. Hancock (eds.), *Asia's High Tech Regions*, Stanford, CA: Stanford Business Press.

World Bank (1999), *World Development Report 1999/2000: Entering the 21st Century*, Oxford: Oxford University Press.

Yoon, Jung Rho (1990), 'Hankook eui Sanup Baljun gwa Gugka: Bandoche Sanup eul Joongsim euro, (Industrial Development of Korea and the State: With Special Reference to the Semiconductor Industry), in Hankook Sahwesa Yunguhwe (ed.), *Hyundai Hankook ui Saangsanryuk gwa Gwahak Gisul* (Productive Capacity, Science and Technology of Contemporary Korea), Seoul: Moonhak gwa Jisungsa

3. High-Tech Industrialisation and Local Capability Formation in South Korea

Jang-Sup Shin

1. INTRODUCTION

South Korea (henceforth Korea) was one of the successful developing countries in the latter half of the twentieth century, rapidly transforming itself into a high-tech economy from a low-tech one. In the beginning of its industrialisation, its export items were predominantly raw materials and light industrial goods, and the share of heavy and chemical industries (including the electronics industry) to the total export was a mere 10.4 per cent in 1965. However, the share jumped to 54.4 per cent in 1990 and further to 74.0 per cent in 2000.[1] In less than three decades, Korea's economy turned into one principally led by high-tech industries.

This chapter attempts to explore important factors behind this transition by focusing on policies and institutions that Korea employed and maintained. Rather than drawing implications for 'universal' policies or generally 'good' institutions, it investigates configurations between those policies and institutions, which worked in the context of Korea's history, geopolitics and characteristics of political leadership. This is because developmental processes and institutions are diverse across countries and, I believe, a 'lower-level' analysis would provide other developing countries with more useful insight for designing and/or managing their own policies and institutions.

Our discussion begins with the overall framework of policies and institutions for industrialisation in Korea (Section 2). Some characteristic features of Korea's transition towards high-tech industrialisation are described in comparison with those of Singapore and Taiwan (Section 3). It then examines how Korea's policies and institutions interacted with each other in acquiring advanced technologies and forming local technological and managerial capability (Section 4). The chapter concludes by drawing some implications for current developing countries on the path of high-tech industrialisation.

2. OVERALL POLICIES AND INSTITUTIONS FOR INDUSTRIALISATION IN KOREA

Korea's overall development programme is often characterised as an export-oriented industrialisation. According to this view, the country was able to grow rapidly because the export orientation of the economy brought the incentive structure of Korean industries in line with the world market. The export market also provided the Korean economy with bigger market potential to realise higher productivity (See Ranis and Fei 1975; Krueger 1980; Bhagwati 1987; World Bank 1987).

The export promotion was certainly a new policy direction in Korea introduced by the authoritarian government, which came to power after a military coup in 1961 and embarked on an ambitious industrialisation programme. The previous government in the 1950s had practised a simple import-substitution policy, with foreign currency and food supplies relying predominantly on generous aids from the USA. As US aids dissipated, export of manufactured products became the only viable option for Korea to finance its industrialisation, given its small domestic market and lack of natural resources or primary products to export. This imperative to export manufactured goods operated similarly in other East Asian newly industrialising economies (NIEs) such as Taiwan, Singapore and Hong Kong.

While it is natural to pay attention to this change in economic policy introduced with the regime change, what is less well-recognised in the neo-liberal literature on the Korean success story is that, within this framework of export promotion, Korea pursued simultaneously a strong import-substitution strategy. Constructing an 'independent economy' was explicitly stated as the top priority on the state's economic policy agenda. Different Five-Year-Plan documents, various issues of the Economic White Paper, and speeches of President Park Chung Hee all laid great emphasis on the independent economy or the increase in self-sufficiency.

For this purpose, the Korean government employed systemic import-substitution policies through tariffs, subsidies, regulations on foreign direct investment (FDI) and technology transfer, and so on (Lee 1991; Chang 1993, 1994). Among East Asian NIEs, this emphasis on import substitution was the strongest in Korea. Therefore, the country's overall policy for industrialisation can be better termed as 'export-promotion-cum-import-substitution policy (henceforth EP-cum-IS)', rather than as a straightforward export-oriented policy.[2]

In this policy framework, the main thrust lay in establishing *internationally competitive local companies*. The Korean government anticipated that these local companies would constitute the backbone of the 'independent economy' and promotional measures were directed at them. In policy formulation and implementation, there was a clear distinction between local and foreign companies, and Korea's reliance on FDI was therefore among the lowest in developing countries (See further discussion in Section 4).

However, in contrast to cases of simple import-substitution policy practised in other parts of the world, in which the performance of local firms bears only a remote relationship with their export activities, nurturing local firms' competitiveness in the export-oriented environment constrained the import-substitution policy in the following ways.

First, as pointed out above, export earnings from manufactured goods were needed to finance industrialisation due to the small domestic market and the lack of exportable natural resources. Therefore, it was often the case that the imperative of export promotion over-ruled that of import substitution, at least in the short run. Especially in the earlier stage of industrialisation, some industries or products, such as garments and electronic parts, were promoted mainly for their abilities to earn foreign currency and thereby ease the balance of payments (BOP) constraint, though they contributed to the import substitution of other industries indirectly by supplying foreign currencies as required.

Second, despite strict regulations on imports in general (mainly to encourage import substitution and partly to ease the BOP constraints), those related to export activities received exceptional treatment. Capital and intermediate inputs used in export activities could be imported without tariffs and quotas applied to imports for other purposes (Lee 1991; Chang 1993; Kim 1993). This was because the Korean government gave top priority to making export industries competitive in the international market, which meant that import substitution was pursued only to the extent that it did not seriously harm export competitiveness.

Of course, there may have been a possible slowdown of the local capital goods sector, which is related to 'the make–import decision', due to this priority of export.[3] However, the slowdown did not seem to be serious in Korea. The Korean economy achieved international competitiveness in final assembly industries first and was later able to move to gain competitiveness in the capital goods and parts industries.

To make this EP-cum-IS policy work, Korea built its own institutions for economic development, that is, the 'state banks–*chaebols*' nexus, which depended on a close cooperation among major local actors. In this system, the state was at the top. The Korean state drafted a comprehensive industrial policy for export promotion and import substitution. Moreover, it nationalised commercial banks and subordinated their lending decisions to its industrial policy. The state also mobilised a vast amount of foreign loans, as an alternative to FDI, by guaranteeing their repayments by domestic banks and firms. Most foreign debts, which jumped from 10.7 per cent of gross national product (GNP) in 1965 to 44.4 per cent of GNP in 1980 and again to 52.1 per cent in 1985 at its peak, was guaranteed by the state and distributed to the *chaebols*, the family-owned and diversified business groups, through the banking system.

Despite the omnipresence of the state in the Korean system, the state entrusted local entrepreneurs to undertake major industrial projects. It relied on public enterprises only when public ownership was deemed necessary for some reasons, such as when local private entrepreneurs were simply not available (the integrated steel industry), where 'natural monopoly' could be easily established (telecommunications industry), or in industries which functioned as basic supply industries for other downstream industries (some parts of the chemical industry and the steel industry). Outside these industries, financial resources, mobilised through the banking system or by the state directly, were mostly channelled to the *chaebols*. The banking system provided the *chaebols* with 'patient money' under the auspices of industrial policy and state ownership. In addition, the *chaebol* structure itself functioned as another credit-creation system through internal transactions and loan guarantee.

The initial period of industrialisation in Korea in the 1960s was led by light industries such as garments, textiles, wigs, plywood and so on, where Korea's comparative advantage of low labour cost could be utilised in competing in the international market. In the electronics industry, a newly emerged technologically dynamic industry in the latter half of the twentieth century, Korea also took part in the assembly of consumer electronics goods, electronic parts and components where its low-cost labour could bring competitive advantage. This pattern of industrialisation beginning from light industries was common to many developing countries and also to East Asian NIEs.

What seems to be particular in Korea, conditioned by its EP-cum-IS policy, was a two-pronged strategy in developing light industries: on the one hand, it applied strong import-substitution measures to some light industries for which domestic demand existed or was soon to grow, such as garments, textiles and consumer electronics; on the other hand, it simply promoted exports of other light industries for which domestic demand hardly existed, such as wigs, electronic parts and components.

This two-pronged strategy was most pronounced in the electronics industry. The Korean government banned the imports of finished consumer electronic products such as radios, monochrome TV sets and so on, and secured the domestic market of these products exclusively for local producers (L. Kim 1980, p. 4; Mody 1990, p. 297). Although most parts and equipment had to be imported, the government attempted import substitution in the assembly of these electronic products in the beginning and anticipated that this would later be extended gradually to import substitution of parts and equipment through backward linkage effects.

However, other segments of the electronics industry were promoted as export items mainly for foreign currency earning without linkage with domestic demand. For instance, electronic calculators, tape recorders and electronic digital watches were initially assembled exclusively for the export market. Most production equipment and intermediate materials for these items were imported and they were again exported after being assembled by the low-cost labour in Korea. The insulation from domestic demand and supply was most visible in electronic parts and components such as semiconductors, capacitors, resistors and so on, for which the ratio of export to total production remained around 82 per cent even until 1972 (Suh 1975, p. 112, Table 6). Semiconductors were the most extreme case. In 1971, nearly 100 per cent of materials and equipment were imported and 95.9 per cent of assembled products were exported (Shin 1996, Table 4.1).

This was because the electronics industry showed a marked trend towards spatial separation of production processes and thereby led the spread of the global production networks (GPNs) from the 1960s. For latecomers like Korea, taking part in the expanding GPNs was an easier way to start production of electronic goods because it was enough for it simply to provide low-cost labour while it received initial capital and technologies from multinational corporations (MNCs). Other East Asian NIEs also began to participate in the GPNs in a similar way in the 1960s (see Henderson 1989, pp. 49–76). Nevertheless, among them, the degree of protection of the domestic market was strongest in Korea.

Thus seen, the initial stage of Korean industrialisation can be characterised as the process of creating a dual structure in the economy, which consisted of industries (or products) having a linkage with domestic demand and being promoted under import-substitution policy from the beginning, and those of 'enclave' connected to GPNs with little linkage with domestic supply and demand. The world economy later evolved rapidly towards the direction of further spread of GPNs, and economic growth through creating 'enclaves' gained more importance in developmental strategies. Korea's transition towards a high-tech economy also benefited greatly from this spread of GPNs. However, compared with those of other East Asian NIEs, it was more in the direction of increasing linkages between initial enclave industries (items) and those under import substitution due to its EP-cum-IS policy, as will be discussed below (Section 3).

3. TRANSITION TOWARDS HIGH-TECH INDUSTRIES

The initial industrialisation beginning with light industries was very successful and Korea could have continued its industrial development gradually along with changes in its comparative advantage and by liberalising the economy. This was in fact the advice given to the Korean government by international organisations like the World Bank, and most foreign and domestic economists. This was also the developmental path largely adopted by Taiwan in the late 1960s when it shifted its economic policy regime towards a more liberalised one (see Table 3.1).

However, the Korean state precipitated itself into a vigorous drive towards heavy and chemical industrialisation (henceforth HCI). The state drafted an ambitious HCI programme and encouraged or sometimes even forced selected firms to enter some designated HC industries by providing them with an array of incentives in the form of tax benefits, subsidised loans, preferential access to limited raw materials and so on. In the case of the petro-chemical and steel industries, which would serve as basic supply industries for other HC industries, the state established them directly as public enterprises.

There were certainly important economic considerations in initiating the HCI programme. First, the balance of payment (BOP) problem was worsening throughout the 1960s despite the impressive export growth contributed by light industries because the export growth exceeded that of imports of raw materials, parts, equipment and technologies from abroad.[4] The Korean government regarded the HCI as inevitable for ameliorating its chronic BOP problem.[5] Second, the Korean economy faced the increasing need to create new markets and develop new products due to the rising protectionism in the USA as well as the increasing catching-up pressure from its followers in Southeast Asian countries.[6]

It might have been possible for Korea to focus more on upgrading existing industries and gradually adding new industries to survive competitive pressure and BOP constraints, as Taiwan or Singapore did, rather than forcefully mobilising national resources for establishing new large-scale HC industries in a hurry. It seems that the Korean state's preference for 'independence' played an important role here. For political leaders in Korea, having its own competitive HC industries, just as Japan

Table 3.1 Korea's Road Map for High-tech Industrialisation

Plans	Objectives	Policy Directions
1st FYP (1962–66)	Breaking the vicious circle of poverty Establishing foundations for self-sustaining economic development	Securing energy supply sources Correcting structural imbalances Expanding basic industries and infrastructure Effective mobilisation of idle resources Improving balance of payments position
2nd FYP (1967–71)	Modernising industrial structure Promoting self-sustaining economic development	Laying the foundation for industrialisation Improving balance of payments position Creating employment Raising farm household income
3rd FYP (1972–76)	Realising a self-reliant economy Harmonising growth, stability and equity Achieving balanced regional development	Promoting heavy and chemical industries Improving science, technology, and human resources Efficient spatial distribution of industries Self-sufficiency in food staples
4th FYP (1977–81)	Achieving self-sustaining economy Promoting technology and improving efficiency	Industrial restructuring and promoting international competitiveness Self-sufficiency in investment capital Expanding investment for science and technology Improving economic management and institutions
5th FYP (1982–86)	Improving technology Establishing foundations for price stability and self-sustaining economy Restructuring government's economic functions	Increasing competitiveness in heavy industries Improving financial institutions Eradicating inflation-oriented economic behaviour Overcoming energy constrains

Plans	Objectives	Policy Directions
6th FYP (1987–91)	Industrial restructuring and improving technology Promoting creative potential and initiative	Solidifying foundation for price stability Improving technology Realising balance of payments surplus and reducing foreign debts Promoting market economic system

Source: Adapted from Song (1990, Table 8.1).

had achieved in the 1950s and 1960s, was regarded as an essential element in establishing an independent economy. There was also a security concern in the HCI drive. The lukewarm US response to North Korea's partial military provocations in 1968,[7] and President Richard Nixon's commitment to the gradual withdrawal of the US force in the South made it urgent for Korea to build HC industries to support its own military industry.[8] 'Self-reliance' meant to Korean leaders, especially to President Park, becoming more independent from foreign resources and technologies as well as becoming more independent from a foreign military presence and the consequent interference of the USA in domestic affairs.[9]

High-tech industrialisation in Korea was basically the process of establishing HC industries, which also included the electronics industry.[10] And the Korean state was the mastermind and the director in this transformation. Although the *chaebols* had already been established since the 1950s, they did not grow enough to launch large-scale projects by themselves. In particular, their capability to mobilise capital for these projects was severely constrained, since the domestic capital market was underdeveloped and they did not have international recognition to raise money from the international financial market. The state was the principal agent serving as a financier of HCI because it owned commercial banks, could mobilise financial resources through the tax system, and was the only one that could draw foreign loans at that time. With its nearly omnipotent power over the *chaebols*, the Korean state even forced some reluctant ones to undertake new heavy industries. In this sense, the growth of major *chaebols* in Korea was mainly the result of the state's initiative in the HCI.

As mentioned in Section 2, the Korean state attempted to develop HC industries through private entrepreneurs as far as it could. Owing to its initiative and support for the HC industries, the public share of fixed capital formation has therefore been the lowest among the East Asian NIEs (Shin and Chang 2003, Figure 2.9). Major investments in HCI came from the *chaebols* mostly with loans provided by the banking system, and ultimately by the state. Major flagship industries of Korea, such as shipbuilding, steel, automobile, chemicals and electronics, took off in the 1970s with the HCI drive.

One characteristic feature in Korea's move towards high-tech industrialisation was that the degree of integration among local firms was strengthened greatly, though this kind of integration was also found in Singapore and Taiwan to some extent in their high-tech industrialisation. As mentioned above, many HC industries were developed as backward linkage industries for existing assembly industries whose equipment, parts and materials had previously been imported from abroad. Replacing previously imported items with domestically produced ones is the process of import substitution and that of increasing the degree of integration among local firms.

A pronounced example in this regard is again the electronics industry. Like Taiwan and Singapore, Korea attracted MNC assembly businesses of electronic parts and components by setting up 'free trade zones' in the 1960s and the early 1970s and initially developed them as 'enclave' industries. However, Korea later applied vigorous import-substitution policies on these enclave items by allowing their imports only if 'the user provided evidence of inferior or insufficient local supply' (L. Kim 1980, p. 11), and eventually displaced many MNCs with local enterprises, mostly *chaebol*-affiliated local integrated electronics firms which started their businesses be assembling consumer electronic goods. Once the attraction of low-cost labour dissipated with economic growth and consequent wage increase, free trade zones lost their attraction and many of them were simply closed in the 1970s.

This is in stark contrast to cases in Singapore or Taiwan. Singapore developed by making the whole country as a (an almost) free trade zone and upgrading its 'complementary assets' so that MNCs could remain and expand in the zone. The upgrading of its industries was carried out by upgrading MNC investments, not by replacing them with local ones. Similarly, Taiwan upgraded its free trade zones into science parks and developed its electronics industry largely by providing MNCs with an attractive environment to increase their investments or form joint ventures with local firms, though the growth of local firms was stronger than that of firms in Singapore. Integration of previous enclave items certainly progressed in Taiwan and Singapore as their industries became more complex with economic development, but its degree seems to have been far weaker than that in Korea.

Table 3.2 Major R&D Indicators in Korea

(Billion won, per cent)

	1976	1981	1986	1988	1990
R&D expenditures	60.90	293.13	1,606.90	2,454.10	3,349.80
Funds from government	39.18	121.73	374.30	522.90	651.00
Funds from private sources	21.72	171.40	1,232.50	1,931.20	2,698.80
Government: private (%)	64:36	42:58	23:77	20:80	19:81
R&D/sales (%)	0.36	0.67	n.a.	1.61	n.a.
R&D/GNP (%)	0.44	0.65	1.77	1.94	1.95

Source: Shin (1996).

Another characteristic feature of high-tech industrialisation in Korea was the swift growth of the *chaebols*. Although they were children of the state-led HCI in the 1970s, they rapidly began initiating new large-scale projects from the 1980s. For instance, their foray into the semiconductor industry in the 1980s can be better understood as a result of oligopolistic competition among themselves in spite of the initial scepticism and reluctance of the government to support it (Yoon 1990; Shin 1996). The pace of the *chaebols*' expansion was partly reflected in the phenomenal growth of research and development (R&D) expenditure from the private sector, which increased 128-fold from 21.7 billion won ($24.6 million, $1 = 850 won) in 1976 to 2,698.8 billion won ($3,175.0 million) in 1990. The public sector share of R&D dropped accordingly from 64 per cent in 1976 to 19 per cent in 1990, similar to such decline in Japan in the 1990s (Table 3.2). In the 1980s, the *chaebols* securely established themselves as the major promoters of high-risk projects in Korea although their financing was still heavily dependent on banks which were under the state's control. This also contrasts with the cases of Singapore and Taiwan whose states maintained a leading role in high-risk projects even up to the 1990s.

4. TECHNOLOGY ACQUISITION AND LOCAL CAPABILITY FORMATION IN KOREA

Although Korea vigorously pursued import-substitution policies, the technologies it employed were mostly imported from abroad as it was a latecomer in the industrialisation process. In order for imported technologies to be applied to the actual production process, they should above all be learnt. They also need to be assimilated to local conditions and can be further improved by initiatives of users. Learning, assimilation of and further improvements in imported technologies are an integral part of technology transfer and its success hinges on building adequate local technological capability to do so.

This section deals with three major routes for technology acquisition and local capability formation in Korea:

1. FDI and local capability formation;
2. import of capital goods and 'learning by doing' at the firm level; and
3. expansion of public education.

I shall particularly emphasise the importance of 'learning by doing' when the country pursues a strong import-substitution strategy in building its high-tech industries.

FDI and Local Capability Formation in Korea

The Korean government clearly preferred, if the situation allowed, to use technology and capital under 'national' management, rather than relying on MNCs, and attempted to develop local capabilities which would eventually substitute foreign capabilities, as pointed out above (Section 2). This was not because the government was unaware of the various benefits of FDI, such as investment augmentation, employment creation, effect of industrial upgrading, technology transfer, creation of foreign market and so on. Rather, it was mainly because it perceived that

> [T]here always exists a possibility that [the] objective [of the stable profit increase in parents firms in home countries] will conflict with the overall economic policy objectives of host countries, such as economic growth, price stabilisation, reduction in unemployment, maintaining current account balance, development of science and technology, and so on.(EPB 1981, pp. 13–14)

However, the Korean government did not succumb to the naïve

'dependency view' that was popular in many developing countries prior to the early 1980s. For it is believed that

> [Most of the problems [arising from FDI] can be solved by host countries' effort to increase their own "bargaining power" over foreign firms' and '[m]any successful developing countries through open-door policies are characterised by their governments' strong effort to "squeeze" foreign firms to act according to their national interest ...] (EPB 1981, pp. 64-5)

The government's regulations over FDI had the clear objective of 'squeezing' foreign firms and gaining access to advanced technologies was the most important consideration in doing so.[11] Many policy measures were therefore deployed in order to maximise technology transfer from FDI.

First, there were measures to ensure that the 'right' kinds of technology were acquired on the 'right' terms. The technology that was to be brought in by the investing MNCs was carefully screened and checked to ascertain whether it was not overly obsolete or whether the royalties charged on the local subsidiaries, if any, were not excessive.[12] Second, those investors that were more willing to transfer technologies were selected over the others that were not, unless they were too far behind in terms of technology. Third, local content requirements were quite strictly imposed, in order to maximise technological spillovers to local suppliers.

Last but not least, the Korean government rarely allowed 100 per cent foreign ownership and encouraged/forced joint ventures in the belief that technology transfer is likely to be more certain and thorough when it occurs directly through joint-venture partners than when relying on 'spillovers' from wholly-owned foreign subsidiaries.[13] A government document states:

> [I]t is possible to acquire foreign technologies through technology licensing. However, the main reason why we need joint venture is that foreign firms often do not offer their technologies in the form of licensing, and, even when they agree to technology licensing arrangements, it is impossible to expect that they will provide core technologies unless they take part in the projects as equity partners. Especially in the heavy and chemical industries and electronics industry, where the support from foreign technologies is critical, this need [to extract advanced technologies through joint ventures] becomes more important. (EPB 1981, p. 26)

Despite its overall restrictive nature, Korea's FDI policy was carried out flexibly because the imperative of export promotion posed constraints on it, on the one hand, and because FDI policy was a matter of negotiation with foreign firms, on the other.

Given the existing BOP constraints, the regulations were applied more loosely to FDIs with high export-earning capabilities. The exceptional allowance of 100 per cent foreign ownership in light industries, especially in the assembly industries in free-trade zones, until the early 1970s is a good example of the flexibility of this policy.

If some foreign firms had great negotiating power thanks to their possession of critical technologies or their special access to certain markets, the Korean government gave them special treatment. For instance, the government initially attempted to set up joint ventures with American semiconductor firms such as Fairchild and Motorola when they were building assembly plants in East Asia in the late 1960s. However, it had to give in to these MNCs' demand for 100 per cent FDI as the first exception to the Foreign Capital Inducement Act. This was because, when the country had neither technologies nor markets for this high-tech industry to be used as bargaining chips with transnational corporations (TNCs), allowing an exception was better than strictly applying its FDI regulations, so that it can at least earn hard currency and create employment (Shin 1996, Ch. 8).

The government also relaxed regulations concerning FDI in the electronics industry in 1975 'for the purpose of advancing the technological level' when it stepped up its effort to promote import substitution in the electronics industry. This was because the government understood that, in this technologically dynamic industry where technological frontiers were rapidly changing, the Korean economy would gain more by allowing freer flow of investment from MNCs rather than by strictly restricting them or taking time for negotiation to extract more favourable terms of investment. A similar step was taken in 1983 when the government revised its regulation on foreign capital introduction and designated the advanced electronics sector as 'completely open to FDI' (Shin 1996, pp. 127–8).

Similarly, the local content regulations were applied rather flexibly. Although the regulations in this regard appeared to be stricter than in most other developing countries, once again the targets for localisation were set realistically, so that they would not seriously damage the export competitiveness of the country.

There is no doubt that the contribution of FDI to technology transfer in the Korean economy was significant, as the government regarded it as a major benefit of FDI. Among foreign technologies – disembodied technologies through technology licensing, excluding embodied technologies in the form of importing capital goods – in Korea during

Table 3.3 Foreign Firms' Contribution to Technology Transfer: Cases of Technology Import

Year	Total cases of technology import	Cases of technology imported by foreign firms	Share of foreign firms (%)
1970	85	28	32.9
1971	45	13	28.9
1972	51	17	33.3
1973	64	24	37.5
1974	87	28	32.2
1975	96	24	25.0
1976	127	37	29.1
1977	164	24	14.6
1978	286	54	18.9
Total	1,005	249	24.7

Note: Technology imports here does not include the import of capital goods.

Source: Adapted from EPB (1981).

1970–78, 24.7 per cent came in through foreign-invested companies (Table 3.3). If we consider the fact that these figures were generally higher in more higher-technology industries such as chemicals, oil refinery and electronics, the contribution of FDI to technology transfer could be higher than the average figures suggest.

However, FDI in technology transfer in Korea pales into insignificance when compared with what was carried out by local firms. As Table 3.3 shows, 75.3 per cent of foreign (disembodied) technologies were introduced by local firms during 1970–78. Considering that 100 per cent wholly-owned foreign subsidiaries constituted less than 5 per cent of foreign-invested firms in Korea in the early 1980s and most of them were joint ventures with local firms having majority shares, there should have been local partners' contributions in technology imports by foreign-invested firms. Moreover, as will be discussed in detail later (See Section 4), the import of capital goods, which embodied advanced technologies whose use on production lines generated 'learning by doing', seems to have been the major channel of technology transfer to Korea during its

HCI drive.

How, then, could the 'squeezing' or 'extraction' of MNCs work in Korea? MNCs certainly have options to shift to other countries if they feel overly squeezed by the local governments or local partners. This is why international organisations and liberal economists currently view government regulations over MNCs negatively and call for liberalisation of FDI policy, including abolition of restriction on maximum foreign shares, local content regulations and so on.

To begin with, it should be noted that the global environment for FDI was different. When Korea embarked on HCI in the 1970s, developing countries in general introduced FDI with much caution and MNCs also accepted it as a norm to operate under stringent government regulations. It is mainly from the 1980s that advanced countries opted for substantial liberalisation of cross-border investments and the 'competitive opening' of national economies towards FDIs spread to developing countries. According to the Korean government's own assessment, its foreign investment regime in the 1970s was neither too restrictive nor too liberal when compared with those in other developing countries (EPB 1981, pp. 28–9). The global environment is leniency to government regulations on MNCs should have been a factor that made MNCs more ready to accept the Korean government's policy measures for 'squeezing' them.

However, what seems to be more important is that, even after being squeezed, MNCs could still envisage lucrative profit potential from investing in Korea. One major reason for this would be that the Korean economy had shown high growth potential and offered MNCs a relatively good business environment, including well-trained labour, a pro-business government attitude and so on, thereby offering them the prospect of gaining from further economic growth. As long as high economic growth potential promises a positive-sum game between MNCs and local governments, 'being squeezed' or 'being extracted' could be permitted by MNCs whatever the actual methods may be.

Moreover, most foreign-invested firms coming to Korea with high technologies occupy monopolistic or oligopolistic positions in the domestic market thanks to their superiority or uniqueness in their technologies, brand names, marketing networks and so on. Reflecting this, profit rates of foreign-invested firms were consistently higher than those of local firms in the 1970s (EBP 1981, Table 3.3.1).

Another important factor that made MNCs comply with the regulations of the Korean government was that it was in their own interest to pre-empt the markets for their own products before local players could grow and challenge them. Since Korea was vigorously promoting its own local firms with successful records, they might have perceived that it would be

Table 3.4 The Composition of Technology Transfer to Korea (US$ million, %)

	Foreign Direct Investment	Foreign Licensing	Import of Capital Goods	Total*
1962–66	45.4	0.8	316.0	362.2
	(12.5%)	(0.2%)	(87.2%)	(100%)
1967–71	218.6	16.3	2,541.0	2,775.9
	(7.9%)	(0.6%)	(91.5%)	(100%)
1972–76	879.4	96.6	8,841.0	9,817.0
	(9.0%)	(1.0%)	(90.1%)	(100%)
1977–81	720.6	451.4	27,978.0	29,150.0
	(2.5%)	(1.5%)	(96.0%)	(100%)
1982–86	1,767.7	1,184.9	50,978.0	53,930.6
	(3.3%)	(2.2%)	(94.5%)	(100%)
1987–91	5,635.9	4,359.4	120,952.0	130,947.3
	(4.3%)	(3.3%)	(92.4%)	(100%)
1992–93	1,938.8	1,797	67,152.0	70,887.8
	(2.7%)	(2.5%)	(94.7%)	(100%)
Total	11,206.4	7,906.4	278,758.0	297,870.8
	(3.8%)	(2.7%)	(93.6%)	(100%)

Note: * This does not mean that there are no other forms of technology transfer other than the three listed in this table.

Source: Adapted from Kim (1997, Table 2-1).

a matter of time before local players entered the markets for their products and regarded it as advantageous to abide by government regulations or form joint ventures with local partners. In this respect, it can be said that the Korean case showed that a developing country has greated scope for regulating MNCs if there is a stronger (potential) presence of local players.

The Import of Capital Goods and 'Learning by Doing' at the Firm Level

The predominant form of technology transfer in Korea was the import of capital goods. During 1962–93, $278.7 billion of capital goods were imported while the amount of inward FDI was $11.2 billion and the expenditure on foreign technology licensing was $7.9 billion (Table 3.4). Although these three forms of technology transfer do not exhaust possible ways of technology transfer and import of capital goods also includes that

Table 3.5 An International Comparison of Structure of Technology Transfer (%)

	Period	Korea	Brazil	India	Argentina	Mexico
FDI/GDP	1967	1.7	4.0	3.0	10.4	7.3
	1977–79	3.2	6.4	2.1	4.7	5.6
PFL/GNP	1970–71	0.04	0.20	n.a.	n.a.	n.a.
	1977–79	0.17	0.33	n.a.	n.a.	0.23
MKG/I	1965	13.0	4.6	10.3	5.3	14.5
	1977–79	27.2	8.4	5.6	8.6	11.8

Notes:
FDI: stock of foreign direct investment.
GDP: gross domestic product.
PFL: payments on foreign licensing.
GNP: gross national product.
MKG: import of capital goods.
I: investment.

Source: Adapted from Westphal et al. (1984, Table 6.3)

carried out by foreign-invested firms in Korea, the figures would suffice to show the significant role played by the import of capital goods in Korea's technology transfer. In addition, the import of capital goods accelerated with the launch of HCI in the 1970s. The value of capital goods imported during 1972–81 ($36.8 billion) was 12.8 times that during 1962–71 ($2.5 billion).

If we compare Korea's figures with those of other developing countries, the country's reliance on the import of capital goods for technology transfer becomes more striking. As seen in Table 3.5, during 1977–79, which was towards the end of Korea's HCI drive in the 1970s, the ratio of payments for disembodied technologies to GNP for Korea was 0.17 per cent–a significant increase on that in the early 1970s, that is, 0.04 per cent in 1970–71, but still much lower than that of Brazil (0.33 per cent) or Mexico (0.23 per cent). On the other hand, the ratio of capital goods imports to investment was 27.2 per cent during 1977–79, far higher than that of Brazil (8.4 per cent), India (5.6 per cent), or Mexico (11.8 per cent).

A main reason for this heavy reliance on the import of capital goods was Korea's push for import substitution and building an 'independent' economy. As argued above, the priority of promoting local firms precluded FDI from possibly playing a major role in technology transfer.

The HCI drive also necessitated large-scale imports of equipment and machinery.

According to Keynesian and Marxian views, capital accumulation is a major means for technological progress, even in advanced countries. First, this is because technological progress is 'embodied' in new equipment. 'The use of more capital per worker ... inevitably entails the introduction of superior techniques which require "inventiveness" of some kind' and 'most, though not all, technical innovations which are capable of raising the productivity of labour require the use of more capital per man – more elaborate equipment and/or more mechanical power', as Kaldor (1957, p. 595) emphasises.

Second, capital accumulation provides opportunities for learning, whereby technological progress occurs. This aspect of the interaction should be distinguished from that of embodiment. Embodied technological progress is, by definition, realised when new equipment is *made*, whereas technological progress through learning is realised when the equipment is *used*. In addition, learning may be embodied again in new equipment, or remain disembodied. However, what is important is the fact that, whether learning is embodied or not, it occurs in the process of working with the equipment which is the result of physical investment. This is, in short, the process of 'learning by doing' (Arrow 1962).[14]

Third, new plant and equipment provide the economy with flexibility for better resource allocation. As Nelson (1964, p. 593) points out, '[t]he productive movement of labour from one sector or industry to another is limited by the rate at which new plant is being built in the industry to which labour is moving'. Causation runs in both directions. Differential technological progress across sectors necessitates reallocation of resources. On the other hand, adequate reallocation of resources across sectors facilitates technological progress at the economy level.

We can expect basically similar interaction between capital accumulation and technological progress in developing countries. New capital investment embodies new knowledge, thus providing opportunities for learning. Structural change of developing countries becomes possible only when accompanied by capital investment.

However, the process of embodied technological progress and learning by doing is quite different in developing countries and this seems to have significant implications for their technological progress. In a developed country, learning of the embodied technologies in itself is really a technological progress at the economy level because technological progress has already been achieved when new machines were made. This is a process of diffusion in a very strict sense.[15] In developing countries, however, the learning of embodied technologies, when capital goods are imported from developed countries, is generally the first step in technological progress. In contrast to the situation in developed countries, the learning of embodied technologies itself, which are completely new to them, is critically important for technological progress in developing countries. The other form of technological progress, that is, improvements upon embodied knowledge occurs only after a substantial period of simple learning.

The learning of embodied technologies is often a difficult task because there is usually a big gap between these embodied technologies and the existing level of knowledge in developing countries. This gap certainly opens the possibility that the process of capital accumulation may be isolated from technological progress in developing countries if there is no subsequent learning. However, technological progress can take place by leaps and bounds with capital accumulation if there is adequate subsequent learning.

It seems that Korea was quite successful in achieving rapid technological progress by maximising the import of capital goods, that is, the inflow of embodied knowledge from advanced countries, and learning and improving on it. Here, the 'EP-cum-IS' policy and the state banks–*chaebols* nexus played an important role.

First, one basic technological characteristic of HC industries is the importance of 'economies of scale'. Without achieving minimum efficient scale of investment, it is impossible for firms to attain competitiveness. In order to make incipient HC industries in Korea *internationally competitive*, economies of scale had to be realised at the international level. Of course, the exact calculation of efficient scale was almost impossible.[16] However, the emphasis on economies of scale and achieving competitiveness in the international market generally led to the pursuit of establishing bigger and bigger plants.

A conspicuous example in this regard is POSCO, the state-owned integrated iron and steel company. The earlier stage of its establishment may be described as a struggle to establish bigger and bigger plants. Difficult negotiations with Western and Japanese counterparts were all related to the issue of size, and POSCO eventually succeeded in establishing as large plants as possible.[17] Another example is Hyundai Heavy Industries (HHI), currently the largest shipbuilder in the world, which started its operation by building the then largest shipyard in the world. The government's push for economies of scale was applied to many other industries. As a result, 'by the end of the 1970s, Korea had the largest textile plant, the largest plywood plant ... the largest cement plant, and the largest heavy machinery plant in the world' (Kim 1993, p. 367).

Owing to the small domestic market and the lack of adequate local technological capability in the beginning, many HC industries, which continued to build ever-larger plants during the 1970s, suffered from extremely low-capacity utilisation at the outset.[18] It caused huge resource allocation problems and financial disturbances in the economy. This was the main criticism that neo-liberal economists have concerning the Korean HCI programme.[19] However, this supposed resource misallocation served to create 'technical imbalance' to overcome, and 'put enormous pressure on both the government and private entrepreneurs to upgrade technological capability for survival' (Kim 1993, p. 362). The fact that the Korean economy turned into one primarily led by HC industries from the late 1980s shows that it was able to overcome this technical imbalance and accumulated high-tech capability in the process.

Second, there was another synergy effect of combining EP with IS. The fact that IS industries were also compelled to export forced them to introduce as much new equipment and technologies as possible, and to keep up with the changes in frontier technologies in order to gain and maintain a competitive edge in the international market. This would have contributed to an increase in the total volume of technology transfer from advanced countries, thus enhancing the potential for technological learning.

This trend was conspicuous especially in the process industries. Y.B. Kim (1980, pp. 226–31) attributes the rapid development of the chemical textile industry to its 'inherent characteristics ... which allow technological innovations to be continuously introduced in order to achieve quality and productivity improvement [to maintain export competitiveness]'. Similar cases could be found in the steel and cement industries. Whenever POSCO expanded its capacity, there arose the so-called 'controversy of localisation of steel-making equipment'. However, POSCO did prefer a gradual upgrade of localisation to a rapid one in order not to compromise its competitiveness (POSCO 1989, pp. 250–52). The cement industry, a typical IS industry, maintained continuous foreign technological assistance rather than turning entirely to local sources of technology, largely because it tried to remain close to the global frontier in process technology (Westphal et al. 1984, p. 197).

The policies of the Korean government also reflect this trend. Despite strict regulations on overall imports, those related to export activities have received exceptional treatment. Capital and intermediate inputs used in export activities could be imported without tariffs and quotas applied to imports for other purposes. The share of capital goods imports to gross domestic investment in Korea was thus one of the highest among major developing countries. This was mainly because the Korean government realised that, without importing advanced capital goods, it was impossible to gain and maintain international competitiveness of its HC industries.

Third, the state banks–*chaebols* nexus also functioned as a mechanism to facilitate technological learning at the firm level. For technological progress actually to occur, continued financing for import of capital goods should be maintained, while learning of and improvement upon the technologies embodied in those capital goods should take place. The state banks–*chaebols* nexus helped increase the flow of capital goods imports by augmenting investment funds. 'Patient money' from the banking sector enabled the corporate sector to have long-term views and adequate time for technological learning to occur in large-scale investments with a long gestation period.

The diversified structure of the *chaebols* also had advantages in these long-term large-scale projects by exploiting 'economies of scope'. Through connected shareholding and internal transactions, the *chaebols* could increase investment funds without actually putting in new money. The structure of the business group also ensured that other resources such as technologies, engineering skills and marketing capabilities were utilised through intra-group transfer among member firms (for details, see Shin and Chang 2003, Ch. 2).

In this respect, the Korean case shows the importance of human capital development at the firm level. The major pressure to acquire, learn and improve upon advanced technologies was placed on the *chaebol* firms, and, for their own survival, they invested vigorously in reducing the time taken for learning imported technologies and their own human capital development. FDI-led industrialisation certainly requires this kind of technological learning at the firm level. Without building and upgrading this 'complementary' capability on the side of FDI recipients, MNCs do not have the incentive to stay or increase their investments in those countries.

However, the intensity and breadth of technological learning seem to be stronger and wider in the case of industrialisation with local actors under the pressure of international competition. Local players not only have to acquire and learn peripheral technologies, but also develop and master 'core' technologies, as they are in direct competition with forerunners. On the contrary, most important technologies in the FDI-driven growth are provided by MNCs and recipient countries can concentrate on working with 'peripheral' technologies, many of which were also transferred from MNCs.

This difference is partly reflected in R&D investment. Compared with Singapore or Taiwan, equally successful East Asian catching-up economies that have relied more on FDI than did Korea, Korea's capability to invest in R&D is more conspicuous. From the 1980s, Korea constantly recorded a higher ratio of R&D expenditure as a proportion of GDP. The difference was bigger when it came to private-sector R&D, given the relative lack of risk-taking capability in the private sectors of Taiwan and Singapore (See Figure 2.7 in the chapter by Shin and Chu in this volume).

Expansion of Public Education

In many studies on Korea, the country's relatively high level of public education at the initial stage of economic development is often pointed out as an important factor accounting for its subsequent industrial success (for instance, see Ranis 1995; Booth 1999). As the concept of 'learning to learn' signifies, a country is certainly better in learning imported technologies if it is already equipped with higher learning capability. Some international comparative studies indeed confirm that Korea's level of social development, especially that of educational development, was far higher than that of other developing countries. For instance, Harbinson and Meyers (1964) show that '[w]hen the per capita income of Korea reached $107, its educational level was equivalent to that of countries with a per capita GNP of $380'.[20]

However, it would be misleading to emphasise only this initial condition as the major factor contributing to Korea's successful local capability formation. What is distinctive in Korea was a further and rapid human educational development *in the process of* economic growth. As Tables 3.6 and 3.7 show, Korea's initial level of educational attainment was in fact not markedly higher than those of other developing countries. For instance, many developing countries in East Asia (for instance, the Philippines and Malaysia) and Latin America (for instance, Brazil and Chile) reached similar or even higher levels of primary school enrolment compared with that of Korea in 1960, which was 94.0 per cent. Korea's secondary school enrolment ratio in 1960 (27.0 per cent) may look higher than those of other developing countries, but was not significantly higher than that of the Philippines (26.0 per cent) or Chile (24.0 per cent). In terms of tertiary school enrolment ratio, Korea was only a medium-level performer in 1960.

What distinguishes Korea, as well as other East Asian NIEs, from other developing countries seems to be the rapid expansion in post-primary education. The secondary school enrolment ratio in Korea reached 78 per cent in 1980 while it was only increased to 48 per cent in Malaysia, 64 per cent in the Philippines, 33.5 per cent in Brazil and 52.5 per cent in Chile. Tertiary school enrolment expanded especially in the 1980s and

Table 3.6 A Comparison of Human Resource Development in Developing Countries: Primary School Enrolment (per cent, gross enrolment)

	1960	1970	1980	1992
Korea	94.0	103.0	110.0	103.0
Taiwan	..	..	101.0	101.0
Singapore	111.0	105.0	108.0	103.0
Indonesia	71.0	80.0	107.0	114.0
Thailand	83.0	83.0	99.0	97.0
Malaysia	96.0	87.0	93.0	98.0
Philippines	95.0	108.0	112.0	110.0
Brazil	95.0	..	97.8	108.6
Mexico	80.0	104.0	120.4	114.2
Chile	109.0	107.0	109.0	99.0

Source: World Development Indicators, Statistical Yearbook 1995, the Republic of China (Taiwan).

Table 3.7 A Comparison of Human Resource Development in Developing Countries: Secondary and Tertiary School Enrolment (per cent, gross enrolment)

	Secondary				Tertiary			
	1960	1970	1980	1992	1960	1970	1980	1992
Korea	27.0	42	78	91	5	7.4	14.7	43.6
Taiwan	..	..	80.3	95.6	..	..	17.9	42
Singapore	32.0	46	60	67	6	6.1	7.8	23.1
Indonesia	6.0	16	29	44	1	2.5	3.6	9.3
Thailand	13.0	17	29	37	2	3	14.7	18.7
Malaysia	19.0	34	48	58	1	..	4.1	9.2
Philippines	26.0	46	64	76	13	16.8	24.4	26.5
Brazil	11.0	26	33.5	42.8	2	4.7	11.1	10.9
Mexico	11.0	22	48.6	53.9	3	..	14.3	13.6
Chile	24.0	39	52.5	69.4	4	9.1	12.3	24.2

Source: See Table 3.5.

reached 43.6 per cent in 1992 while the corresponding figures were 9.2 per cent for Malaysia, 26.5 per cent for the Philippines, 10.9 per cent for Brazil and 24.2 per cent for Chile.

Similar to the case of 'learning by doing' through the import of capital goods, this expansion in public education *with* economic growth should be given due attention. As the import of higher-end capital goods forced the country to accelerate its learning effort, the move towards higher-tech industrialisation increasingly required engineers, scientists and managers who were equipped with higher education. It is not convincing to say that the opportunity for high growth and transition to a high-tech economy was realised in Korea because it had already expanded public education sufficiently. Rather, it would be more plausible to say that Korea's initiative towards high-tech industrialisation, which required rapid local capability formation, was facilitated by concurrent expansion in public education.

5. CONCLUSIONS

The Korean way of making the transition to a high-tech economy was 'nationalistic'. It minimised the role of FDI and developed its high-tech industries mainly relying on local firms in its pursuit of building an 'independent' economy, although it had to draw heavily on foreign loans and import foreign technologies due to its lack of resources and capabilities in the initial stage. This 'unconventional' path of transition was sustained by policies of 'EP-cum-IS' and institutions of the 'state banks–*chaebols*' nexus. I have put emphasis on the configuration among these policies and institutions: possible inefficiencies of the simple IS process, which might be generated under the wall of protection and subsidies, were substantially reduced because it was combined with EP, which forced new industries to be launched exploiting economies of scale and using advanced equipment, thereby exerting strong pressure for technological learning. The state banks–*chaebols* nexus provided firms involved in high-tech industries with structure and resources to sustain their long-term large-scale projects.

What can current developing countries attempting to foster growth of high-tech industries learn from the Korean experience? One big difference between these developing countries and Korea lies in the fact that globalisation has progressed rapidly since then and the global environment has subsequently changed markedly. For instance, with the coming of the World Trade Organization regime, many government regulations that Korea applied to its high-tech industrialisation – such as tariff protection, preferential subsidies for local industries and restrictions on FDI – are difficult to enforce now, though not impossible. The rapid progress of GPN also widened the scope for developing countries to take part in high-tech industries by forming complementary relations with

MNCs, rather than trying to compete directly with them. The intellectual climate also underwent a change and FDI-driven industrialisation is now regarded as a norm, rather than an exception, among scholars, international organisations and governments.

However, GPN is widespread mostly in light industries in which production processes can be segmented and moved easily to other locations. In many HC industries, such as chemicals and steel, whose production processes are tightly integrated geographically and competitiveness comes from integration, economies of scale still matters and the possibility of latecomers from small segmented production processes taking part in the industries is limited. When a developing country considers entering these industries, the Korean experience still has relevance.

I have also emphasised the importance of local capability formation *during* high-tech industrialisation, not before it. Technology transfer from foreign-invested firms, the importance of which was much less than in other countries, was sought mainly through working with foreign firms in the form of joint ventures. The import of capital goods and the subsequent learning process, which I have argued is the most important form of local capability formation in Korea, was also what happens in the process of working with machines in the firm. I have also pointed out the importance of *expansion* of public education along with high-tech industrialisation.

It will surely require further empirical investigations to determine the relative importance between the initial level of local capability and what is created in the process of high-tech industrialisation. However, it is far-fetched to say that Korea possessed/created the necessary capabilities before it commenced its high-tech industrialisation. The Korean case seems to foster a more dynamic view of local capability formation.

NOTES

1. Song (1990, Table 7.1) and BOK website.
2. We use the term 'export promotion' and 'import substitution' literally, that is, the latter is a strategy to substitute imports with domestic production while the former is a strategy to boost exports through various policy measures. We object to the definition of export promotion widely employed in the neoclassical trade literature, which says that export promotion is a policy setting price incentives in such a fashion as to ensure that the home market does not become more lucrative than the foreign market (for this view, see Bhagwati 1987).
3. There is a trade-off between promoting the export of finished goods and promoting the local capital goods sector. If a developing country is to make competitive products in the world market, it is necessary to import capital goods from advanced countries. However, this may retard the development of the local capital goods industry. On the other hand, if exporters are forced to use local capital goods, the

competitiveness of exporters may be harmed. For a discussion of this make–import decision, see Fransman (1985).

4. From 1961 to 1968 when export-oriented industrialisation based on light industrial goods was well under way, total imports almost doubled total exports (Cole and Lyman 1971, p. 128, table 6.4). This BOP problem continued until the mid-1980s, resulting in heavy accumulation of foreign debts.
5. For instance, Suh Suk Joon, then Deputy Vice Minister of the Economic Planning Board (EPB) recalled that 'the HCI policy was necessary [to solve the BOP problem] because the ingrained nature of the economy was such that the dependence on foreign supplies of intermediate goods was too high, the value-added productivity was too low, and the upgrading of export industries was in an unsatisfactory state' (quoted in Yoo 1990, p.25).
6. The US protectionism in the early 1970s was mostly directed at labour-intensive products, especially at textiles which had been the single most important export item of Korea. The bilateral textile agreement with the US in 1973 was a 'heavy shock' to the Korean business community and the government. Moreover, the so-called second-tier NICs in Southeast Asian countries such as Thailand, Indonesia and Malaysia were beginning their catching-up process in light industries (Yoo 1990, pp. 20-21).
7. North Korea dispatched a group of specially trained soldiers to Seoul in an attempt to take President Park's life in January 1968, and a few days later it seized the USS *Pueblo*, a US navy vessel and its 83-man crew in the open sea (Yoo 1990, p. 19).
8. The Korean state in fact extracted concessions regarding assistance in building its military industry from the US as compensation for the withdrawal of its troops (Nolan 1986, p. 62).
9. President Park explicitly expressed this kind of nationalist vision in one of his speeches: 'Economic resurgence is an integral part of the nationalistic vision of a more independent Korea to become more independent of the United States aid and control and, as an economically stronger and independent entity, more able to deal with North Korea' (Park 1963, quoted in Mason et al. 1980, p. 46).
10. The six strategic industries chosen by the Korean HCI programme were steel, non-ferrous metal, machinery, shipbuilding, electronics and chemicals. According to Kim Ki-Hyung, the Science and Technology Minister during 1967–71, President Park initially paid no attention to the electronics industry, regarding it as a simple 'light industry'. However, the industry was included in the HCI programme thanks to persuasion by technocrats (Joong Ang Daily 1998, pp. 151–6).
11. According to a survey by the Korean government in 1979, local firms rank technology transfer as the most important reason for establishing joint ventures with TNCs. The respondents rank the factors in the following order: (1) technology transfer (35 per cent), (2) securing raw materials and export markets (28 per cent), (3) utilising foreign partners' experience and facilities (21 per cent), (4) supplementing capital (13 per cent), and (5) benefits from the government (2 per cent). If we interpret 'utilising foreign partners' experience and facilities' as technology transfer in the broad sense, technology transfer was the over-riding reason (56 per cent) for setting up joint ventures with foreign firms. See EPB (1981, p. 27, Table 2-1-2).
12. For instance, the Korean government gave priority to technologies that develop intermediate products for capital goods industries and bring about diffusion effects to other sectors. It was also stipulated that royalties should be no more than 3 per cent of sales and contract duration no more than three years for the manufacturing sector (Kim 1985).
13. Consequently, as of the mid-1980s, only 5 per cent of TNC subsidiaries in Korea were wholly owned, whereas the corresponding figures were 50 per cent for Mexico and 60 per cent for Brazil, countries which were often believed to have had much greater 'anti-foreign' policy orientation than that of Korea (Evans 1987, p. 208).
14. While acknowledging the two aspects, Arrow (1962) incorporated only embodied

technical progress in his growth model.

15. Of course, even this kind of diffusion may be regarded as technological progress at the economy level. As more firms adopt new machines, their productivity will rise and thereby technological progress in the economy will occur. However, as far as the content of technology is concerned, diffusion itself cannot be technological progress if there is no improvement in the embodied knowledge or their usage.
16. For a general account of this problem, see Streeten (1959).
17. When the idea of setting up POSCO was first discussed in the 1960s, there was a considerable gap between the Korean government and its foreign collaborators in terms of the size of the plant. Korea International Steel Associates (KISA), an international consortium formed by Western steel-makers, agreed reluctantly to give support in building a plant with a capacity of 0.6 million tonnes in 1968, which was criticised by the International Bank for Reconstruction and Development as too big. By contrast, the Korean government thought that the plant should be producing at least 2 million tonnes, which was the production capacity of the newest furnaces at that time. When POSCO secured collaboration from the Japanese steel makers and banks after the deal with KISA had been aborted, it obtained an agreement from the Japanese that the capacity of the plant should be expanded to 3 million tonnes from the initial 1.03 tonnes within the next three to five years. Moreover, when POSCO was going to build the fourth plant, its president Park Tae Joon persuaded the Japanese that it should have a capacity of 9.1 million tonnes, not the 2.6 million figure that was the basis of the first plan received from Nippon Steel's technical cooperation department. POSCO grew into one of the most efficient steel makers in the world within 10 years of its foundation. There have been many factors contributing to this achievement. Among them, its continuous emphasis on the economies of scale is understood as an important factor (Enos & Park 1988; Amsden 1989; and POSCO 1989).
18. This was serious especially in heavy machinery industries. Capacity utilisation in the fabricated metal products and machinery sector during 1976–79 was a mere 60.6 per cent. See J.H. Kim (1990, p. 34, Table 19).
19. See Y.C. Park (1986), J.H. Kim (1990) and Yoo (1990).
20. Quoted in Kim (1997, p. 62). Adelman and Morris (1967) also find that 'of 74 developing countries, Korea was 60th in 1961 per capita income but 14th in a composite indicator of socio-cultural development' (quoted in Jones and Sakong 1980, pp. 1–2).

REFERENCES

Amsden, A.H. (1989), *Asia's Next Giant*, New York: Oxford University Press.

Arrow, K. (1962), 'Economic Implications of Learning by Doing', *Review of Economic Studies* 29, 155–70.

Booth, A. (1999), 'Initial Conditions and Miraculous Growth: Why is South East Asia Different from Taiwan and South Korea?', *World Development*, 27/2, 302–21.

Bhagwati, J. (1987), 'Outward Orientation: Trade Issues', in V. Corbo, M. Khan and M. Goldstein (eds.), *Growth-Oriented Structural Adjustment*, Washington DC: IMF and World Bank.

Cole, D.C. and Lyman, P.N. (1971), *Korean Development: The Interplay of Politics and Economics*, Cambridge Mass.: Harvard University Press.

Chang, H.J. (1993), 'The Political Economy of Industrial Policy in Korea', *Cambridge Journal of Economics*, 16/2.

Chang, H.J. (1994), *The Political Economy of Industrial Policy*, London: Macmillan.

Enos, J.L. and Park, W.H. (1988), *The Adoption and Diffusion of Imported Technology: The Case of Korea*, New York: Croom Helm.
EPB (Economic Planning Board) (1981), *Oegoogin Tooja Baeksuh* (White Paper on Foreign Investment) (in Korean), Seoul: Government of Korea.
Evans, P. (1987), 'Class, State, and Dependence in East Asia: Lessons for Latin Americanists', in F. Deyo (ed.), *The Political Economy of the New Asian Industrialism*, Ithaca, NY: Cornell University Press.
Fransman, M. (1985), 'Conceptualising Technical Change in the Third World in the 80s: An Interpretive Survey', *Journal of Development Studies*, 21/4.
Harbinson, F. and Meyers, C.A. (1964), *Education, Manpower, and Economic Growth*, New York: McGraw-Hill.
Henderson, J. (1989), *The Globalisation of High Technology Production*, London: Routledge.
Jones, L.P. and Sakong, I. (1980), *Government, Business, and Entrepreneurship in Economic Development: The Korean Case*, Cambridge, Mass.: Harvard University Press.
Joong Ang Daily (1998), *Shilok Park Chung Hee* (Chronicles of Park Chung Hee), Seoul: JoongAng M&B (in Korean).
Kaldor, N. (1957), 'A Model of Economic Growth', *Economic Journal*, 67, 591–624.
Kim, J.H. (1990), 'Korean Industrial Policy in the 1970s: The Heavy and Chemical Industry Drive', KDI Working Paper, no. 9015, Seoul: Korea Development Institute.
Kim, L. (1980), 'Stages of Development of Industrial Technology in a LDC: A Model', *Research Policy*, 9/3, 254–77.
Kim, L. (1985), 'Technology Transfer and R&D in Korea: National Policies and the U.S.–Korea Link', *Korea's Economy*, 1/1.
Kim, L. (1993), 'National System of Industrial Innovation: Dynamics of Capability Building in Korea', in R. Nelson (ed.), *National Innovation Systems: A Comparative Analysis*, Oxford: Oxford University Press.
Kim, L. (1997), *Imitation to Innovation: The Dynamics of Korea's Technological Learning*, Boston: Harvard Business School Press.
Kim, Y.B. (1980), 'The Growth and Structural Change of the Textile Industry', in C.K. Park (ed.), *Macroeconomic and Industrial Development in Korea*, Seoul: Korea Development Institute.
Krueger, A. (1980), 'Trade Policy as an Input to Development', *American Economic Review*, 70, Papers and Proceedings.
Lee, H.K. (1994), *Oegoogin Jikjup Tooja wa Tooja Jungchek* (Foreign Direct Investment and Investment Policy) (in Korean), Seoul: Korea Development Institute.
Lee, K.T. (1991), *Sanup Jungchaek ui Iron gwa Hyunsil* (Theory and Reality of Industrial Policy) (in Korean), Seoul: KIET.
Mody, A. (1990), 'Institutions and Dynamic Comparative Advantage: The Electronics Industry in South Korea and Taiwan', *Cambridge Journal of Economics*, 14, 291–314.
Nelson, R. (1964), 'Aggregate Production Function and Medium-Range Growth Projection', *American Economic Journal*, September.
Nolan, J.E. (1986), *Military Industry in Taiwan and South Korea*, New York: St. Martin's Press.
Park, Y.C. (1986), 'Foreign Debt, Balance of Payments, and Growth Prospects:

The Case of the Republic of Korea', *World Development*, 14/8, 1019–58.

POSCO (1989), 'Pochul 20 Nyunsa' (20 Years' History of POSCO), POSCO: Pohang.

Ranis, G. (1995), 'Another Look at the East Asian Miracle', *World Bank Economic Review*, 9/3, 509–34.

Ranis, G. and Fei, J. (1975), 'A Model of Growth and Employment in the Open Dualistic Economy: The Cases of Korea and Taiwan', in F. Stewart (ed.), *Employment, Income Distribution and Development*, London: Frank Cass.

Shin, J.S. (1996), *The Economics of the Latecomers: Catching-Up, Technology Transfer and Institutions in Germany, Japan and South Korea*, London and New York: Routledge.

Shin, J.S. and Chang, H.J. (2003), *Restructuring Korea Inc.*, London and New York: Routledge.

Song, B.N. (1990) *The Rise of the Korean Economy*, Oxford University Press.

Streeten, P. (1959), 'Unbalanced Growth', *Oxford Economic Papers*, 11, 167–90.

Suh, S.C. (1975), 'Development of a New Industry through Exports: The Electronics Industry in Korea', in Wontack, Hong and A.O. Krueger (eds), *Trade and Development in Korea*, Seoul: Korea Development Institute.

Westphal, L., Rhee, Y. and Pursell, G. (1984), 'Sources of Technological Capability in South Korea', in M. Fransman and M. Kings (eds.), *Technological Capability in the Third World*, Basingstoke: Macmillan.

World Bank (1987), *World Development Report*, New York: Oxford University Press.

Yoo, J.H. (1990), 'The Industrial Policy of the 1970s and the Evolution of the Manufacturing Sector in Korea', KDI Working Paper, no. 9017, Seoul: Korea Development Institute.

Yoon, J.R. (1990), 'Hankook eui Sanup Baljun gwa Gugka: Bandoche Sanup eul Joongsim euro' (Industrial Development of Korea and the State: With Special Reference to the Semiconductor Industry), in Hankook Sahwesa Yunguhwe (ed.), *Hyundai Hankook ui Saangsanryuk gwa Gwahak Gisul* (Productive Capacity, Science and Technology of Contemporary Korea), Seoul: Moonhak gwa Jisungsa.

4. The Political Economy of Taiwan's High-Tech Industrialisation: The 'Developmental State' and its Mutinous Mutation

Yun-Peng Chu

1. THE 'MIRACULOUS' COMEBACK OF THE MANUFACTURING SECTOR IN THE 1990s

The manufacturing production index in Taiwan had followed a continuously growing trend since the 1950s. However, as seen in Figure 4.1, signs of slowing down and peaking off emerged in the late 1980s. At that time, in view of the sharp appreciation of the New Taiwan Dollar (NTD) and the rising labour cost, many people painted a rather pessimistic picture of Taiwan's manufacturing sector: 'The days of growth in manufacturing are gone, and attention must be shifted to services instead'.[1]

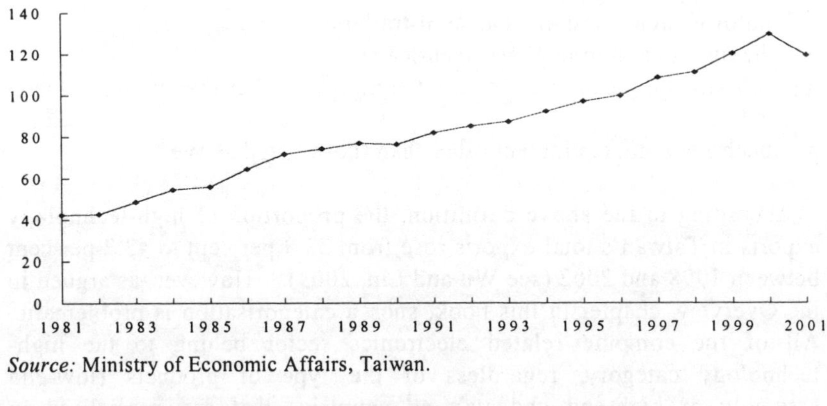

Source: Ministry of Economic Affairs, Taiwan.

Figure 4.1 Index of Manufacturing Production in Taiwan

However, to the surprise of many, only a few years later there emerged an obvious reverse in the peaking-off trend. As shown in Figure 4.1, between 1992 and 1995, manufacturing production started to rebound, and grew at an increasing rate until 2000. What has contributed to this 'miraculous' comeback in the 1990s? Will the trend continue into the future? These are the two major questions addressed in this chapter. This section along with Sections 2–4 discusses mainly the first issue, while the analysis in Sections 5–8 focuses more on the second question. Finally, Section 9 contains the concluding remarks.

A quick answer to the first question is that it was made possible by a surge in technology-intensive exports, which survived the currency appreciation and labour cost hikes. What is meant by 'technology intensive'? There is a two-digit definition given by the Organisation for Economic Cooperation and Development (OECD) (see DSTI, OECD, 2004), and the sectors that belong to the 'high-technology industry' are as follows:

1. aircraft and spacecraft;
2. pharmaceuticals;
3. office, accounting and computing machinery;
4. radio, television and communications equipment; and
5. medical, precision and optical instruments.

'Medium–high-technology industries', on the other hand, include:

1. electrical machinery and apparatus other than those listed in the high-technology category;
2. motor vehicles, trailers and semi-trailers;
3. chemicals excluding pharmaceuticals;
4. railroad equipment and transport equipment other than motor vehicles; and
5. machinery and equipment other than those listed above.

According to the above definition, the proportion of high-technology exports in Taiwan's total exports rose from 38.9 per cent to 43.2 per cent between 1998 and 2002 (see Wu and Lin, 2003). However, as argued in the Overview chapter in this book, such a categorisation is problematic. All of the computer-related electronics sector belong to the high-technology category, regardless of the type of products (low-end assembly or high-end gadgets); so countries that are specialised in electronics exports would display an obviously exaggerated degree of high-technological intensity.[2] For this reason, in that chapter we design and use an alternative index using the proportion of employees with college-or-above education in a sector, to gauge its technological content, and the results for the eight economies considered in this book seem to match other supporting evidence.

Here we shall also use that index to show how the structure of Taiwan's exports had changed during the 1990s. Specifically we divided the HS (the Harmonized Commodity Description and Coding System) four-digit 227 industries in the manufacturing sector into three divisions according to their labour intensity, namely the high skilled (with the highest proportion of employees with college or above education), the medium skilled and the low skilled. In terms of the value contributed by each of the three groups to the total exports of Taiwan, we found that the group with high-skilled labour intensity accounted for 37.8 per cent of total exports in 1993 and the proportion reached 52.7 per cent in 2000. On the other hand, the proportion of low-skilled labour-intensive manufacturing in total exports declined continuously year by year, from 26.7 per cent in 1993 to 12.9 per cent in 2000 (see Figure 4.2).

In the case of South Korea (see Figure 4.3), the high-skilled labour-intensive group accounted for 32.6 per cent of total exports in 1993 and the proportion continued to grow, though slowly, reaching 55.6 per cent in 2000. However, in 1995, the export proportion of the high-skilled labour-intensive group (438.9 per cent) was still lower than that of the medium-skilled labour-intensive group (43.9 per cent). The former did not surpass the latter until 1996. In the case of the PRC (see Figure 4.4) the low-skilled labour-intensive group accounted for 66.7 per cent of total exports in 1993 and the proportion declined to 45.9 per cent in 2000, which was still much higher than that of the other two groups. In fact, the high-skilled labour-intensive group in the PRC accounted for only 23.7 per cent of the total exports in 2000.

From these 227 industries, we further chose those whose export value exceeded US$100 million in 2000, and those which experienced the largest changes in Taiwan's market share in four major markets (US, EU, Japan and the PRC) during 1993–2000. Figures 4.5, 4.6 and 4.7 show the trend of market shares of these major exports from 1993 to 2000. Those with the highest growth rate in market share, that is, growth rate of 100 per cent or above from 1993 to 2000 are shown in Figure 4.5; those with the second-highest growth (50–100 per cent) are given in Figure 4.6, and finally those with the largest decline (with minus growth rate of 60 per cent or higher) are displayed in Figure 4.7.

In the case of Taiwan, most of the products in the high-growth category belong to either the medium-skilled (dashed line) or high-skilled (solid line) labour-intensive industries. In contrast, industries

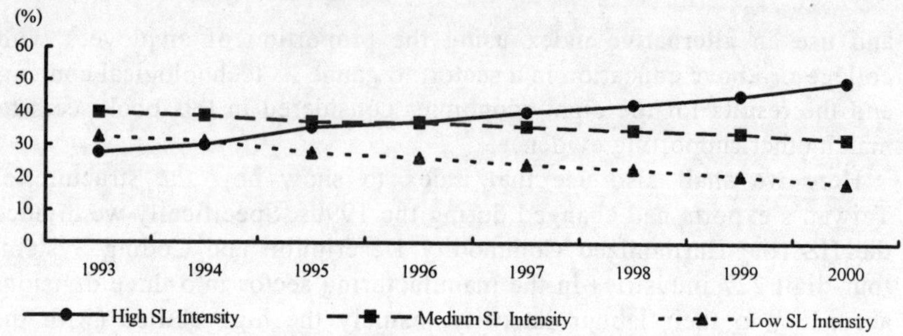

Figure 4.2 Categorisation of Exports by Skilled-Labour (SL) Intensities: Taiwan

Figure 4.3 Categorisation of Exports by Skilled-Labour (SL) Intensities: South Korea

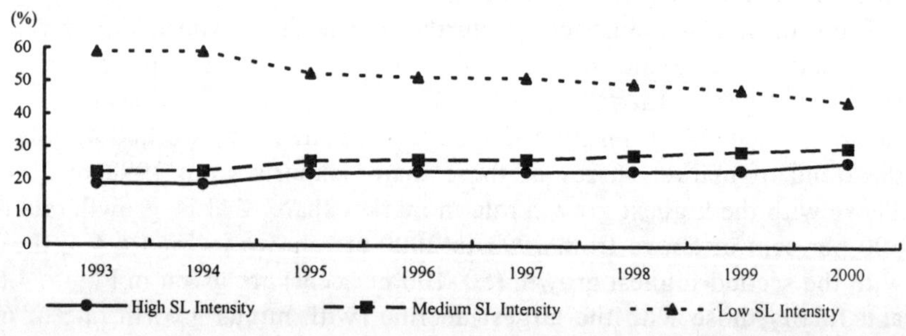

Figure 4.4 Categorisation of Exports by Skilled-Labour (SL) Intensities: Mainland China

experiencing the largest declines in market share were predominantly low-skilled labour intensive.

The 'star industries' belonging to the high-growth group in Figures 4.5 and 4.6 include data-processing equipment, steel surface treatment, steel rolling, secondary steel processing, miscellaneous scientific and optical instruments, glass fibre products, secondary aluminium processing, ship building and repairing, aircraft parts (Figure 4.5), data storage media manufacturing, electronic tube manufacturing, secondary copper processing, semiconductors manufacturing, steel wire manufacturing, paint/lacquer/dye and related products, TV and VCR manufacturing, and petro-chemical materials (Figure 4.6).

In Figure 4.7, industries with obvious declines in their market shares belong mostly to lower-end consumer products. These include leather shoes and other leather products, preserved foods, meatpacking, frozen food, jewellery and related products, watches and clocks, wood furniture and fixtures and miscellaneous wood products, lighting fixtures, and food and beverage production equipment.

Figures 4.8 and 4.9 show the four-country (the PRC, Korea, Taiwan and Japan) comparison of the growth rate of exports of the same category of products as in Figures 4.5 and 4.7 (remember the category is chosen from Taiwan's perspective only).

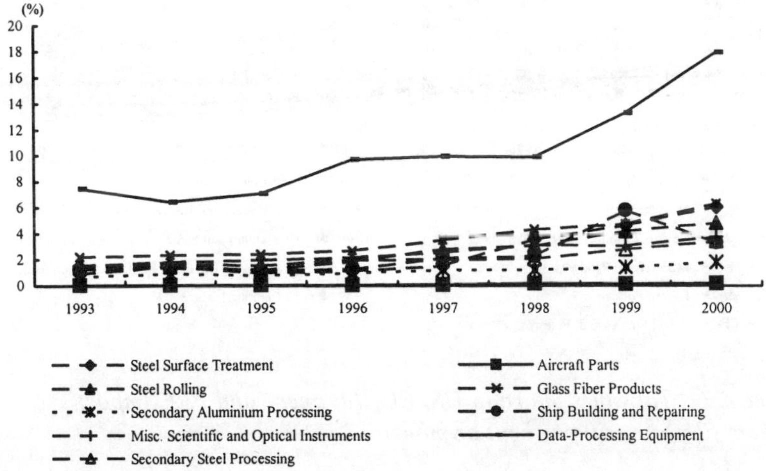

Figure 4.5 Taiwan's Share in US, EU, Japanese and PRC Import Markets (up by 100 per cent or more)

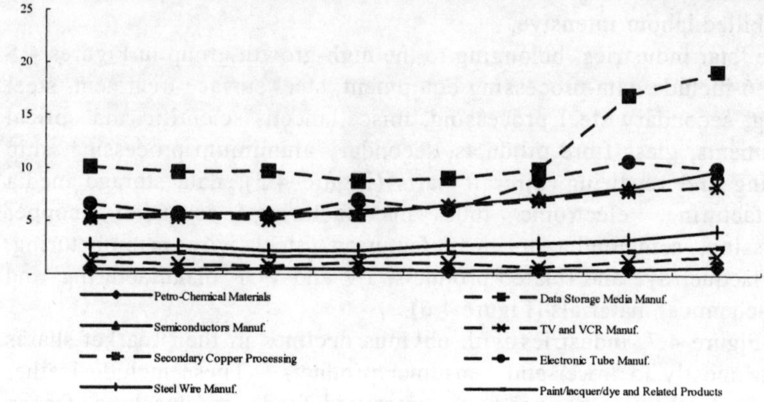

Figure 4.6 Taiwan's Share in US, EU, Japanese and PRC Import Markets (up by 50–100 per cent)

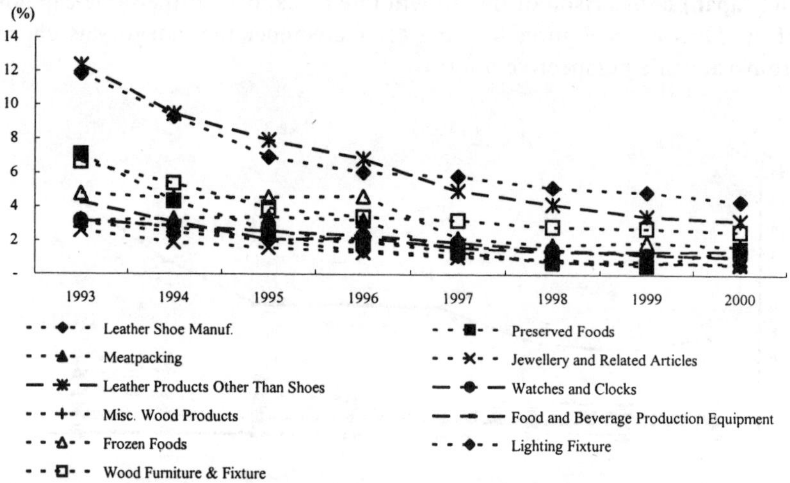

Figure 4.7 Taiwan's Share in US, EU, Japanese and PRC Import Markets (down by 60 per cent or more)

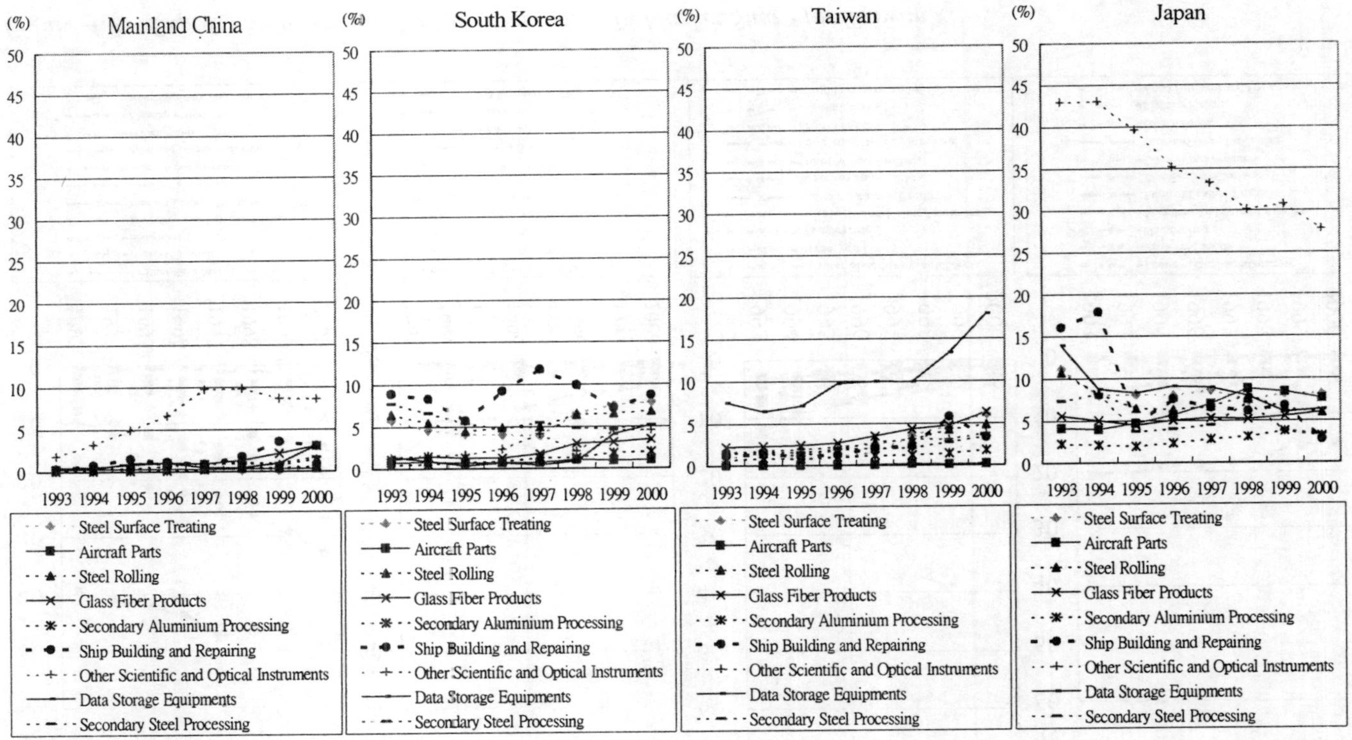

Figure 4.8 Exports Enjoying the Largest Increases in Market Share for Taiwan

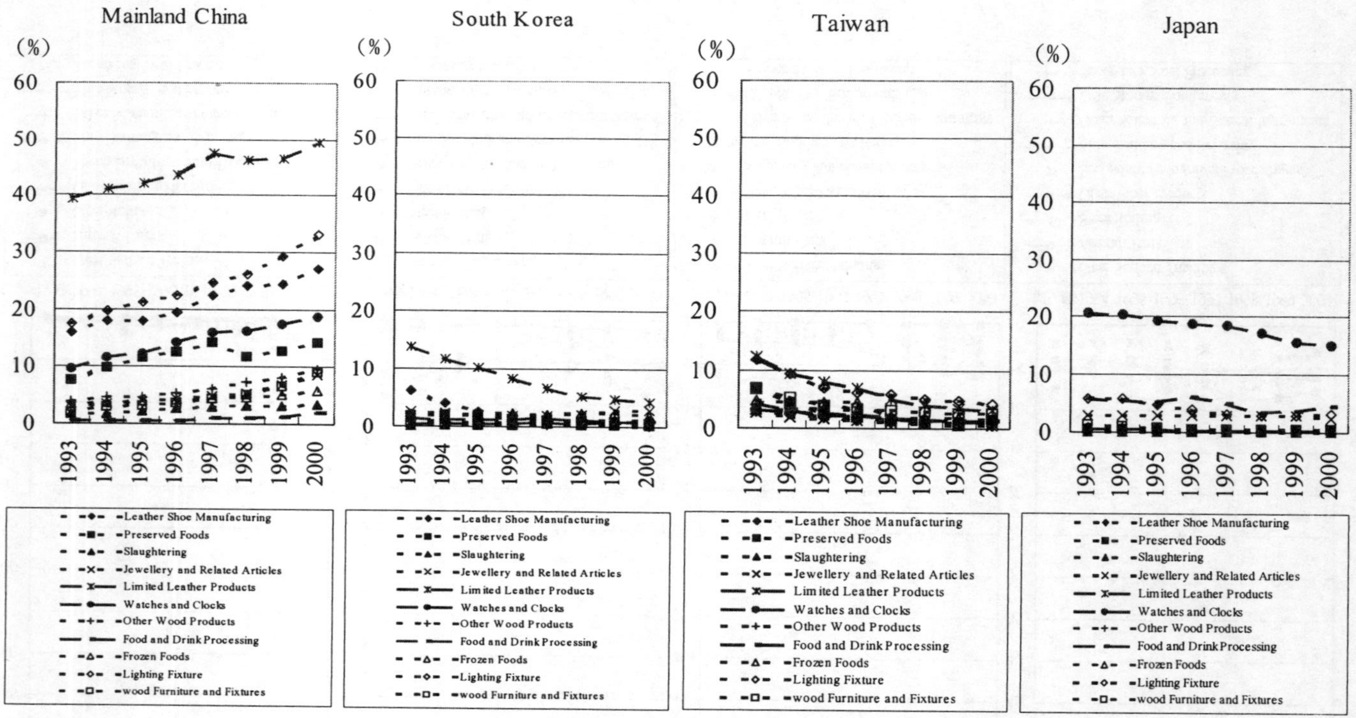

Figure 4.9 Exports Suffering the Largest Decreases in Market Share for Taiwan

It is clear that for most of the products in Figure 4.8, the market shares of the PRC's exports remained small.[3] The pattern in Figure 4.9 is the opposite: all these lower-end consumer products are increasing their production in the PRC, while their scale in Japan, South Korea and Taiwan has shrunk.

2. MAIN CAUSES OF THE COMEBACK: ACTIVE VENTURE CAPITAL, BOOMING STOCK MARKET AND DIVIDENDS TO EMPLOYEES

Section 1 offered a brief answer to the question of why manufacturing made a miraculous comeback in Taiwan during the 1990s. This section will deal with other contributing factors in greater detail.

One possible cause of the rise in skilled-labour intensity of exports was the exodus of capital from Taiwan to other parts of the world with lower labour costs (mainly the PRC), while the remaining (surviving) ones upgraded their equipment and manpower. If so, it was in essence the market, particularly the global movement of capital that held the key to the success in adjustment.

However, the pattern of survival is far from uniform. Although the overall picture pointed to the rise in skilled-labour intensity in Taiwan's exports, from Figure 4.5 it is obvious that the successful grabbing of world export market share is far from uniform among all medium- to high-skilled labour-intensive industries. Why were some so successful, and others less so? Were government policies at work?

If we use the two-digit grouping instead of the four-digit one employed in Figures 4.2–9, the picture of Taiwan's export performance could not be more obvious. By 2003, Taiwan's total exports amounted to US$130.6 billion, of which the largest five product groups in descending order are electronics, information and communication products (hereafter EIC products; 31.4 per cent), base metals and their products (9.9 per cent), precision instruments and miscellaneous equipments not hitherto listed (9.5 per cent), plastics and related products (6.9 per cent), and machinery (6.8 per cent) (see Table 4.1). Among these five, between 1989 and 2003, it was the EIC products sector that registered the highest rate of growth, leaving all other sectors far behind. No doubt this sector was the core of the comeback story of Taiwan's manufacturing in the 1990s.

What has contributed to the miraculous rise in manufacturing exports, EIC products in particular? The history of its development and that of the artificial fibre industry (a bulk part of Taiwan's chemical industry) are presented in greater detail in Chen et al. (2001b) and Chen et al. (200la), respectively. Suffice it is to say that both international

Table 4.1 Major Manufactured Exports of Taiwan: 15 Categories, 1989–2003

	Chemicals		Plastics and Related Products		Leather and Fur Products		Wood and Related Products		Textiles Other Than Garments		Garments		Footwear, Handicrafts and Related Products	
	US$m	Per cent	US$m	Per cent	US$m	Per cent	US$m	Per cent	US$m	Per cent	US$m	Per cent	US$m	Per cent
1989	1,058.3	1.60	4,335.9	6.54	1,493.8	2.25	1,162.0	1.75	6,408.7	7.55	3,946.9	5.95	4,483.7	6.76
1990	1,282.1	1.91	4,429.7	6.59	1,315.9	1.96	962.0	1.43	7,093.9	8.47	3,190.3	4.75	4,116.2	6.12
1991	1,569.8	2.06	5,171.2	6.79	1,262.7	1.66	921.5	1.21	8,478.6	8.86	3,518.6	4.62	4,423.7	5.81
1992	1,769.3	2.17	5,454.9	6.70	1,173.0	1.44	856.1	1.05	8,713.4	8.52	3,128.5	3.84	4,326.6	5.31
1993	1,937.4	2.28	5,693.0	6.69	1,053.2	1.24	785.2	0.92	9,271.4	8.78	2,768.0	3.25	3,335.8	3.92
1994	2,404.1	2.58	6,402.7	6.88	1,287.2	1.38	753.5	0.81	11,461.3	10.09	2,538.1	2.73	2,242.2	2.41
1995	3,238.2	2.90	8,074.7	7.23	1,446.9	1.30	741.0	0.66	13,272.0	9.77	2,350.0	2.10	1,876.0	1.68
1996	3,257.3	2.81	7,713.8	6.65	1,445.9	1.25	686.9	0.59	13,381.9	9.50	2,286.1	1.97	1,656.9	1.43
1997	3,277.6	2.68	7,716.3	6.32	1,456.3	1.19	653.3	0.54	14,209.8	9.62	2,450.7	2.01	1,402.1	1.15
1998	2,857.0	2.58	6,870.1	6.21	1,286.4	1.16	446.7	0.40	12,208.1	9.26	2,337.8	2.11	1,043.6	0.94
1999	3,266.9	2.69	7,524.3	6.19	1,196.0	0.98	413.6	0.34	12,052.8	8.32	2,119.9	1.74	978.2	0.80
2000	4,044.0	2.73	9,055.8	6.11	1,181.5	0.80	397.2	0.27	13,003.7	7.31	2,213.4	1.49	920.5	0.62
2001	4,137.9	3.37	7,992.8	6.51	999.1	0.81	325.3	0.26	10,861.4	7.31	1,768.7	1.44	740.9	0.60
2002	4,667.0	3.57	8,798.6	6.74	956.5	0.73	300.8	0.23	10,606.7	6.63	1,543.0	1.18	651.9	0.50
2003	5,661.4	3.93	9,975.4	6.92	924.9	0.64	298.0	0.21	10,410.6	5.81	1,466.9	1.02	605.1	0.42

	Non-metallic Materials and Products		Base Metals and Their Products		Electronic, Information and Communication Products		Machinery		Electrical Machinery Products		Household Electrical Appliances		Precision Instruments and Miscellaneous Equipment Not Hitherto Listed	
	US$m	Per cent	US$m	Per cent	US$m	Per cent	US$m	Per cent	US$m	Per cent	US$m	Per cent	US$m	Per cent
1989	1,142.5	1.72	5,192.2	7.83	12,562.1	18.95	4,668.1	7.04	2,191.4	3.31	1,096.1	1.65	1,341.6	2.02
1990	1,103.8	1.64	5,215.2	7.76	12,748.9	18.97	5,776.9	8.59	2,207.7	3.28	903.5	1.34	1,494.0	2.22
1991	1,162.0	1.53	5,805.9	7.62	13,772.1	18.08	6,794.7	8.92	2,538.6	3.33	1,006.0	1.32	2,187.5	2.87
1992	1,208.5	1.48	6,464.5	7.93	15,105.7	18.54	7,533.0	9.25	2,783.6	3.42	1,034.8	1.27	3,233.2	3.97
1993	1,084.0	1.27	7,133.1	8.38	16,698.7	19.62	7,045.6	8.28	3,217.9	3.78	874.9	1.03	5,637.6	6.63
1994	1,041.8	1.12	8,074.9	8.68	19,124.7	20.55	7,207.5	7.75	3,446.6	3.70	825.7	0.89	7,205.6	7.74
1995	1,089.4	0.98	10,026.8	8.98	26,157.2	23.43	8,349.7	7.48	3,998.6	3.58	875.8	0.78	9,447.8	8.46
1996	1,032.5	0.89	10,245.7	8.84	29,177.5	25.17	9,482.3	8.18	4,224.5	3.64	890.1	0.77	9,952.8	8.58
1997	1,129.1	0.92	11,530.6	9.45	32,465.7	26.59	9,651.5	7.91	4,767.5	3.91	828.1	0.68	11,278.0	9.24
1998	927.2	0.84	10,870.1	9.83	30,658.4	27.72	7,807.3	7.06	4,351.4	3.93	663.8	0.60	11,793.1	10.66
1999	935.3	0.77	11,606.9	9.55	36,974.2	30.41	7,921.1	6.51	4,601.1	3.78	690.9	0.57	13,974.6	11.49
2000	1,034.8	0.70	13,518.6	9.11	51,235.6	34.54	9,665.5	6.52	5,371.1	3.62	650.3	0.44	15,639.1	10.54
2001	883.5	0.72	11,330.9	9.22	39,269.2	31.96	8,348.1	6.79	4,665.1	3.80	574.8	0.47	13,994.3	11.39
2002	922.1	0.71	12,542.7	9.60	41,877.0	32.07	9,258.4	7.09	5,898.7	4.52	555.6	0.43	13,043.6	9.99
2003	974.6	0.68	14,330.3	9.94	45,215.1	31.36	9,864.9	6.84	6,012.3	4.17	526.8	0.37	13,733.2	9.53

Note: Percentages are in proportion to total exports.

Source: Ministry of Finance, Taiwan, R.O.C.

environment and government policies mattered. For the EIC sector, the former study points mainly to the following facts:

1. International movement of capital from the US and Japan initiated the first electronic products assembly industries in Taiwan, and enabled the local engineers and businessmen to learn about the trade. Many engineers who used to work there became employed by newly established local electronics companies who first served as original equipment manufactures (OEM) suppliers to the brand-name owners in such products as radios, calculators, cassette tape recorders, and the like.
2. The new IBM policies of adopting an open system in personal computer manufacturing, that is, the development of the market for 'IBM-compatibles', made possible the shift from Taiwan's mushrooming production of printed circuit boards or PCBs for the illegal (and later stopped by the government) video game machines to IBM compatible PCs.
3. The world demand for computer-related electronic products expanded greatly during the 1990s, thanks to the so-called 'digital revolution'.

All the above three facts have in common was the rising demand for consumer-end electronic products. This is important. As Agrawal et al. (2000) point out, sequencing is vital. According to them, in both Taiwan and South Korea, there was:

> The autonomous pressure that the first generation of exporters put on subsequent entrants ... [The exporters] imposed a strict and credible constraint on domestic import-competing input suppliers ... subjecting the downstream producers to the pressures of the global competition first, and, through this, creating a vested interest in efficiency that puts enough pressure on upstream producers to attain efficiency or go out of business.[4]

For Taiwan's EIC sector, the rising world demand for consumer electronic products first enabled the downstream sector to expand. Then, a local supply network of electronic components emerged autonomously, and all of these newly emerged middle- or up-stream suppliers were subject to the pressure from their downstream customers as the latter was in turn subject to the competition in the international market. A virtuous cycle then developed.

Government policies also contributed noticeably to enhancing the technology of integrated circuit (IC) manufacturing,[5] which later gave rise to the emergence of the foundry service sector in Taiwan. Other related policies included the following:

1. Establishment of the semi-official Industrial Technology Research Institute (ITRI),[6] which was aimed at promoting technologies for use by (and diffusion among) industries.
2. Establishment of the science-based industrial parks which attracted overseas Taiwanese talent to return from the advanced countries to start up business, and at the same time provided the needed infrastructure (hardware), administrative conveniences (software), high-quality manpower supply (major science and technology or S&T oriented university in the vicinity).
3. Granting tax and other incentives to the targeted 'emerging' industries.[7]
4. Promotion of R&D activities in general by granting tax incentives and by directly subsidising research on the development of targeted new products and key components. Under such policies, the total granted tax exemptions/benefits as a percentage of total tax revenue (for all levels of government) reached 1.2, 4.3, 5.4, 3.1 and 4.5 per cent in 1994, 2000, 2001, 2002 and 2003, respectively (see Table 4.2).
5. De-regulation regarding the entry requirement for the listing of high-tech[8] companies, which could then readily draw a large amount of capital from the booming capital market.[9]
6. Promotion of venture-capital activities including tax incentives.
7. Allowing high-tech as well as general companies to give stock dividends to employees and give technical shares to engineers in return for the core technologies they possessed.
8. The promotion of S&T education in general.

Which of the above policies[10] were more important than the others? Items 1–4 were probably important, but certainly not unique to Taiwan. Most other NIEs tried broadly similar formulas although with varying degrees of intensity. Nevertheless, these were probably not necessary conditions. The benefit from the establishment of science-based industrial parks, for example, could not by itself have been necessary, as numerous firms mushroomed in areas outside these parks and became equally successful. In fact, the total number of electronics manufacturing factories in Taoyuan (a one-hour drive from Taipei), which housed only the traditional type of industrial parks with far fewer amenities and subsidies than those enjoyed by factories in the science-based industrial parks, has been even greater than that of Hsinchu (a two-hour drive from Taipei), where the latter was located. The same situation prevails in terms of total employment in that sector.

In comparison, items 5–7 were probably more important and pivotal.

Table 4.2 Tax Exemptions as Percentage of Tax Revenue

Calendar Year	Tax Exemptions Awarded by the Statute for Upgrading Industries (NT$ million)	Tax Revenue of All Levels of Government (NT$ million)	Tax Exemptions as Percentage of Tax Revenue (%)
1994	12,843	1,116,849	1.15
1995	13,985	1,155,201	1.21
1996	17,947	1,132,556	1.58
1997	29,694	1,309,646	2.27
1998	42,598	1,314,044	3.24
1999	51,487	1,279,080	4.03
2000	55,785	1,296,272	4.30
2001	65,308	1,200,278	5.44
2002	36,264	1,190,874	3.05
2003	54,910	1,220,116	4.50

Source: Ministry of Finance, Taiwan, R.O.C.

Taiwan first established its stock market as early as 1962. The size of the stock market rose continuously over the years, but it was during 1988–91 when Taiwan experienced a general asset inflation that the stock market clearly boomed (see, e.g., Chu, 1994). At the height of the 'bubble', the value of listed companies reached NT$6,174 billion or 156.75 per cent of GDP in 1989. This was the best place to draw investment funds from. Given the general high level of prices of most stocks (high price-earning ratios), the cost of capital became negligible.

The Taiwanese electronics companies seized the opportunities to expand. Contrary to the situation in matured developed economies, at that time companies in Taiwan that re-invested the majority of their dividends in new expansions were more welcome in the market, because the newly issued stocks could be sold at prices much higher than their face value. Thus, there emerged a 'virtuous' circle:

1. With the government's blessing, venture-capital companies mushroomed, so there was a large sum of funds to be invested in newly established companies or in ones that planned to expand their operations.
2. With the lowered entry barrier to being listed in the stock market,

many firms became listed, so the venture capital had a profitable exit, which enabled them to invest in new projects.
3. In the booming stock market, the supply of stocks was almost always short of demand, so the cost of capital was very low for firms that wished to expand further.
4. As it was fashionable to issue dividends in stocks to the employees, many talented people were attracted to the electronics sector as the stocks they received as dividends could be sold at high prices, which brought them an income many times that of their regular salaries.

Of course, the virtuous cycle would have ceased to work if the companies could not sell in the international market. Fortunately, this was not the case in the 1990s. The digital bubble did not really burst until 2001, by which time Taiwanese firms had become world-class competitors at least in the areas of OEM and ODM (original design manufactures) manufacturing, and could deal with the blow.

The stock market in Taiwan also burst, at a much earlier date, in 1989. However, before long it regained its strength and continued to provide a cheap and reliable source of funds for investment from computer-related electronics companies. From the viewpoint of these companies, the stock market, which displayed a bullish pattern in most years during the 1990s, generally gave the green light to investment in new equipment whenever it was deemed necessary, in spite of occasional corrections.

The last (eighth) policy item listed above, namely the promotion of S&T education, could also be important. Here too, many other NIEs tried to do the same, but there was a development in Taiwan that could be regarded as quite unique: the big emphasis that the government put on technical training. During most of the 1990s, about 70 per cent of students graduating from K9 were forced[11] to attend vocational high schools, which taught practical knowledge of occupation-specific skills. If they opted for advanced learning instead of getting a job, they could attend the 'technical' colleges or universities, which again featured a curriculum that was meant to be practical. In this way, Taiwan trained a huge number of high-school and college graduates who were trained in skills directly related to the operation of various industries.[12]

The above touches upon the specific case pertaining to the development of computer-related electronics manufacturing. However, many of the factors presented actually followed a long tradition of active government industrial policies, and were part of a much larger picture, which is summarised in Chu (2001a). This is not to say that the government did not respect the market. The international market clearly has had the last say in almost all policy matters. Many domestic markets were established and nurtured, and market signals were in most cases highly respected and carefully monitored. So the market gradually played a larger role in the determination of allocation of resources, while the government gradually played a lesser role. However, during the process, the basic pattern and perceived mandate of the government has been the same: most of the time it tried to do more than the current market would allow, that is, it tried to anticipate what would happen next and did the 'preparations' accordingly, even though the scope of such preparatory tasks became smaller over time because more and more markets matured and the need for additional preparatory actions by the state became more limited.[13] This was the situation most of the time from the 1950s to the 1990s, as we shall explain in greater detail below.

3. THE 'DEVELOPMENTAL STATE' MODEL OF INDUSTRIAL AND TECHNOLOGY POLICIES PRIOR TO 1988

After 1945 when Japan returned Taiwan to China, and especially after 1949 when the Nationalists retreated to Taiwan following their defeat on the Mainland, the Taiwanese government mostly consisted of bureaucrats who had had experience in the administration of the central government in China during the war years (1937–45, Sino-Japanese War). A sizeable group of them, mostly technocrats with an engineering or technical background, formerly worked for the National Resources Committee, which supervised the operations of state enterprises in such industries as petro-chemicals, power generation, sugar-making, paper, cement, coal mining, steel, machinery, rubber, fertilisers, glass and textiles. The resources under their control were staggeringly high, given the extensive involvement of the state in industrial production at that time (see Lin and Chu, 2001). While other parts of the government were often accused of corruption and ineffectiveness, these technocrats were considered relatively clean and able. One group of them followed the Nationalist retreat to Taiwan, and they constituted the chief bureaucratic manpower in charge of economic development for decades to come.[14]

Under the authoritarian rule which characterised the political scene from the 1950s until the late 1980s, political competition among parties was limited, but that among bureaucrats who tried to impress their leaders never stopped as in other governments. The periods of rule by Chiang Kai-Shek (CKS) (1949–75) and his son, Chiang Ching-Kuo (CCK) (1978–88) as well as the earlier years of the Lee regime (1988–2000) saw the faithful commitment of the leaders to economic development along with the requirement that bureaucrats be free from corruption. One bureaucrat after another competed to perform accordingly.

However, sometimes the bureaucrats performed so well that they became popular among the general public. This could represent a potential political challenge, or could be perceived as such, and the leaders would bend the rule of 'meritocracy' in manpower allocation among ministries as a result. One clear example of this was what happened to Li Kuo-Ting (LKT), who became immensely popular during his term as Minister of Economic Affairs (1965–69). In 1965, the then heir apparent to CKS, Premier Chen, who installed people from the National Resources Committee to key positions to promote economic development in the first place, died of an illness. The next potential heir was Chiang's son Chiang Ching-Kuo. However, CCK was still relatively less well-known and experienced in economic affairs. He was promoted to the vice-premiership in 1969, in preparation for taking over administrative power. In the same year, the then popular LKT was removed from the Ministry of Economic Affairs and appointed as Minister of Finance, which he had often criticised as being too conservative in matters of development promotion (meaning resistance to tax exemptions, cheap loans and the like) when serving as the Minister of Economic Affairs.[15] Sun Yun-Chuan (SYC) who was generally regarded as Chiang's own choice became the Minister of Economic Affairs.[16]

In 1976, by which time CCK had become the Premier, a dispute erupted. CCK was unhappy with the way LKT's deputy handled the price of salt, which was under the jurisdiction of the Ministry of Finance, and removed him from office without prior consultation with LKT. Citing reasons of health, LKT resigned, and was appointed as Minister without Portfolio (see Wang, 1995). LKT, who was accustomed to gathering information from industrialists in Taiwan and experts from overseas, perceived the importance of advanced technologies in the coming age. He took a personal interest in S&T-related matters, which was a field relatively far away from politics. His superior, happy with the arrangement, supported him and let him 'play' in that arena.

During his term of office, LKT established a research committee on applied technologies to promote and coordinate the activities among different ministries. Given his prestige, he soon de facto took on many of the responsibilities of the National Science Council (see Chang, 1987). In 1978, at his suggestion, the first national S&T conference was held, which brought forth a comprehensive S&T Development Programme, the first of its kind in Taiwan, which became a roadmap for S&T development for years to come. In 1979, by which time SYC had become the Premier,[17] a Science and Technology Advisory Group was established under the Premier, and several foreign experts were appointed as advisors (see Li, 1997). In the same year, the Institute for Information Industry (III) was established to promote software-related technologies.[18] From then on, the S&T budget, the bulk of which went to the Ministry of Economic Affairs,[19] the National Science Council, and various government-sponsored research institutes including the Academia Sinica, has been substantially increased, in comparison with other items on the budget. The powerful figure of LKT, with the support of both CCK and SYC, was one of the most important reasons behind this.

SYC himself has been a figure at least as important as LKT in technology promotion during his terms as the Minister of Economic Affairs and as the Premier. He noted the Korean experience of a technology research institute supported by the government that was free to hire Korean experts coming back from abroad with market-competitive salaries, and decided to establish the Industrial Technology Research Institute (ITRI), which after some wrestling with the Legislature was approved in 1973 (Yang, 1989, Ch. 11). The development of IC technology was also initiated by SYC at the suggestion of Dr Pan Wen-Yuan who was the head of the research department at RCA (in the US) and returned to Taiwan in 1974 for a visit. SYC accepted the suggestion and two years later the ERSO (Electronics Research and Service Organisation) of ITRI sent a group of people to RCA to learn about semiconductor technology. Robert Tsao, who is currently the head of the famous semiconductor producer UMC (United Microelectronics Corp.), was a member of that team. He helped to establish a 'model factory' at ITRI following the team's return, and that factory was subsequently spun off to form UMC, Tsao being the vice general manager at that time at the age of 33 (*Digital Times Biweekly*, 1 April 2004).

Morris Chang, the head of TSMC (Taiwan Semiconductor Manufacturing Company Limited), who is also very well known, is believed to have been approached in 1982 by a group of people including SYC, who urged him to return to Taiwan to help develop the IC industry (Yang, 1998).[20] Chang could not oblige at that time, but finally came back in 1985 to head the ITRI at the invitation of the then Premier Yu, who succeeded the ailing SYC in 1984. Then in 1987, Chang headed a group that set up the TSMC, which was the second semiconductor firm spun off from the ITRI.[21]

In short, what mattered most was the close cooperation among SYC (and Premier Yu in later stages) and LKT as well as several other important figures such as Dr Hsu, head of the National Science Council, which made many things possible, including the initiation of IC projects, the setting up of the science-based industrial parks, inviting able entrepreneurs to return from abroad, the setting up of comprehensive S&T development and manpower plans, the increase in the S&T (and R&D) budget, and the reasonably successful implementation of the budget.

During that period, political competition was not very intensive at the higher level, unlike the situation in 1969 when CCK suddenly removed LKT from the post of Minister of Economic Affairs. During that time, SYC was completely trusted by CCK, who was ailing; LKT was popular but himself aging and in poor health. There was little to fight for politically. It was actually the rule of meritocracy at work. When such a rule prevailed, technocrats behaved impeccably, which in turn resulted in sensible policies closer to the social optimum (see the argument concerning optimal government intervention versus non-intervention in Chu et al. 2001).

It should be emphasised here that these stories about the key decision-makers were not meant to imply that they were exceptionally talented and able to accurately predict the future course of events. In fact, there were no such things as secret formulas or omnipotent bureaucrats. If one examines the detailed chronology of incidents leading to the decision making and implementation of various technology-related policies, one finds the sequence simple and logical. Excessive insightfulness was not required of the bureaucrats:

1. They were expected by their superiors to promote economic development.
2. They were inquisitive, eager to learn from the world, eager to listen and then to digest the collected information.
3. They enjoyed the support of their superiors and were thus reasonably free from unduly political intervention.[22]
4. They did not engage in fierce infighting with each other, because the power ladder had already stabilised.
5. They were free from corruption (another requirement from the top).

These characteristics remind us of the course of a 'developmental state' in economic growth.[23] Many authors now attribute the existence of such a mechanism to the particular circumstances of late development and nationalism in response to war and imperialism.[24] This could certainly be true, and similar factors were at work in Taiwan as well, as documented in numerous studies.[25] However, in Taiwan, there is a special condition deserving attention. The government took over Taiwan from the Japanese in 1945, and in the process confiscated all assets owned by the Japanese authorities, companies or citizens. As the

Japanese were dominant in economic power during the colonial period, so was the government in the post-War period.[26] The government's assets would dwarf those of the biggest enterprises even today, after years of 'privatisation' measures. It has been the biggest shareholder and landowner, and controls the largest enterprises of the nation.[27]

Specifically, even by as late as 2003, the government still controlled a large number of enterprises, as Table 4.3 indicates, ranging from power, telecommunications, water supply, petro-chemicals and steel, to banking and insurance, and transportation. The total sales of state-owned enterprises (with the state's share exceeding half) amounted to NT$ 2,774.9 billion, while those of the state-controlled ones (with the state's share less than half but de facto controlling) amounted to NT$ 453.8 billion, both in 2003. This was much larger than the first largest private enterprise in Taiwan, FOXCONN, whose sales were NT$ 328 billion in the same year. Through its ownership of Taiwan Sugar Co., which established itself by confiscating the major sugar companies in colonial times, the government has been the largest owner of arable land in Taiwan.

Given its size and the accompanying influence, the government has been the key to economic development for better or for worse. By the late 1980s, the records show that it was mainly for the better, thus deserving the status of a developmental state as indicated above.[28] The situation would be very different in later periods, as the subsequent analysis will demonstrate.

4. THE POLITICAL ECONOMY BETWEEN 1988 AND 2000: A TRANSITION HEADING ULTIMATELY FOR POPULIST POLITICS

After 1988, when CCK died of illness and Lee Teng-Hui (LTH) became the acting president, the major mechanism of the development-related policy-making process remained largely the same for a few years. Nevertheless, important changes were under way.

First, since 1987, a major political trend has emerged, demanding more political and economic and liberalisation (see Chu, 2001b). The Nationalist Party's traditional power core dominated by Mainlanders who followed the government to Taiwan in 1949 began to be eroded. Entry barriers to various domestic market-oriented industries or services such as insurance, banking, security brokerage houses, newspapers, inland transportation and so on were lifted one after another. The US demand to let American insurance companies enter Taiwan was used as a trigger by legislators to open up all markets in the financial sector, and

Table 4.3 Revenue of State-owned and -controlled Enterprises in 2003

State Owned		State Controlled	
Name	Revenue (NT$ billion)	Name	Revenue (NT$ billion)
Taiwan Railway Administration	19.86	China Petrochemical Development Corp.	23.65
Aerospace Industrial Development Corp.	9.41	China Steel Corp.	160.00
RSEA Engineering Corp.	31.15	Taiwan Fertilizer Co.	7.20
National Health Insurance	356.01	Mega Financial Holding	22.96
Central Deposit Insurance Corp.	4.25	Farmers Bank of China	14.40
Labour Insurance Bureau	213.73	Central Reinsurance Corp.	11.57
Taiwan Sugar Corp.	36.75	Yang Ming Marine Transport Corp.	71.11
Taiwan Water Supply Corp.	23.58	Bank of Kaohsiung	6.15
China Shipbuilding Corp.	14.63	Chang Hwa Bank	43.61
Export–Import Bank of the Republic of China	3.69	Hua Nan Commercial Bank	10.51
Central Trust Bureau	82.10	First Commercial Bank	10.47
Tang Eng Iron Works Co. Ltd.	18.43	Taiwan Business Bank	30.32
Land Bank of Taiwan	53.56	Taiwan Fire & Marine Insurance Co. Ltd.	8.31
Taiwan Cooperative Bank	56.44	Taiwan Navigation Co.	2.44
Bank of Taiwan	77.16	Taiwan Life Insurance Co.	31.08
Taiwan Tobacco & Liquor Corp	69.50		
Chinese Petroleum Corp.	464.49		
Chunghwa Post Co	367.77		
Taiwan Power Co.	339.65	Sub-total	453.78
Chunghwa Telecom Co.	181.35		
Central Bank of China	330.94		
Others	20.45		
Sub-total	2774.90		
Total: NT$ 3228.68 billion			

Source: Ministry of Audit (National Audit Office), Taiwan, R.O.C.

subsequently all other domestic markets. Regulations barring the outflow of capital were largely relaxed by the government as an emergency measure to remove market pressure for the appreciation of the NTD, which ultimately were ineffective.

In the political arena, because of the elections the competition between the Nationalists and the main opposition, the Democratic Progressive Party (DPP) became more important, as the Legislative Yuan reforms gradually removed the non-elected members. In the process, LTH himself promoted direct election of the president, rocking the then existing constitutional design. These occurrences gave rise to populist politics. In 1996, LTH himself became the first directly elected president in the history of the Republic of China (Taiwan's official name).

Most of these events did not, however, affect the S&T policies, although the decision makers were gradually replaced. Lee Yuan-Tze, a Nobel laureate in chemistry and a 'Taiwanese',[29] was invited by President LTH to come back to Taiwan in 1994 to head Academia Sinica, the largest non-military research institute directly under the government. His influence in S&T policies gradually increased. Although towards the end of the LTH administration, he and LTH were reputedly not always on very good terms, these occurrences, even if true, would most likely have been minor incidences, and in general Lee trusted and respected LTH in S&T policies.

With this change in the allocation of decision-making power surrounding S&T policies, the tradition of active promotion of S&T and of technology-intensive industries remained intact. Like SYC and LKT, Lee also enjoyed both high popular reputation and political power, and like them was therefore able to ensure that the S&T policies were free from the possible influence of potential rent-seekers. With his knowledge of developments in the west, he was able to introduce many new and advanced technologies into Taiwan.

Let us then review again the five conditions given above for a reasonably sensible policy-making environment under which the bureaucrats in charge operated: the mission of promoting development, eagerness for updated information, freedom from politics, little infighting, and little corruption. All five remained true to a large extent, although for different reasons.

First, economic development was still desired by the government, but it has been increasingly obvious that short-term political considerations were more important. However, given the high prestige of LYT, such considerations seldom reached the arena of S&T policy making or budget allocation, although sometimes it did influence the deployment of personnel in such a way that it could adversely affect bureaucratic morale.

The second (eagerness to acquire new information) and third (freedom from daily politics) conditions were largely still true, and for the same reasons. The fifth condition (little corruption) also remained intact. As for the fourth condition (little infighting), it was again true, but there was an added aspect to the story already told. Given LYT's prestige and popularity, it occurred to some people that he could have taken the post of premier or even decided to run for the presidency. However, because by the time LYT returned from the US, LTH was already firmly in power, and would soon become the directly elected president enjoying both popular support and power, he could not have become a direct threat to LTH. So the trust and respect was sustained. Moreover, it was reported that LTH once asked LYT to be the premier but he declined. Such a move further cleared him for possible political entanglements.[30] As a result, like LKT (in his later years) before him, LYT became a popular figure but posed no threat to the regime.

However, some problems were brewing among the various organisations implementing the S&T budget, which was allowed to increase at an annual rate of 12 per cent in spite of the general worsening of the government financial situation.[31] The budget allotment process and review mechanism of S&T programmes in Taiwan are as follows. First, the Executive Yuan set aside a rough figure for the total amount of the annual S&T budget according to 'administrative priorities' (in fact political priorities). Then the vice premier coordinates the allocation of this budget among various ministries and councils. Finally, the various ministries and councils draw up their respective mid-term development projects following administrative priorities given the rough preliminary budget.

In terms of budgetary review, after a self-review conducted by the various ministries and councils, all budgets are submitted to the National Science Council (NSC) for primary review (the NSC budget is reviewed by the Science and Technology Advisory Group or STAG). The purpose is to avoid the repetition of programmes or projects and decide on their priority. After primary reviews, the NSC makes a detailed examination of the budgets, focusing on issues such as the rationale behind fund allocation and estimate, the appropriate distribution between basic, applied and technical development expenditures, and building a consensus on new programme starts. Finally, the budgets are submitted to the Council Meeting of the NSC for final review, and then to the Executive Yuan for approval. Although LYT has been an important member of the Council Meeting, and he has personally repeatedly emphasised the importance of true rather than 'business-as-usual' kind of 'formality' reviews, sclerosis has developed within some organisations which seek to honour long-existing programmes without really setting new priorities in response to the changing environment. However, because the total amount of the budget has increased rapidly relative to almost all other government budget categories, new and worthy programmes have had a good chance of being funded, although some wastage and extravagant spending, that is, the 'white elephants', cannot be ruled out.[32]

If things largely remained the same in the sphere of S&T policies, they underwent major changes in the general political environment facing the private enterprises in the second half of LTH's rule, particularly after 1995. In the first half of LTH's term, that is from the late 1980s to the early 1990s, cross-Strait relationships improved greatly and several direct talks were held. Nevertheless, all that changed after LTH's high-profile visit to Cornell University in the US in early 1995. Initially, the US administration was against the visit, but Taiwan's lobbying was powerful enough to generate a sweepstake resolution of support in Congress, and so the visit was realised.

Some time later, cross-Strait talks came to a standstill, and a 'missile crisis' erupted in late 1995 when the PRC trained its missiles around targets near Taiwan. Consequently, the US sent an aircraft carrier to nearby waters. In 1996, LTH began his term as a directly elected president, and soon after his inauguration, he imposed a 'No Haste, Be Patient' policy towards Taiwanese investment in the PRC, by tightening up the provisions regulating such investment, and at the same time promoted investment in areas other than the PRC. The new regulation was effective with regard to the big enterprises, mostly listed companies, but was not really enforceable among the small and medium-sized enterprises (SMEs), given currency convertibility under high freedom of capital flows. Nevertheless, the policy became a source of inconvenience in the globalisation scheme of the private enterprises. This was where political ideology intervened in the business decisions of enterprises. In 1999, LTH's announcement that the cross-Strait relationship was a 'country-to-country' one sent further shock waves to the business community as it attracted fierce adverse responses from the PRC. Here one sees the onset of the end of a pro-development state free from populist politics in Taiwan.

5. THE POLITICAL ECONOMY OF DEVELOPMENT IN THE POST-2000 ERA: STRONGER PRIVATE ENTERPRISES VS. POPULIST POLITICS

In 2000, the opposition DPP party, which had been pro-'Taiwan

Independence'[33] won the presidential election. In theory, this should have had profound effects on the choice of S&T policy makers and possible change in the S&T policies. However, nothing happened. Just before the presidential election day, LYT and a group of well- known scholars and businessmen held a press conference openly endorsing the DPP candidate Chen Shuei-Bian (CSB). Given this controversial and risk-taking move which later proved to be on the side of the winner,[34] LYT's say in the S&T circle became even greater.

Most of the other heads of important S&T organisations such as the NSC, the Minister without Portfolio in charge of S&T policies, and the Minister of Education were politically much less powerful than he was, and some of them were perhaps even appointed on his recommendation. So if it can be said that he gradually became an important figure of S&T policies during 1994–2000, the second half of President LTH's term, he has definitely been *the* key figure after 2000. His influence over the Council of Agriculture, the Council for Economic Development, and the Ministry of Economic Affairs was not as large, but had increased over time. It can be said that the various S&T budget and programmes that received his blessing have been much respected by the regime since 2000. So once again, the key mechanism for and attitude towards S&T policy making did not really change much after 2000. This has been a plus for Taiwan's economy and businessmen.

However, upheavals have occurred in the general environment. Some changes concerning cross-Strait relations had already occurred in the later years of the LTH term as noted above, but they intensified after 2000. Meanwhile, after a decade of intensive development and globalisation, many private enterprises became powerful economically by the year 2000. These two factors would have profound effects on the high-tech as well as general economic development in Taiwan.

The second development will be discussed first. Prior to the 1990s, Taiwan had had only a few large private enterprises that did not rely on the local market for their development and survival, and could somehow determine the direction of Taiwan's industrial development in the sectors they belonged to independent of what the government thought or did. During the 1990s, however, with the rise of high-tech, particularly computer-related industries, Taiwan saw the emergence of such an array of industries. The main players in wafer fabrication such as TSMC and UMC, although still enjoying the many tax privileges available to high-tech industries in general, had their own plans of development which were tuned to the global IC development. They had their own large research departments, which no longer needed much help from the government.

Moreover, most of the Taiwanese firms in the computer-related sectors were OEM or ODM manufacturers. Their scale of production had been quite large in the area of LCDs, notebook computers, DRAMs/DDRs, mobile phones, CDs and digital cameras.[35] However, their survival and development have mainly depended on how their customer companies (the brand-name owners) fared in the world market, rather than on government policies or initiatives. They must be cost-competitive to survive in the world market and that has chiefly been the result of years of own experience and research, rather than government assistance, although they would still benefit from the general fiscal, infrastructure and financial environments of the economy as a whole.

An accompanying development is intensive globalisation and the rise of the PRC (Mainland China) as a main production base for manufacturers. Taiwanese businessmen had invested there for a long time, dating back at least to the late 1970s, when Deng Xiao-Ping initiated the policy of 'reform and open'. Investment on a larger scale took place after 1987, when the Taiwanese government officially allowed residents to 'visit relatives' in Mainland China, and when the NTD experienced sharp appreciation amidst an asset inflation.[36] While many labour-intensive Taiwanese firms migrated to the PRC permanently, still more seized the opportunity to enlarge their scale of production by moving the manufacturing of lower-end products westwards, while initiating that of higher-end products in Taiwan.

That move not only enhanced the scale of production of Taiwanese firms, which became increasingly global in the process, but also shifted the balance in the structure of Taiwan's exports. The triangular pattern of trade, namely Taiwan importing a large amount of raw materials and key components from Japan, assembling them into finished products to be exported to the US (and Europe), gradually evolved into a tetragonal pattern, with Taiwan increasingly becoming a major supplier of raw materials (mainly artificial fibre and other chemical materials, and steel) and key components (mainly computer or telecommunication related) to the PRC, which then assembled them into finished products for export to the US, Europe and Japan.

During the process, the economic relationship between the two sides of the Taiwan Strait became considerably closer than before. According to estimates made by the PRC, by March 2004, Taiwanese businessmen's (realised) direct investment in the PRC had accumulated (from 1979) to US$24.3 billion. Meanwhile, by 2002, the PRC had become the largest market for Taiwan's exports – its share in total exports of Taiwan amounted to 22.56 per cent in that year.[37] The frequency of passenger travel across the Taiwan Strait rose from a trickle before 1987 to

2,871,813 person/times in 2003. It is a fact that the two sides became economically integrated. In the early 2000s, when the domestic economy suffered from a recession and a slow recovery (more on this later), it was generally believed that the double-digit growth of Taiwan's exports had been the only remaining 'engine of growth' for Taiwan. And there is little doubt that the rapid expansion of the PRC's economy contributed to the high growth of Taiwan's exports. So Taiwan's economic growth rate would probably have been even lower without it, other things being equal.

This trend of increasing *economic* integration has, however, been contrary to the direction of cross-Strait *political* relations. When the pro-independence DPP came to power in 2000, initial announcements made by the then newly elected CSB hinted at sticking to the policy of maintaining the status quo, that is, not formally declaring the independence of Taiwan. Nevertheless, the subsequent actions taken by his administration as well as the announcements made by them have become increasingly nationalistic, and anti-PRC.

For the majority of Taiwanese big enterprises, such a situation caught them in the crossfire. They would be suspected of 'helping the enemy' or 'hollowing out Taiwan, going to the Mainland' if they invested in the PRC;[38] while in Mainland China they could be suspected of supporting the pro-independence DPP government and under the threat of adverse actions taken against them. The old trust between the businessmen and the government in Taiwan subsided, and it became harder for the administrators to acquire industrial information from the former, who sometimes had to 'hide' their operations from the government.[39]

Not only that, but the status of development-related agencies inside the administration suffered a general collapse. The Ministry of Economic Affairs used to be a leading ministry whose decisions were highly respected by the other agencies; however, after 2001, its head became increasingly less powerful inside the cabinet and later became almost voiceless.[40]

In terms of actual policies implemented by the government, several of them were particularly disturbing to the businessmen:

1. A shock rocked the business community in late 2000, when the administration decided to reverse the long-made decision of building a fourth nuclear power plant, which was already under construction. That decision, which was later again reversed, costing the public coffer an estimated $1 billion, was attributable to ideological rather than practical reasons, according to most analysts.
2. In early 2001, to allay the fears of businessmen, a 'National

Economic Development Conference' was held. Although the conclusions of the conference called for relaxing the 'No Haste, Be Patient' policy and allowing direct passenger flights between the two sides of the Strait, they were not really implemented by the administration.[41] Some minor changes in the regulations took place, but in reality the influence of the Mainland Chinese Affairs Council increased in the process.

3. In August 2002, CSB made an open announcement that there was 'one country on either side of the Taiwan Strait'. Many analysts said this was at odds with the long-held tacit understanding across the Taiwan Strait that as long as the 'one China in the remote future' principle is adhered to, the status quo can be sustained. Like the 'country-to-country' announcement made by LTH in 1999, this new interpretation created another shock wave in the cross-Strait relationship.
4. At the end of 2003, when the campaigning process was about to be unfolded for the upcoming presidential election in 2004, CSB announced that he would promote the scheme of writing a new constitution by way of a referendum. Reputedly under pressure from the US, the scheme was changed to a referendum on weapon procurement and cross-Strait negotiations. This move put the cross-Strait relationship under further stress.
5. More and more voices in the mass media, most with the government's blessing, openly endorsed the formal announcement of Taiwan's independence, and the mood of Taiwanese nationalism ran high.
6. After his inauguration as a second-term president,[42] CSB sent contrasting signals regarding the cross-Strait relationship. Sometimes he said he was for peaceful negotiations not precluding some form of unification in the future, but at other times would speak the same language as the Taiwanese Independence fundamentalists. There were also moves to change the guidelines for school textbooks on history and geography into something more consistent with Taiwanese nationalism.

As these events and many others displaying the rising influence of populist politics were unfolding, Taiwan's economy suffered a recession during 2001–03. Although that recession could also be attributable to the international environment, the lack of business and consumer confidence was widely suspected to have played a role. The situation is summarised in Table 4.4 below. The table shows the following:

1. Private investment fell to low levels: in constant 1996 dollars: it

dropped from NT$2,282 billion in 2000, when the CSB government was in the process of taking over political power, to NT$1,812, NT$1,773 and NT$1,737 billion in 2001, 2002 and 2003, respectively; with annual growth rates of –20.6, –2.15 and –2.03 per cent for these years, respectively.[43]

2. Inward foreign direct investment also fell sharply.
3. The unemployment rate rose from below 3 per cent prior to 2001 to around 4–5 per cent by 2002–04, while income inequalities rose.
4. Government finance deteriorated, with the total debt of all levels of government rising from 22.0 per cent at the end of 2000 to 35.6 per cent by 2003. Some scholars, after adding some allegedly 'hidden' items including anticipated shortfalls in upcoming social insurance payments, claim that the real figure is two to three times larger.
5. In 2001, Taiwan suffered the first negative growth in GDP since the 1950s. Manufacturing output suffered a 5.73 per cent fall in the same year.

It seems that within a very short period of time, most of the principles behind a state operating in a 'developmental' manner were disappearing, with the narrow sphere of S&T policies being an exception. Johnson (1987, p. 145) describes a structural model of the East Asian high-growth system as consisting of the following elements:

1. stable rule by a political-bureaucratic elite not acceding to political demands that would undermine economic growth;
2. cooperation between public and private sectors under the overall guidance of a pilot planning agency;
3. heavy and continuing investment in education for everyone, combined with policies to ensure equitable distribution of wealth; and
4. a government that understands the need to use and respect methods of economic interventions using the price mechanism.

All four elements have basically either strayed away or been put in reverse gear. It seems that the 'developmental state' has mutated into something different. Although S&T policies have remained largely the same, the overall political environment has become increasingly unfriendly to the globalisation trend. However, the state remained very big economically, as it continued to be the largest shareholder, the largest landowner, and still controlled the largest enterprises. During the developmental stage, such a structure enabled it to act as a powerful engine for growth, but when it became unfriendly to developmentalism, the same structure became instead a powerful threat to growth. This is

Table 4.4 Investment and General Economic Conditions

	Gross fixed capital formation at 1996 prices		Approved inward foreign direct investment		Real GDP annual growth %	GNP per capita US$ (for the last year in period or for the designated year)	Manufacturing gross output annual growth %
	NT$ million	Annual growth %	US$ million	Annual growth %			
1971–80 av.	336,535	11.53	215.92	19.16	9.83	2,344	13.65
1981–90 av.	691,167	6.82	1053.32	22.81	7.96	8,111	7.57
1991–2000 av.	1,727,134	8.51	3131.47	19.43	6.42	14,188	5.38
2001	1,811,585	–20.61	5128.529	–32.59	–2.18	12,876	–5.73
2002	1,773,022	–2.13	3271.747	–36.20	3.59	12,916	6.26
2003	1,737,456	–2.01	3575.656	9.29	3.31	13,156	5.76

	Tax revenue as % of GNP	Unemploymeт Rate	Government budget balance in NT$ million (for the last year in period or for the designated year)	Public debt as % of GDP % (for the last year in period or for the designated year)	Ratio of income share of the highest 20% to that of the lowest 20% among income recipients	Stock price index (annual average for the last year in period or for the designated year)
1971–80 av.	7.67	1.61	–4,303	–	–	547
1981–90 av.	12.91	2.11	–94,269	–	6.61	6775
1991–2000 av.	15.88	2.17	–434,854	27.37	5.79	7847
2001	13.19	4.57	–635,126	32.00	6.01	4907
2002	12.34	5.17	–420,584	33.13	5.87	5226
2003	12.14	4.99	–275,936	35.59	5.67	5162

Source: Directorate, General of Budget, Accounting and Statistics (DGBAS), Taiwan, R.O.C.

the great risk underlining any societies following or intending to follow the model of the 'developmental state'.

What remains to be seen in the future is whether the first trend mentioned in this section, that is, the emergence of large-scale globally oriented Taiwanese enterprises continuously enjoying support from the S&T policies, would be strong enough to sustain changes in the political atmosphere which has been contrary to the economic trend and constituted a major source of deterrence to Taiwan's increasingly strong international economic power. To be sure, most important economic master plans conceived by the government such as building Taiwan into a 'regional operation centre', a 'regional financial centre', a 'regional centre of high value-added manufacturing' and the like, were left in a vacuum, given the intensifying cross-Strait hostility.

If the former trend had a decisive victory in spite of the latter, the future of Taiwan's high-tech enterprises would remain optimistic, because of the continued emphasis on S&T and the increasing independence of the globally oriented enterprises resulting from government policies. If the latter trend were strengthened instead of softened and had a decisive victory in the future, more and more of the globally oriented enterprises would probably have to reassess their corporate strategies and to consider relocating the core of operations elsewhere on a more permanent basis even if reluctantly, despite the continued increases in the S&T budget.

It has been very fortunate that Taiwan has entered a stage of populist politics with the mood of Taiwanese nationalism and anti-globalisation as well as anti-PRC feelings being stirred up by politicians, at a time when many of its enterprises have already become large and global enough to take care of themselves. The pro-active policies of the SYC and LKT eras are no longer needed. However, it remains to be seen whether it would still be too early for Taiwan to enter such a stage as far as economic development in general and high-tech development in particular are concerned: even if many Taiwanese firms have matured, they might not be strong enough to withstand the political wave of anti-globalisation and anti-PRC sentiments, should these become very strong.

6. OTHER CHALLENGES TO TAIWAN'S HIGH-TECH DEVELOPMENT IN THE FUTURE

Against the background of the general environment of political economy described above, we discuss in this section the strong and weak factors that are directly related to Taiwan's high-tech industrialisation. Most of the analysis will be performed by reference to innovation activities and related data.[44]

Distribution of R&D Manpower

Taiwan has an ample supply of researchers. For every 10,000 employees, there were 63.6 researchers in 2001 and the number rose to 68.1 in 2002. This ratio of R&D employees to the number of overall employees was higher than many other countries such as Canada (61.1 in 1999), the UK (55.2 in 1998) and Germany (66.7 in 2001); and much higher than the PRC (10.2 in 2001).

However, most of these high-calibre personnel are not working in the enterprises. Table 4.5 shows the allocation of R&D human resources.

It is true that in 2002, 62.6 per cent of R&D talent was employed by business enterprises, 20.39 per cent by the government, 16.19 per cent by universities or colleges, and the remaining 0.86 per cent worked for private non-profit-making institutes. But for R&D employees with a PhD degree, the percentage of people working for universities or colleges was 68 per cent (9.71 per cent divided by 14.23 per cent in Table 4.5), and only 13 per cent of these worked for business enterprises. Doctoral researchers still preferred to work in universities or research institutions and became increasingly so. The situation has been especially serious during recent years: as higher education capacities increased sharply in Taiwan, the demand for academic researchers with a PhD degree increased tremendously.

R&D Expenditure

Just as there has been an ample supply of R&D personnel, the R&D expenditure of Taiwan also increased considerably: from NT$125,031 million in 1995 to NT$224,428 million in 2002, indicating a 79.5 per cent growth in seven years. In terms of the proportion of R&D expenditure in GDP, Taiwan reached 2.3 per cent in 2002, ahead of many other NIEs but lagging behind Korea, which achieved almost 3 per cent in this regard in 1997 (see Table 4.6). According to the government's 'Knowledge-based Economy Development Programme', the proportion of R&D expenditure in GDP will be further increased to 3 per cent by 2007.

Table 4.7 displays R&D expenditures by sector of performance. The main R&D spender was business enterprises, followed by the government (24.8 per cent), and higher education institutions (12.3 per cent). Compared with that of most other countries, the ratio of government R&D expenditures is too high, implying an overdependence on public research institutions.

Table 4.5 Total R&D Personnel by Sector of Employment and Formal Qualification

	2000	2001	2002	2002 (%)
Business Enterprise Sector				
Scientific Fields				
1. Second-stage tertiary education (ISCED 6)	2,349	2,411	2,631	1.8
2. First-s tertiary education - theoretical (ISCED 5A)	44,313	46,554	53,612	35.7
3. Sub-total university level degrees (ISCED 5A + 6)	**46,662**	**48,965**	**56,243**	**37.4**
4. First-stage tertiary education - practical (ISCED 5B)	31,838	30,588	29,064	19.3
5. Post-Secondary non-tertiary education (ISCED 4)	..	..	..	
6. Upper secondary education (ISCED 3)	11,346	9,264	8,646	5.8
7. Others	..	..	..	
8. Total business enterprises	**89,846**	**88,818**	**93,952**	**62.6**
Government Sector				
Scientific Fields				
9. Second-stage tertiary education (ISCED 6)	3,428	3,510	3,874	2.58
10. First-stage tertiary education - theoretical (ISCED 5A)	13,965	14,860	17,543	11.68
11. Sub-total university level degrees (ISCED 5A + 6)	**17,393**	**18,370**	**21,417**	**14.26**
12. First-stage tertiary education - practical (ISCED 5B)	4,509	4,359	5,230	3.48
13. Post-secondary non-tertiary education (ISCED 4)	..	..	..	
14. Upper secondary education (ISCED 3)	3,238	3,117	3,981	2.65
15. Others	..	..	..	
16. Total government	**25,140**	**25,846**	**30,628**	**20.39**
Higher Education Sector				
Scientific Fields				
17. Second-stage tertiary education (ISCED 6)	12,556	13,350	14,588	9.71
18. First-stage tertiary education - theoretical (ISCED 5A)	7,735	8,123	8,670	5.77
19. Sub-total university level degrees (ISCED 5A + 6)	20,291	21,473	23,258	15.48
20. First-stage tertiary education - practical (ISCED 5B)	885	934	854	0.57
21. Post-secondary non-tertiary education (ISCED 4)	..	..	..	
22. Upper secondary education (ISCED 3)	134	162	211	0.14
23. Others	..	..	..	
24. Total higher education	21,310	22,569	24,323	16.19

	2000	2001	2002	2002 (%)
Private Non-profit-making Sector				
Scientific Fields				
25. Second-stage tertiary education (ISCED 6)	301	289	276	0.18
26. First-stage tertiary education – theoretical (ISCED 5A)	713	604	720	0.48
27. Sub-total university level degrees (ISCED 5A + 6)	**1,014**	**893**	**996**	**0.66**
28. First-stage tertiary education – practical (ISCED 5B)	166	143	181	0.12
29. Post-secondary non-tertiary education (ISCED 4)	..	..	..	
30. Upper secondary education (ISCED 3)	146	140	120	0.08
31. Others	..	..	..	
32. Total private non-profit-making	**1,326**	**1,176**	**1,297**	**0.86**
R&D Personnel Total				
Scientific Fields				
33. Second-stage tertiary education (ISCED 6)	18,634	19,560	21,369	14.23
34. First-stage tertiary education – theoretical (ISCED 5A)	66,726	70,141	80,545	53.63
35. Sub-total university level degrees (ISCED 5A + 6)	**85,360**	**89,701**	**101,914**	**67.85**
36. First-stage tertiary education – practical (ISCED 5B)	37,398	36,024	35,329	23.52
37. Post-secondary non-tertiary education (ISCED 4)	..	..	..	
38. Upper secondary education (ISCED 3)	14,864	12,683	12,958	8.63
39. Others	..	..	..	
40. Total R&D personnel	**137,622**	**138,409**	**150,200**	**100.00**

Notes:
1. Second-stage tertiary education (ISCED 6): mainly those with a PhD or equivalent degree.
2. First-stage tertiary education (ISCED 5): Masters, Bachelors as well as graduates of junior colleges.
3. ISCED 5A: theoretically based tertiary programmes: not belonging to ISCED 5B.
4. ISCED 5B: practically oriented occupation-specific programmes.
5. Post-secondary, non-tertiary education (ISCED 4): holders of post-secondary non-tertiary diplomas other than ISCED 3 (not applicable to Taiwan).
6. Upper secondary education (ISCED 3): mainly high school or equivalent.

Source: Indicators of Science and Technology, National Science Council, Taiwan, R.O.C. (2003).

Table 4.6 Comparison of R&D Expenditure in GDP among Major Countries

Country	Proportion of R&D expenditure in GDP (per cent)							
	1995	1996	1997	1998	1999	2000	2001	2002
Sweden	3.59	3.74	3.85	–	3.65	–	4.27	–
Finland	2.30	2.53	2.73	2.88	3.23	3.40	3.40	–
Japan	2.94	2.99	3.12	2.94	2.94	2.98	3.09	–
South Korea	2.68	2.79	2.89	2.55	2.47	2.65	2.96	–
US	2.61	2.67	2.71	2.60	2.65	2.72	2.82	2.82
Germany	2.30	2.30	2.31	2.31	2.44	2.49	2.49	2.50
France	2.34	2.32	2.23	2.17	2.18	2.18	2.20	–
Denmark	1.91	1.94	1.99	2.06	2.19	–	–	–
Taiwan	**1.78**	**1.8**	**1.88**	**1.98**	**2.05**	**2.05**	**2.16**	**2.30**
Netherlands	2.07	2.11	2.13	1.94	2.02	1.94	–	–
Canada	1.57	1.55	1.54	1.79	1.81	1.87	1.94	1.85
UK	2.02	1.95	1.87	1.80	1.88	1.85	1.90	–
Australia	1.65	1.68	1.71	1.51	–	1.53	–	–
Ireland	1.28	1.42	1.48	1.25	1.22	1.15	1.17	–
Italy	1.01	1.02	1.00	1.07	1.04	1.07	–	–
Spain	0.85	0.87	0.86	0.89	0.88	0.94	0.96	–
Portugal	0.59	0.58	0.65	0.69	0.75	0.79	0.83	0.78

Sources: OECD, *Main Science and Technology Indicators*, 2000, 2001, 2003; and National Science Council, Taiwan, R.O.C., *Indicators of Science and Technology*, 2001 and 2003.

Table 4.7 R&D Expenditures by Sector of Performance

	Business Enterprise	Government*	Higher Education	Private Non-profit
Taiwan (2002)	62.2	24.8	12.3	0.7
France (2001)	62.4	17.7	18.5	1.4
Germany (2002)	70.0	13.5	16.5	..**
Japan (2002)	73.7	9.5	14.5	2.3
Korea*** (2002)	76.2	12.4	10.4	1.1
PRC (2002)	60.4	29.7	9.8	..**
United Kingdom (2001)	67.4	9.7	21.4	1.4
United States (2002)	72.9	7.6	14.9	4.6

Notes:
* : Including institutions funded or established by government agencies, as well as public enterprises.
** : Included elsewhere.
*** : Excluding R&D in the social sciences and humanities.

Source: OECD, *Main Science and Technology Indicators*, 2003,

With regard to the R&D expenditure in different industries (see Table 4.8), the proportion of manufacturing R&D expenditure in total expenditure was over 80 per cent in most major countries, among which Taiwan took the lead with 92 per cent. Moreover, we can see the focus of development of each individual country from the ratio of its R&D expenditure to total R&D expenditure. In the case of Taiwan, the R&D expenditure was relatively concentrated in such ICT sectors as electronic communications equipment, computers, and office automation equipment.

Changes in the Value-added Ratio, Its Contribution to Industrial Growth, and Increases in Total Factor Productivity

In order to understand the factors contributing to the enhancement of manufacturing value added in Taiwan in the last decade, we employed the data obtained from the Factory Surveys conducted by MOEA to answer the following questions: (i) How much of the growth in manufacturing GNP (value added) was derived from the expansion of sales, with value-added ratio being held constant? (ii) How much of it was derived from the increase in the value-added ratio, with sales volume held constant?

Table 4.8 *Structure of Business Enterprises' R&D Expenditure (BERD) of Major Countries, 1995 vs. 2001 (per cent)*

	Service Sector		Manufacturing Sector		Hi-tech Manufacturing		Electronics		Office Machinery & Computers		Aerospace		Instruments		Pharmaceuticals	
	1995	2001	1995	2001	1995	2001	1995	2001	1995	2001	1995	2001	1995	2001	1995	2001
Taiwan, R.O.C.	2.9	7.4	97.1	92.0	51.0	66.0	15.8	18.2	33.8	43.7	0.1	0.7	1.5	2.3	1.2	1.1
Japan	0.2	7.2	96.2	90.6	37.8	40.5	9.0	13.0	17.5	15.3	0.7	0.8	3.8	4.3	6.8	7.1
Germany	3.5	8.4	94.6	90.9	32.7	27.9	3.9	1.9	10.0	10.7	8.1	3.6	6.0	4.9	4.6	6.8
France[3]	7.1	10.6	88.3	85.0	49.2	43.1	2.7	1.3	11.0	12.9	13.2	10.3	10.3	6.4	12.0	12.1
Sweden	10.0	11.5	87.5	87.4	47.5	55.1	1.4	0.8	19.9	28.9	5.1	2.7	6.9	4.8	14.3	17.9
Finland	9.0	12.4	88.0	84.6	42.3	56.4	2.1	0.2	31.2	47.5	0.1	0.3	3.9	2.4	5.1	6.0
South Korea	7.6	12.6	83.3	82.8	37.0	51.4	1.8	7.8	31.6	36.2	1.5	3.8	0.7	1.4	1.4	2.2
Belgium	14.2	13.7	84.1	82.9	32.9	41.7	0.2	0.3	15.3	17.5	1.5	1.8	2.3	1.2	13.6	20.9
UK[3]	18.0	16.6	78.5	80.2	41.2	48.1	1.6	0.9	6.6	8.5	9.7	10.2	3.3	4.0	19.9	24.6
Netherlands[3]	11.5	19.7	82.5	75.9	34.4	39.5	12.1	25.7	11.1	0.3	2.7	0.1	1.7	4.5	6.8	8.9
Italy	10.7	19.9	86.4	79.4	45.3	41.6	4.6	1.1	19.9	18.3	8.7	10.2	2.5	4.0	9.6	8.1
Ireland[2]	9.5	24.6	89.8	75.0	50.9	51.5	5.3	5.1	23.9	30.6	0.4	0.4	5.1	5.0	16.2	10.5
Canada	32.2	26.4	61.4	69.8	42.5	49.4	4.2	4.1	22.5	33.7	8.9	7.1	1.3	2.0	5.5	5.6
USA[3]	21.1	34.4	78.9	64.9	47.9	39.4	6.7	5.2	11.6	12.9	12.8	5.2	9.1	9.6	7.7	6.5
Denmark[2]	31.3	38.9	67.9	64.1	33.2	34.0	0.9	0.8	6.3	4.0	0.0	0.0	6.0	6.1	20.0	23.0
Australia[3]	22.3	39.9	62.4	50.4	17.1	21.4	1.6	1.9	7.4	9.9	0.3	0.1	2.1	2.7	5.7	6.8
Norway[1]	40.5	48.0	40.5	48.0	25.7	21.4	1.5	1.0	15.8	13.5	0.4	0.4	1.8	2.1	–	6.3

Notes: 1: 1998; 2: 1999; 3: 2000.

Sources: OECD, *Main Science and Technology Indicators*, 2004; National Science Council, *Indicators of Science and Technology R.O.C.*, 1995 and 2003.

(iii) How much of it was derived from the structural change in production among various industries?[45]

Specifically, at any period of time, the real value added of all (private) manufacturing industries can be expressed as:

$$
\begin{aligned}
&RVA \\
&= \Sigma_i \ (var_i) \ \cdot \ (s_i) \ \cdot \ (TRS) \\
&= \Sigma_{i \neq k}(var_i - var_k) \cdot \ (s_i) \ \cdot \ (TRS) \ + var_k \cdot TRS \qquad (4.1)
\end{aligned}
$$

where

var_i	=	ith industry's value-added ratio (value-added over sales);
s_i	=	ratio of ith industry's sales in total sales;
TRS	=	total real sales;
i	=	1, 2, …, n; and
k	=	a selected 'benchmark' industry (its *var* is close to average *var* of all industries).

Taking the differences of equation (4.1) yields:

$$
\begin{aligned}
\Delta\ (RVA) = \ & \Sigma_i s_i \cdot TRS \cdot \Delta(var_i) && \text{[value-added ratio effects]} \\
& + \Sigma_{i \neq k}(var_i - var_k) \cdot (TRS) \cdot \Delta(s_i) && \text{[production structure effects]} \\
& + \Sigma_i var_i \cdot s_i \cdot \Delta(TRS) && \text{[total sales effects]} \\
& + \text{non-linearity residual.} && (4.2)
\end{aligned}
$$

We applied the above formula to the data obtained from the factory surveys conducted by MOEA. The years in which a factory survey was conducted during the 1990s belonged to either of the two groups: 1992–95 and 1997–99, as there was no such survey in 1996. We then conducted the following three analyses.

1. Applying the formula to the change in real value added of the private manufacturing between 1992 and 1995.
2. Applying the same formula to the same variable for the period between 1997 and 1999.
3. Integrating and merging the industries of different categories in the above two periods by means of the least common factor across the two periods. Then we take the difference between the first three-year average in the first period (1992–94) and the three-year average of the second period (1997–99), and analyse the contributing factors (see Table 4.9).

Table 4.9 Sources of Increases in Value Added of Manufacturing Industries (private enterprises)

	1992–99		1992–95		1997–99	
	NT$ m in 1996 prices	per cent	NT$ m in 1996 prices	per cent	NT$ m in 1996 prices	per cent
Total Change	848,186.47	100.00	488,391.48	100.00	499,921.93	100.00
Value-added Ratio Effects	261,884.27	30.88	198,367.55	40.62	72,017.52	14.41
Share Effects	5,537.72	0.65	- 2,113.82	- 0.43	- 12,083.03	- 2.42
Output Effects	575,033.57	67.80	294,525. 60	60.31	442,717.05	88.56
Non-linearity Residual	5,730.91	0.67	- 2,387.85	- 0.50	- 2,729.61	- 0.55

Source: Computed by author.

As shown in Table 4.9, when analysing the period as a whole (1992–99), about two-thirds of the value-added growth in the manufacturing was derived from sales increase effects and the rest was from value-added ratio effects. The product mix net effects were relatively small after the positive and negative effects offset each other. When analysing the two periods separately, during 1992–95, total sales effects accounted for around 60 per cent of the value-added growth in manufacturing and the other 40 per cent was derived from value-added ratio effects, while during 1997–99, total sales effects accounted for nearly 90 per cent of the value-added growth in manufacturing; and value-added ratio effects only 14 per cent. This is not a very encouraging finding.

Admittedly such a decomposition analysis is made assuming that such activities as vertical (dis-)integration did not occur strongly, and has to be treated with caution. However, the basic finding that Taiwan's manufacturing has been increasingly dependent on large-scale production with falling profit margins seems to be consistent with observations given by industrial specialists. One recent report on the profit margins of major electronic manufacturers in Taiwan, for example, points to a clear trend of declining margins for almost all of them from June 2001 to June

2004. These include the following producers: ASUSTEK (producing mainly motherboards and notebook computers): the profit margin fell from 23.6 to 16.49 per cent; Quanta (notebook computers): from 13.49 to 5.52 per cent; BENQ (peripheral, digital display, mobile phones and input): from 11.36 to 9.15 per cent; Acer (personal computers or PCs and notebook computers): from 10.69 to 4.7 per cent; FOXCONN (printed circuit boards or PCBs, PCs and mobile phones): from 10.43 to 7.05 per cent; COMPAL (notebook computers): from 9.56 to 6.76 per cent; and Inventec Corporation (notebook computers): from 7.26 to 5.39 per cent (see *Chinatimes*, 19 October, 2004).

Venture-capital Activities

Taiwan's venture-capital industry has been very active. At the end of 2002, there were 217 venture-capital companies in Taiwan, which have cumulatively received NT$151 billion in paid-in capital. Over the years the investment cases accumulated to 7,560 in total and the capital invested at home and abroad amounted to NT$145 billion. If the companies that are not named as such but actually engaged in venture-capital business were included in the calculation, the investment amount would have been much bigger.

In terms of the industries that had received investment from the venture-capital businesses, among those 7,560 investment cases accumulated over the years, the semiconductor industry got the highest share, accounting for 16.1 per cent of the total, followed by the electronics industry (15.5 per cent), the information industry (13.7 per cent) and the telecommunications industry (13.3 per cent). Similar patterns hold in terms of investment amount.

The timing of investment by venture capital can be divided into five stages according to the development level: incubating, initial, expanding, maturing and restructuring. In terms of investment amount, the expanding stage got the highest share, accounting for 45.3 per cent of the total amount accumulated over the years, followed then by the initial stage (24.8 per cent), the maturing stage (21.4 per cent), the incubating stage (7.6 per cent) and the restructuring stage (0.9 per cent) (see Table 4.10).

In spite of its glorious past, however, Taiwan's venture capital in recent years had not been as active as before. One factor is the overall political environment as noted above, that prompted many venture capitalists to seek opportunities elsewhere. The other is the lack of high-calibre R&D personnel in industry, which hampered path-breaking advances in technologies.

Table 4.10 Analysis of Investment Structure of Taiwan's Venture Capital (accumulated as of 2002) (NT$ million)

Industry	Stage					Total	%
	Incubat-ing	Initial	Expandin g	Maturing	Restruc-turing		
Semi-conductors	2,200.46	6,851.15	11,690.45	5,530.57	83.06	26,355.69	18.13
Information	1,994.03	4,207.92	9,929.36	6,639.34	252.10	23,022.75	15.83
Electronics	558.81	3,352.01	11,428.67	5,436.01	451.60	21,227.10	14.60
Telecom	1,907.29	5,008.61	10,047.15	3,012.64	223.63	20,199.31	13.89
Opto-electronics	1,291.10	5,659.80	5,857.08	2,184.02	27.00	15,019.00	10.33
Software	567.74	2,694.33	3,815.81	1,319.11	32.96	8,429.96	5.80
Traditional Manufacturing Industry	109.80	370.91	2,402.59	2,207.09	45.65	5,136.03	3.53
Biotech	561.38	1,465.89	1,894.25	955.26	20.00	4,896.77	3.37
Other Major Technology	234.60	501.88	2,872.15	874.05	28.96	4,511.64	3.10
Venture-capital Industry	992.62	1,966.21	576.51	197.94	0.00	3,733.28	2.57
Internet	256.65	1,910.56	1,086.22	441.07	27.10	3,721.61	2.56
Precision Machinery & Automation	86.16	219.61	1,495.11	1,309.17	0.00	63,110.05	2.14
Advanced Materials	14.00	508.98	725.92	470.04	3.00	1,721.95	1.18
Health Care	21.44	324.58	544.15	173.85	0.00	1,064.02	0.73
Technology Services	50.30	239.12	350.44	158.44	0.00	798.30	0.55
Resources Exploitation	0.00	352.53	235.24	57.30	0.00	645.07	0.44
Specialty Chemicals & Pharmaceuticals	64.06	208.25	250.30	43.02	0.00	565.63	0.39
Aerospace	0.00	41.23	354.60	57.90	47.19	500.92	0.34
Advanced Sensing System	41.11	65.90	294.45	50.91	0.00	452.37	0.31
Pollution Control	64.00	103.57	66.67	48.80	0.00	283.04	0.19
Total	11,015.55	36,053.04	65,917.12	31,166.53	1,242.25	145,394.49	100
Per cent	7.58	24.80	45.34	21.44	0.85	100	

Source: Taiwan Venture Capital Association.

An Urgent Need for Enhancing Pasteur's Quadrant in Technology Development

Economists and regional scientists have long identified the contribution of university research to industrial innovation as one of the key factors driving the US economy. Knowledge spillages from the academic sector are thought to have positive effects on related industrial firms, especially for those operating in technology-intensive domains, even before the passage of the Bayh - Dole Act in 1980. Additionally, the ratio of US patents filed by academic institutions rose from 1 per cent in 1980 to 1.5 per cent in 1991, and to 3 per cent in 1998.

According to the US National Research Council, the factors affecting competitive advantages have changed since the 1990s: interaction between researchers and users has become more important than before. Transferring knowledge between industrial firms and university research centres is important since the industry is relying more heavily on these types of relationships to broaden and enhance their existing knowledge base. Therefore, industrial firms transfer their R&D from Pasteur's quadrant (basic research stimulated from applications) to short-term, practical, application research. Universities increase the research on Pasteur's quadrant (basic research stimulated from applications) instead of Bohr's quadrant (basic science research) (see Table 4.11).

For Taiwanese firms, a move from OEM to ODM or even to OBM (original brand manufactures) requires spillages from universities in order to create innovative products. This would inevitably involve an increase in the emphasis on Pasteur's quadrant. It in turn depends on the success of university - industry collaboration as high-calibre R&D personnel have been concentrated in the former, as noted above. To this important topic we now turn.

7. THE PROMOTION OF INDUSTRY - UNIVERSITY COLLABORATION

R&D Focuses of Universities and Industry: Lack of Collaboration

At various important meetings concerning technology development policies, it was often emphasised that in order to overcome the current situation of declining profit margins in the manufacturing of major export commodities, the high-calibre R&D personnel that has been concentrated in the universities, has to be mobilised. That is, university - industry collaboration holds the key to the future development of advanced technologies which Taiwan's industries badly need.

Table 4.11 Stokes's Classification of Research

Quest for Fundamental Understanding? \ Considerations of Use?	No	Yes
Yes	Pure Basic Research (Bohr)	Use-inspired Basic Research (Pasteur)
No		Pure Applied Research (Edison)

Source: Stokes (1997).

Efforts along this line have been made by Taiwanese authorities, but solid results have yet to be shown. Academic R&D personnel performed very well in the number of papers published. A total of 237.6 engineering papers (EI) per million people were published and Taiwan was ranked the seventh globally in 2002. In addition, a total of 480.9 SCI (scientific citation index) papers were published per million people and Taiwan was ranked the 20th in the world.

However, few of these papers had a lot to do with the ongoing needs of Taiwan's industry. Most academicians published the papers mainly to gain promotion or awards.

On the business fronts, industries have not been hesitant in filing patents. Taiwan patents were ranked the fourth in the world among all approved US patents as Table 4.12 indicates. The total number of patents per million of population in Taiwan were ranked second globally and utility patents were ranked third. Ranked by field, the number of semiconductor process technical patents was third, medical equipment was ninth, communication technology was 11th, and biotech and medicine technology were 17th and 18th in the world (see Tsai and Chu, 2004).

At first glance, it seems that Taiwan's performance has been remarkable in the number of patents awarded. Upon closer examination, however, at least two puzzles arise. The first is that Taiwan's patents are not closely associated with scientific discoveries. On average, citations

Table 4.12 Total Number of US Patents for Different Countries

Rank	Country	1990	2001	2000 - 2001 Growth%	1990 - 2001 Growth%
1	USA	52,977	98,663	1.70	5.82
2	Japan	20,743	34,890	5.97	4.84
3	Germany	7,862	11,894	9.91	3.84
4	Taiwan, ROC	861	6,545	12.73	20.25
5	France	3,093	4,456	6.78	3.37
6	UK	3,017	4,356	6.50	3.40
7	Canada	2,087	4,063	3.52	6.24
8	Korea	290	3,763	8.38	26.24
9	Italy	1,498	1,978	0.56	2.56
10	Sweden	885	1,935	11.33	7.37

Source: Taiwan Institute for Economic Research.

in academic papers number about 0.21 times per patent, far below those of the US (4.46), the UK (3.20), Japan (0.99) or South Korea (0.76) in 2002. In fact, most of Taiwan's patents focused on process technology and few were related to scientific basic research.

In addition, although Taiwan held a large number of patents, its balance in trade in technologies has been suffering from a severe deficit. Table 4.13 reports the coverage ratio of technology balance of payments for various countries, and it is clear that by 2000, Taiwan's receipt from exports of technologies was only 10 per cent of its imports. This is consistent with two observations. One is that most of the patents field by Taiwan's firms were for preparatory legal defence, and not for future receipts of royalties. The second is that as a major base of OEM or ODM production, many of Taiwan's technology secrets have been embodied in the process of production and need not be filed as patents. Both point to the same conclusion: Taiwan has not been among the leaders in advanced technologies which could reap sizeable returns in the form of royalties or authorisation in the future.

Measures Taken by the Government

Realising the problem, in recent years the government has installed several programmes to promote university–industry collaboration. These programmes are administered by various organisations as follows.

Table 4.13 Coverage Ratio of Technology Balance of Payments (receipts/payments)

	1997	1998	1999	2000	2001
Taiwan, ROC	0.05	0.02	0.03	0.10	..
Australia	..	0.46	..	..	..
Austria	0.78	0.79	0.87	1.00	..
Belgium	1.30	1.23	1.27	1.33	..
Canada	1.20	1.58	1.43	1.99	..
Czech Republic	0.73	0.56	0.50	..	..
Denmark	..	..	1.57	..	..
Finland	0.19	0.26	..	..	..
France	0.71	0.83	0.87	1.04	1.19
Germany	0.83	0.83	0.75	0.75	0.67
Hungary	0.63	0.44	0.43	..	..
Italy	0.94	0.84	0.79	0.80	0.78
Japan	1.90	2.13	2.34	2.39	
Mexico	0.26	0.31	0.08	0.11	0.10
New Zealand	0.59	..	2.14	..	
Norway	0.66	0.68	0.74	0.82	
Poland	0.48	0.35	0.19	0.17	
Portugal	0.39	0.34	0.38	0.43	0.47
Spain	0.15	0.19	..	..	
Switzerland	2.45	2.23	1.30	1.49	1.67
United Kingdom	1.72	1.81	2.00	2.07	2.22
United States	3.63	3.17	2.93	2.46	2.36

Source: Taiwan Institute for Economic Research.

National Science Council

The NSC has taken the following measures to enhance industry–university collaboration.

1. Large Grants (the average amount is 8 million, the minimum is NT$ 4 million) to industry–university (or research institutions) for collaborative research: Each year the participating enterprise is required to contribute 25–50 per cent to the total budget, not

including the advanced technology licensing fee. In 2002, the total application of such projects amounted to 25, of which 12 were approved. These projects resulted in patent applications for 21 cases, nine of which have been approved. The total budget amounted to NT$ 147 million.

2. Small Grants (between NT$ 0.5 million and NT$ 1.5 million): During 2002, a total of 1,293 applications was received, of which 927 were approved. The total NSC funding amounted to NT$ 344 million.
3. Outstanding Awards for Industry–University Collaboration: These awards were designed to encourage university professors to devote more effort in this aspect.
4. The NSC also encouraged schools to establish technology transfer centres/offices. A total of 20 centres was established, and during 2002, 10 of them received grants of NT$ 2.5–5 million each. Until February 2003, a total of 266 patents was filed, seven of which have been approved.

Ministry of Education

The Ministry also adopted various measures, financial as well as administrative, to promote university–industry collaborations. Six university–industry collaboration centres were established around six main S&T universities (vocational schools, not general universities) in different areas throughout Taiwan.

Ministry of Economic Affairs (MOEA)

The Small and Medium-Sized Enterprises Bureau has helped schools establish incubating centres since 1996. More than 60 such centres were established island-wide by 2002.

Under MOEA's Department of Industrial Technology (DOIT), there are three Science and Technology Special Programmes, applicable to the universities, industry and non-profit research institutions, respectively. The aim of these programmes is to grant funding to those projects that cultivate advanced technologies, including development and application of research results as well as the development of key technologies and components. In order to promote university–industry collaboration, an increasing (though still small) share of the budget has been allocated to the first category, namely that applicable to the universities.

In addition, in 2001 the MOEA promulgated the National Science and Technology Development Plan. MOEA's Industrial Development Bureau (IDB), funded projects to help the industry upgrade their technologies and to develop 'key new products'. These projects belonged to one of the three categories: national, main and general projects. In 2002, NT$68.7 million was allocated to the first, NT$643.1 million to the second, and NT$2,760.7 billion to the third category. For all these projects, particularly those in the national category, university–industry collaboration is encouraged.

The main division of labour with regard to the government's efforts to upgrade industrial technologies is described in Figure 4.10 below.

Although various efforts have been launched, the results have not been significant so far. The basic problem is that Taiwan's universities belong to the 'public sphere' where salaries have been uniform for any given rank of professorship largely independent of performance and of whether the university is public or private. Private universities' tuition and fees have been heavily regulated; and almost all professors enjoy de facto tenure the moment they are employed. Under such a non-market institutional arrangement, most professors are concerned only with academic promotion but not with helping the domestic economy. Thus, changes are needed with respect to the incentive system.

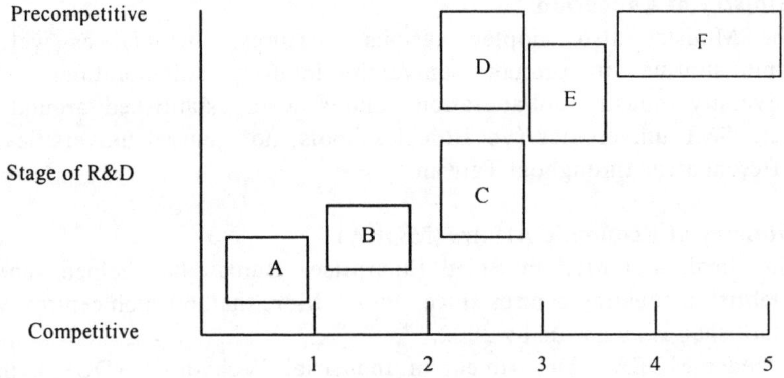

Notes:
A: Small-sized University–Industry Collaboration Projects (NSC);
B: Small Business Innovation Research (DOIT, MOEA);
C: Key New Products Projects (IDB, MOEA);
D: S&T Special Programme Applicable to Industry (DOIT, MOEA);
E: Large-sized University–Industry Collaboration Projects (NSC);
F: S&T Special Programme Applicable to Universities (DOIT, MOEA).

Source: MOEA.

Figure 4.10 Industrial Technology Upgrading: Division of Labour Among Government Agencies

8. THE DECISION-MAKING PROCESS IN S&T POLICIES: LACK OF AN EFFECTIVE FEEDBACK MECHANISM

Currently, the formation mechanism of the national S&T policies is mainly composed of the National Science and Technology Meetings, Science and Technology Advisory Group (STAG) Board Meetings, Science and Technology Meeting of the Executive Yuan, and various Strategic Review Board (SRB) meetings (see Figure 4.11). The conclusions of these meetings, after being approved by the premier, will be implemented. For those conclusions that involve jurisdictions of more than one ministry or council and need Cabinet-level integration and coordination, various Cabinet-level programmes or projects are formulated and sent to related ministries, councils and agencies to develop various implementation plans. In addition, planning activities concerning advanced S&T development include the National Science and Technology Programmes, the Advanced Research Projects (by DOIT), and the 'Fields Priority Planning' (by the NSC).

With such a policy formulation mechanism, the strength has been its flexibility, as all new opinions brought up in the various meetings could find their way to the year-end STAG conference and become official policies. The weakness has been the lack of a systematic check of the performance of the existing programmes. As a result, there could have been too many programmes circulating, wearing thin the resources with or without the accompanying fruits of benefits. This also has something to do with the non-existence of a formal agency in charge of the matter. STAG is not a formal organisation; it has no permanent staff and has to 'borrow' people from various government agencies for a cycle of several years. Knowledge accumulation and systematic evaluation then became a problem.

9. SUMMARIES AND CONCLUSIONS

Taiwan's manufacturing sector appeared to be heading for a fall in the late 1980s as the local currency substantially appreciated and labour costs rose sharply. However, during the 1990s, miraculously the manufacturing sector began growing at an increasingly speed again. Why? Will it continue into the future? These are the two questions to be addressed in this chapter.

A quick answer to the first question is that the phenomenon was made possible by industrial upgrading: the proportion of high-skilled labour-

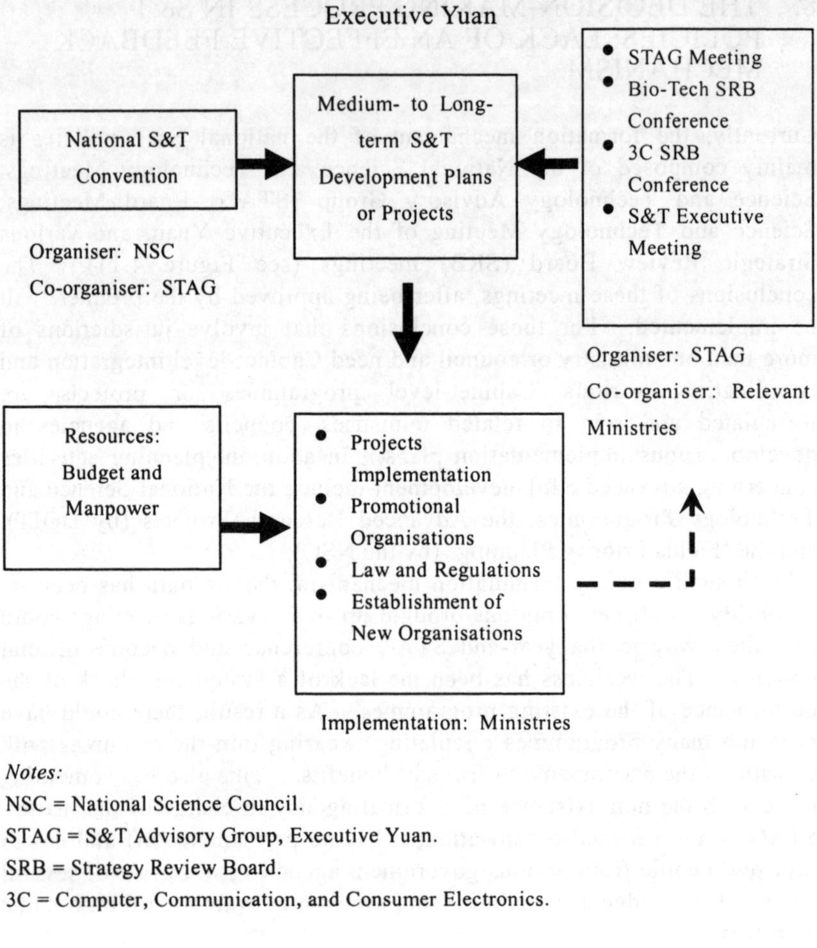

Notes:
NSC = National Science Council.
STAG = S&T Advisory Group, Executive Yuan.
SRB = Strategy Review Board.
3C = Computer, Communication, and Consumer Electronics.

Source: STAG, Taiwan, R.O.C.

Figure 4.11 Existing Flow Chart of the Formation of Science and Technology Policy

intensive products in exports rose while those of the medium- and low-skilled labour-intensive ones both declined. Many industries belonging to the last two groups moved their factories to other places including Mainland China.

Thorough examination of the issue is like peeling an onion layer by layer. The first factor to be noted is the steep growth of the computer-related (otherwise called pan-electronics) sector during the 1990s. This can in turn be attributed to the experiences accumulated earlier from doing assembly work in electronics for multinational companies. Government policies also mattered, including: (i) the establishment of the semi-official ITRI, which was aimed at promoting technologies for use by (and diffusion among) industries; (ii) the establishment of the science-based industrial parks which attracted Taiwanese expertise from abroad to come back from the advanced countries to start up business as well as providing the needed infrastructure (hardware), administrative (software), high-quality manpower supply (major S&T-oriented university in the vicinity) conveniences; (iii) the granting of tax and other incentives to the targeted 'emerging' industries; (iv) the promotion of R&D activities in general by granting tax incentives and by directly subsidising research on the development of targeted new products and key components; (v) the de-regulation regarding the entry requirement for the listing of high-tech companies, which could then readily draw a large amount of capital from the booming capital market; (vi) the promotion of venture-capital activities including tax incentives until recently; (vii) the willingness of high-tech as well as general companies to give stock dividends to employees and technical shares to engineers in return for the core technologies they possessed; and (viii) the promotion of S&T education in general.

While each of these policies had its effects, items (v)–(viii) are judged to be of greater importance. In particular, it is argued that the booming stock market, the active venture capital, and the stock-dividend-to-employees system cleared the way for the computer-related companies to acquire high-grade manpower and large amounts of investment funds at negligible cost. These made possible the operation of 'twin engines', namely capital investment and recruitment of engineers, to propel the electronics sector towards great expansion in the international market.

The main earlier architects of the above-listed policies were LKT and SYC in the 1970s and 1980s, when the foundations for later development were established. It is argued that although they are very capable people, the decision-making process was not at all magical. It contained such simple elements as collecting the latest information from both Taiwan and abroad, digesting it well, and then making a judgement that was in most cases quite obvious at the time. What mattered more was the specific political environment, explained in detail in the text above, which made it possible for them to perform their duties properly. In short, the bureaucrats in charge of S&T and general economic development (i) were expected by the powers that be to promote economic development, (ii) were inquisitive, eager to learn from the world, eager to listen and then to digest the collected information, (iii) had the support of their superiors and were thus reasonably free from political intervention, (iv) did not engage in fierce infighting with each other, because the power ladder had already stabilised, and (v) were free from corruption (another requirement from the top).

These are the conditions consistent with the 'developmental state' hypothesis in the literature. Indeed, these are not conditions readily replicable everywhere in the world, and even in Taiwan the power ladder became a stable 'equilibrium' only after some political struggles in earlier periods. Nevertheless, these conditions were perhaps more important to Taiwan than to the other economies, given the massive concentration of resources in the hands of the government, which has been the largest shareholder and the largest land-owner of the nation, and has controlled enterprises with a total revenue dwarfing any private enterprises in Taiwan. Given its influence in the economy of the country, the government of Taiwan, if it were a successful developmental state, would bring great impetus to growth; while if it became non-developmental, there could be dire consequences for the nation.

In the 1988–2000 era, when LTH succeeded CCK to become president, some personnel changes occurred, notably LYT became the head of Academia Sinica as well as a powerful figure in R&D decision-making. However, the basic nature of the S&T decision making process remained largely intact. R&D expenditure remained a high priority in government budget allocation. Although in this period as in earlier ones, the existence of some 'white elephant' projects cannot be ruled out.

Another thing that remained intact during the entire period of the late 1980s and first half of the 1990s was the liberal economic environment. Taiwan already had a quite open economy and trade was largely free prior to 1987. Since 1987, capital movement has been largely liberalised and many remaining barriers to entry in the domestic market such as banking, insurance, inland transportation, and mass media have been lifted. Visits to Mainland China were also allowed in 1987. As a result, Taiwan's outward direct investment soared with a large proportion going to the booming Mainland Chinese economy. It was also in this period that many Taiwanese firms became internationalised, and made good use of the global division of labour and supply chain network to become dependable OEM or ODM suppliers to brand-name-holding MNC in the West.

A broad summary of the factors contributing to the 'miraculous' comeback of the manufacturing sector in the 1990s is depicted in Figure 4.12 below, which is self-explanatory and dose not need to be elaborated here. This would be the detailed answer to the first question to be addressed in this chapter.

As for the second question, although the period under observation would still be too short to warrant definitive conclusions, one can at least make educated guesses. In the narrow area of S&T policies, which has been under the heavy influence of LYT during the CSB as well as the second half of the LTH terms, similar mechanism and budgetary priorities remained. However, in this area Taiwan is not free from the major challenges ahead:

1. A decomposition of changes in Taiwan's manufacturing value added during the 1990s shows that the sales (or volume) effects dominated while the rise in value-added ratio effects turned out to be small and increasingly so. This has also been reflected in the deteriorating profit margins of Taiwan's chief export items such as notebook computers, mobile phones and even wafer fabrication. The large number of patents (mostly for legal defence purpose) awarded in contrast to the low revenue from technology exports essentially tells the same story. It seems that Taiwan's 'high-tech' industrialisation has reached a bottleneck. There has to be change in developing truly advanced technologies and this involves the use of high-calibre R&D personnel.
2. However, currently the high-calibre R&D personnel have been concentrated in the non-market, non-competitive academic sector and government agencies. The distribution of R&D expenditure though rising as a percentage of GDP reflects the same situation.
3. The situation would become manageable if university–industry collaborations were active. However, these have yet to be realised in spite of initiatives taken by the government. The proper incentive system has not been installed yet, as most university professors are mainly interested in pure academic publications which have little to do with the industrial needs.

The S&T policy arena is not free from challenges, but the general political environment is even more challenging. Here the development of the globally oriented Taiwanese firms would be increasingly at odds with populist politics, which has drifted ever closer towards Taiwanese nationalism. The former needs closer integration with the Mainland Chinese market and economy, while the latter seeks to antagonise or even make an enemy of the PRC.

The drift had already begun in the second half of LTH's term from 1995 to 2000. LTH decided to make bold moves towards enhancing Taiwan's own identity, thus aborting the cross-Strait talks which went

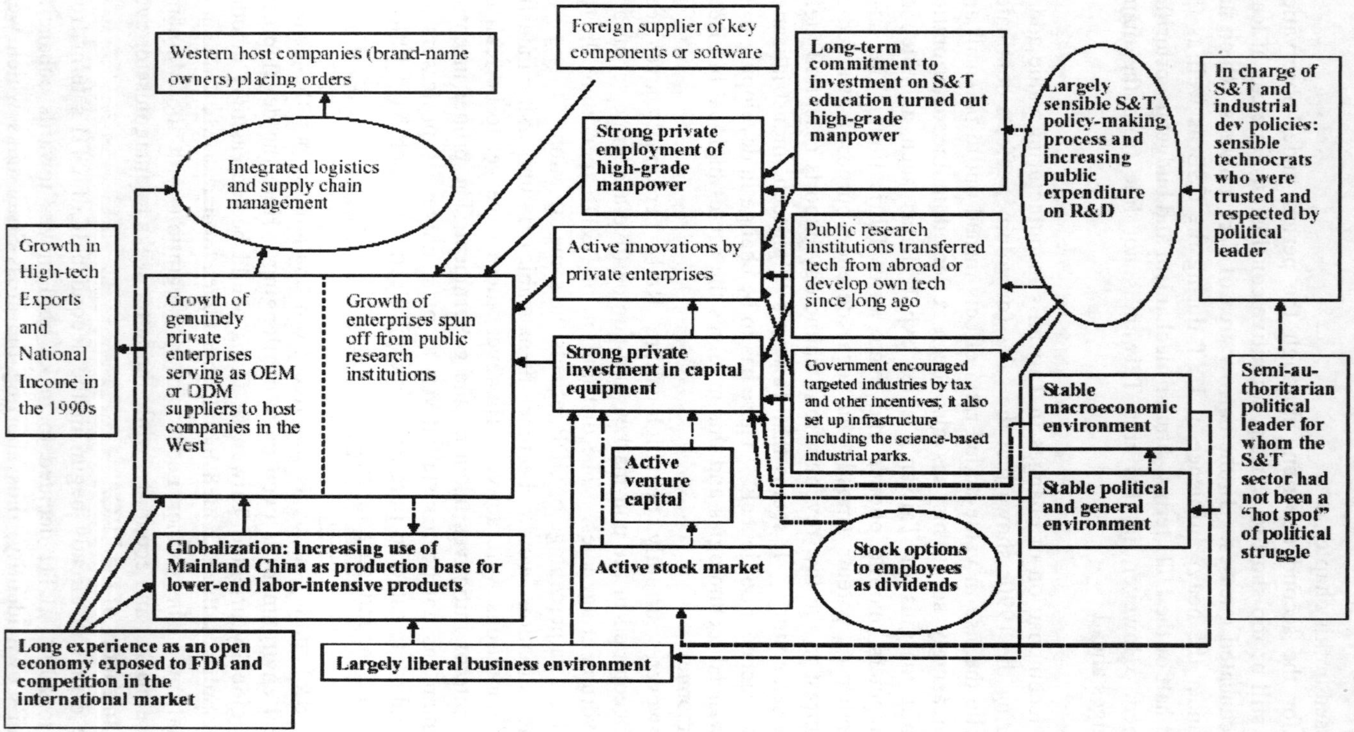

Figure 4.12 Factors Contributing to Taiwan's Growth in High-tech Exports and GDP in the 1990s

Notes:

Explanations of symbols:

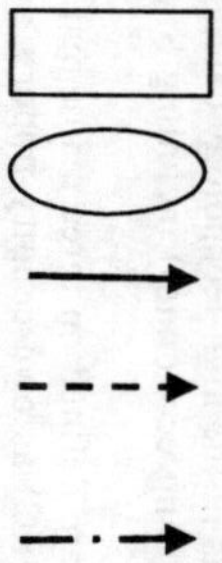

: An entity, a policy measure, or a condition.

: A function or mechanism.

: Supply of goods and services to.

: Contribute to.

: Invest in.

Source: Compiled by author.

well prior to 1995. A 'No Haste, Be Patient' policy was imposed on outward direct investment to Mainland China.

In the post-2000 era, following a major change in political power, the ideology of the new ruling party would run even more contradictorily to the development of increasing globalisation and dependence on the PRC market by Taiwanese enterprises. That is, politically, the two sides of the Taiwan Strait became more hostile, but economically they became closer than ever. The PRC has become not only the largest recipient of Taiwan's direct investment, but also the largest export market for Taiwan.

These two trends going in opposite directions cannot possibly be sustained in the long run. If populist politics subside or become moderate, growth would be likely to be sustained. If on the other hand political ideologies eventually become truly interventionist, the maintenance of respectable economic growth, which has already slowed down during 2001–03 due to both political and economic reasons, could meet increasing challenges in the future.

All country experiences are experiments of economic and political theories. In Taiwan, researchers who are lucky to witness the changes in major political and economic factors are in the middle of a major experiment. The results of the experiment will be obtained only after a longer period of observation and preferably in comparison with other economies. However, one important factor particular to the Taiwanese experience is the strong economic power directly within the grasp of the state. This has made the intention of the state crucial to economic development in general and high-tech industrialisation in particular. When the state was pro-development, it behaved like a powerful 'developmental state'. However, when populist politics turn the state in other directions, the same powerful machine could become a deterrent to economic development that even the collected efforts of and influences by private enterprises could not hope to counter-match. This is the biggest risk embodied in the model of the developmental state: the possibility of a mutation and refusal to fade away even when private enterprises are strong enough to take care of themselves.

NOTES

1. The analysis here about the trend of the manufacturing index and later about the change in composition of Taiwan's exports draws heavily on Chu (2004).
2. For example, in Figure 2.2 of Lall (2003), the Philippines stands out as the economy having the highest proportion of its exports in the high-technology category in 1998, dwarfing 12 other NIEs (newly industrialising economies) including Singapore and Korea. See also Agrawal et al. (2000).
3. The market shares of Japan and the PRC are defined on three major markets instead of four (the US, the EU, Japan and the PRC) as own-country imports are excluded

from the computation.

4. Agrawal et al. (2000, pp. 264–5).
5. The Electronics Research and Service Organisation (ERSO), the electronics branch of the Industrial Technology Research Institute (ITRI, see below) was established in 1974. It acquired IC manufacturing technology from RCA during 1975–79. Later, the United Microelectronics Corp. (UMC) was established in 1980, and was headed and staffed by experts spun off from ERSO, to produce ICs; before that, most operations in both local and multinational firms are limited to the assembly and testing of ICs. Still later, the Taiwan Semiconductor Manufacturing Co. (TSMC) was established in 1987 to provide foundry services to local and international chip design houses and to be headed by the then president of ITRI, Morris Chang; Royal Philips Electronics was invited to take 27.5 per cent of the shares. In the process, ERSO acquired the static random access memory technology from Philips and around 200 staff then spun off to join TSMC. More on this in the text.
6. The government consolidated industrial technology centres into ITRI in 1973 and then expanded its research spheres so that it catered for research areas as machinery, electronics, energy and resources, chemicals and materials.
7. From 1992 to 2000, the government targeted 'ten emerging industries', including communication, information, consumer electronics, aerospace, medical and health care, pollution control, high-grade materials, semi-conductors, special-purpose chemicals and pharmaceuticals, and precision machinery and automation. Since 2001, these have been changed to 18 industries with the following adjustments: (i) the further division of information industry into software and hardware, (ii) high-grade materials were further divided into special alloy, high-grade plastic or rubber, high-grade fibre materials, electronic materials, high-grade ceramic materials, and high-grade composite material, and (iii) bio-tech and pharmaceuticals were added (see Chu, 2002)
8. Here and thereafter the term 'high-tech', unless otherwise noted, means those industries that have been designated by the Taiwanese government as belonging to the important science-and-technology (S&T) category. In 2002, these referred to a wide range of products, most of which belong to the 'emerging industries' targeted by the government.
9. The stock market enjoyed the largest boom ever during 1988–90, amidst an asset inflation triggered by the inflow of short-term capital chasing the sharply appreciating NTD. Both the stock price index and real estate prices hit record highs at that time.
10. According to Lall (2001, Ch. 2), the so-called 'market-stimulating technology policies' fall into three major categories: (1) setting national priorities, (2) providing incentives and signals to private sectors, and (3) generating non-market institutions to underpin the first two. The above list seems to have incorporated all three categories.
11. The government controlled the number of students each school was allowed to accept, at all levels of education, among many other regulations, which is why many people in Taiwan say that there is only one university – the 'University of the Ministry of Education'.
12. Li K.-T., to be discussed in greater detail below, is believed to be the person who initiated the policy. While this policy has its merits, it is not free from shortcomings or side-effects. Teenagers who could not enter regular high schools were labelled as failures in educational attainment, and this could have produced adverse psychological impacts on many people, including the students and their parents.
13. As in other countries, it did not always correctly anticipate what happened. However, given that Taiwan was technologically following in the footsteps of Japan and the US, it would probably not be too hard to anticipate correctly most of the time, under the condition of the existence of reasonably competent bureaucrats who were loyal to their jobs (rather than having a personal interest in wealth).

14. See, e.g., Wade (1990) for accounts of the state mechanisms in Taiwan that are consistent with the 'developmental state' hypothesis. More on this below.
15. At that time, there was a widely circulated saying inside bureaucratic circles that LKT used his influence to install many people of his choice throughout the government agencies related to development. This circle was nicknamed 'The KT Clique'
16. But not directly. The direct successor to LKT was Tao Sheng-Yang, another well-known and highly regarded technocrat, who however, died shortly after he took office.
17. According to Chang (1987), SYC, appointed by CCK as premier in 1978, invited LKT to return to the post of Minister of Economic Affairs but LKT declined for health reasons. SYC, who himself gave a lot of weight to technological development, continued to support the S&T schemes which LKT in his position as Minister without Portfolio in charge of S&T policies promoted.
18. Not all programmes promoted by LKT were indisputably successful. The operation of III is an example. While some have seen III as an important organisation promoting software technology in Taiwan, others have seen it as a 'strange animal', which has had governmental support but has been competing with other private business for various government-granted projects. It is argued that the whole purpose of government sponsorship of a research institute on applied software (or actually on any applied technology) should be to promote those technologies that the private enterprises hesitate to develop for reasons of excessive risk or public good problems. If so, the institute cannot be expected to generate enough revenue to cover its cost, at least in the short run. However, with the tightening of the government budget over the years, the Ministry of Economic Affairs instructed III to soldier on alone. This forced III to participate in those activities that can be immediately marketable, thus competing with private enterprises even more directly. By so doing, the government should have ordered III to be privatised completely, enabling it to compete on a level playing field with the other private firms. However, this has not been done because III has preferred to retain its semi-official status, perhaps under pressure from its employees. The same dilemma is applicable to ITRI as well.
19. A sizeable percentage of the budget allocated to the Ministry of Economic Affairs was actually executed by ITRI and III, which are under the supervision of MOEA.
20. According to LKT, Dr Hsu, the then head of the National Science Council, introduced Morris Chang to high officials in Taiwan (see Yang, 1998).
21. LKT also helped to set up the TSMC. As Minister of Finance, he supervised a development fund, which worked as venture capital to invest in projects that are believed to benefit the Taiwanese economy and to be able to achieve market successes. A total of 48 per cent of TSMC's initial capital was from that fund; the second-largest shareholder was Philips, as noted above (Yang, 1998).
22. The importance of the insulation from rent-seeking activities is also emphasised in, e.g., Hsueh et al. (2001, p. 186).
23. The term in its Japanese application, attributable to Johnson (1982), describes a state mechanism consisting of four elements: (1) the existence of a small, inexpensive, but elite state bureaucracy staffed by the best managerial talent available in the system; (2) there is a political system where the bureaucracy is given sufficient scope to take initiatives and operate effectively; (3) methods of state intervention are market conforming instead of market obstructing; and (4) the existence of a pilot organisation like the Ministry of International Trade and Industry (MITI). See Johnson (1982, pp. 314–20).
24. See, for examples, Woo-Cumings (1999a) and Cumings (1999). The latter also emphasises the importance of world economic conditions in shaping the patterns of responses from the late industrialisers.
25. See, e.g., Gold (1986), Pang (1992), Hsu (1994), Wu (1994) and Kuo (1995).
26. During the Japanese colonial period (1789–1945), according to Yauchihara (1965),

in 1925 arable land owned by the Japanese amounted to 15 per cent of the total; other estimates as cited by Lin (1995) put the figure at 18–20 per cent on the eve of the end of the colonial period. Yauchihara (1965) also estimates that 97 per cent of the shares of all limited liability companies with paid-in capital exceeding 5 million Yen was owned by the Japanese in 1938–41.

27. See Huang (2004) for an analysis of the state–business relationship in Taiwan.
28. Deyo (1987) attributes the relative insulation from business associations, labour unions and foreign capital that has been characteristic of the developmental states in East Asian NIEs to the sequence of establishment of institutions. In these NIEs, 'externally fostered emergence of strong exclusionary states before extensive industrialisation or linkage to foreign capital permitted East Asian governments to impose pre-emptive controls over both emergent social classes and external capital ... Labour unions were controlled from the outset, so subsequent unionisation conformed to ... corporatist principles ... foreign capital confronted states able and predisposed to determine the magnitude, sectoral distribution, and operating conditions of the investment. This sequence clearly differentiates the Asian NICs from their counterparts in Latin America' (Deyo, 1987, p. 240). In Taiwan, this is especially true given that the Kuomingtang (KMT) government took over Taiwan from Japan at the end of the Second World War, and that the government became immensely powerful economically with assets confiscated from the Japanese as noted above, not to mention the then authoritarian-military rule.
29. The term 'Taiwanese' is frequently used in Taiwan to describe the group of ethnic Chinese who migrated to Taiwan several generations ago (mostly before 1900). They constituted the majority. The 'Mainlanders', on the other hand, were people whose families came to Taiwan after 1945.
30. The case of LKT when he was the finance minister is already given above, although the threat in that story in retrospect was probably more perceived or imagined than real.
31. Increased spending included the bail-out of troubled banks following the wave of the Asian financial crisis coupled with the bursting of the real-estate bubble, spending on public health insurance, pensions, and other forms of welfare. Meanwhile, tax revenue as a percentage of GNP fell continuously, becoming a mere 12.3 per cent by 2002. The main reasons for the low percentage are: (a) a tax reform (paid corporate income tax deductible from individual tax returns) was adopted by LTH, which de facto lowered the burden of businessmen; (b) the nominal tax rate on landowners' capital gains was halved, pleasing the landowners and the real-estate developers; and (c) firms that expanded by making new investments in general and high-tech firms in particular continued to enjoy low or zero taxation given the provisions in the Statute for Upgrading Industries, and their share in GDP rose (see discussion above on the actual numbers for this item)
32. Theoretically, zero base budgeting is adopted. The criteria of allocation of S&T resources are as follows: (a) the annual growth rate of the total S&T expenditure is set at 12 per cent; (b) the national S&T programmes take the highest priority, and currently there are nine of them, namely System-On-Chip, Natural Disaster Prevention, Telecommunications, e-Learning, Digital Archive, Biotechnology and Pharmaceuticals, Agricultural Biotechnology, Nanoscience and Nanotechnology, and Genomic Medicine (see http://www.nsc.gov.tw/project.asp); (c) the second priority is given to programmes prioritised by various ministries, councils and agencies which propose the budget; (d) programmes of continuity precede emerging programmes; (e) programmes with better performance precede those with poorer performance; and (f) all other 'competitive' programmes of various ministries, councils and agencies will be prioritised through reviews and usually only those in the top 60 per cent will be supported.
33. The Taiwan independence movement refers to actions aimed at declaring *de jure* independence of Taiwan, in the name of the Republic of Taiwan, and the associated ideological intention to promote Taiwanese nationalism in terms of history and

culture. The current status of Taiwan is *de facto* independence but its constitution installed by the Nationalists holds that the Republic of China, Taiwan's official name, is a Chinese republic that sees China as a unified nation, the division of which into two parts ruled by two different sovereignties is unintended or temporary.

34. These phrases describe the process from a retrospective and matter-of-fact point of view. They do not imply that LYT made political calculations beforehand.
35. Taiwan in 2000 was the world's largest producer of CD-ROMs, DVD-video discs, CD-RW discs, scanners and PCs; the second-largest producer in LEDs, digital cameras and small-size TN/STN-LCDs; and the third-largest producer in large-size TFT-LCDs.
36. Asset inflation both in the real estate and stock market as well as the appreciation of local currency made the sale of industrial land (or any other forms of land property) or that of stocks highly profitable in terms of US dollars, resulting in rises in investment abroad.
37. From Mainland Affairs Council, *Cross-Strait Economic Statistics Monthly*, April 2004. Direct trade with the PRC is illegal in Taiwan, so all trade goes through a third party, mainly Hong Kong. As a result, the trade figures have to be estimated.
38. In fact, among all industries, the electronic equipment (mainly the ICT sector), which has been the largest investor in the PRC, has also been the largest investor in Taiwan. During the 1990s, total direct investment in the latter was about seven times that in the former.
39. A new wafer fabrication plant, He Jian Technology, was established near Shanghai in 2001, and many of the managers working there used to work for the UMC, so it is widely held that it was actually an investment project undertaken by the UMC. Taiwan's Prosecutor's Office opened an investigation of the case in early 2005.
40. In contrast, the Chairman of the Mainland Chinese Affairs Council became the most outspoken person regarding many important economic policy decisions. In theory, that Council should be concerned with affairs related to Mainland China only, but, as explained above, because the PRC has become the largest export market and destination of outward direct investment, it has been *de facto* making important economic policy decisions overshadowing the Ministry of Economic Affairs. Here one sees the strange twisting of the 'embedded autonomy' hypothesis given by Evans (1995). Most of the Ministry of Economic Affairs officials can be said to be embedded in the business sector (but without losing autonomy as Evans emphasises), so they are not likely to make decisions which they know would jeopardise the business sector in general. However, the same cannot be said of the Mainland Chinese Affairs Council, which became autonomous but not embedded.
41. The business communities, including the US and European Chambers of Commerce in Taiwan, repeatedly called for the government to establish direct passenger flights between Taiwan and the PRC, but to no effect.
42. CSB was re-elected in 2004 amidst controversies over a mysterious gunshot incident and other events, see *China Post*, 20 and 21 March 2004.
43. The poor performance also had something to do with the general slackening in the US economy following the 9/11 Attack. The situation became better in 2004 in terms of both real private investment and general economic growth, but neither the absolute level of the former nor that of per capita national income in US dollars returned to their 2000 records.
44. The analysis in this section draws heavily on Tsai and Chu (2004).
45. See Chu (2002) for details on the decomposition analysis.

REFERENCES

Aberbach, Joel D., David Dollar and Kenneth L. Sokoloff (eds) (1994), *The Role of the State in Taiwan's Development*, Armonk, NY: M.E. Sharpe.

Agrawal, Pradeep, Subir V. Gokarn, Meena Mishra, Kirit S. Parikh and Kunal

Sen (2000), *Policy Regimes and Industrial Competitiveness: A Comparative Study of East Asia and India*, London: Macmillan.
Chang, Jun (1987) (in Chinese), *People Who Made the Economic Miracle*, Taipei: Biographic Literature.
Chen, Been-Lon, Tain-Jy Chen and Yun-Peng Chu (2001a), 'The Role of Textiles and Man-made Fiber in the Process of Industrialization: The Case of Taiwan', in Wong and Ng (eds), pp. 283–321.
Chen, Tain-Jy, Been-Lon Chen and Yun-Peng Chu (2001b), 'The Development of Taiwan's Electronics Industry', in Wong and Ng (eds), pp. 245–82.
Chen, Tain-Jy and Joseph S. Lee (eds) (2004), *The New Knowledge Economy of Taiwan*, Cheltenham, UK: Edward Elgar.
Chu, Yun-Peng (1994), 'Taiwan's External Imbalance and Structural Adjustment: A General Equilibrium Analysis', *Asian Economic Journal*, 8, pp. 85–114.
Chu, Yun-Peng (2001a), 'Markets Grew and Matured as a Result of State Actions: The Taiwan Experience', *International Journal of Development Planning Literature*, 1, pp. 204–34.
Chu, Yun-Peng (2001b), 'Liberalization Policies since the 1980s', in Mai and Shih (eds), pp. 89–119.
Chu, Yun-Peng (2002), 'The Development of Newly Emerging Industries and their Knowledge Content: How to Release the Bottlenecks to the Enhancement of Industrial Value-Added', in Lin (ed.), pp. 1–14.
Chu, Yun-Peng (2004), 'The Transformation of Traditional Manufacturing Industries', in Chen and Lee (eds), pp. 103–17.
Chu, Yun-Peng, Been-Lon Chen and Tain-Jy Chen (2001), 'Rethinking the Development Paradigm: Lessons from Taiwan – The Optimal Degree of Government Intervention', in Wong and Ng (eds), pp. 197–244.
Cumings, Bruce (1999), 'Webs with No Spiders, Spiders with No Webs: The Genealogy of the Developmental State', in Woo-Cumings (ed.), pp. 61–92.
Deyo, Frederick (1987a), 'Coalitions, Institutions, and Linkage Sequencing – Toward a Strategic Capacity Model of East Asian Development', in Deyo (ed.), pp. 227–48.
Deyo, Frederick (ed.) (1987b), *The Political Economy of the New Asian Industrialism*, Ithaca, NY: Cornell University Press.
DSTI (Directorate for Science, Technology and Industry), OECD (2004), *STAN Indicators: 1980–2002*, http://www.oecd.org/dataoecd/60/28/21576665.pdf.
Evans, Peter (1995), *Embedded Economy: States and Industrial Transformation*, Princeton, NJ: Princeton University Press.
Gold, Thomas B. (1986), *State and Society in the Taiwan Miracle*, Armonk, NY: M.E. Sharpe.
Hsu, Cheng-Guo (1994), 'Ideological Reflections and the Inception of Economic Development in Taiwan', in Aberbach et al. (eds), pp. 307–20.
Hsueh, Li-Min, Chen-Kuo Hsu and Dwight H. Perkins (2001), *Industrialization and the State: The Changing Role of the Taiwan Government in the Economy, 1945 - 1998*, Cambridge, MA: Harvard Institute for International Development.
Huang, Tsung-Hao (2004) (in Chinese), 'State - Business Relations in Taiwan Since 1949', *Wenti Yu Yanjiu*, 43/4, pp. 35–72.
Johnson, Chalmer (1982), *MITI and the Japanese Miracle*, Stanford, CA: Stanford University Press.
Johnson, Chalmer (1987), 'Political Institutions and Economic Performance: The Government - Business Relationship in Japan, South Korea and Taiwan', in

Deyo (ed.), pp. 136–64.
Kuo, Cheng-Tian (1995), *Global Competitiveness and Industrial Growth in Taiwan and the Philippines*, Pittsburgh, PA: University of Pittsburgh Press.
Lall, Sanjaya (2001), *Competitiveness, Technology and Skills*, Cheltenham, UK: Edward Elgar.
Lall, Sanjaya (2003), 'Foreign Direct Investment, Technology Development and Competitiveness: Issues and Evidence', in Lall and Urata (eds), pp. 12–56.
Lall, Sanjaya and Shujiro Urata (eds) (2003), *Competitiveness, FDI and Technological Activity in East Asia*, Cheltenham, UK: Edward Elgar.
Li, Kwoh-Ting (1997) (in Chinese), *My Taiwan Experience*, Taipei: Commonwealth.
Lin, Chien-Fu (ed.) (2002), *Knowledge Economy and Research and Development: Proceeding of the Seventh Conference in Memory of K.-S. Liang*, Taipei: Department of Economics, National Taiwan University.
Lin, Ching-Yuan and Yun-Peng Chu (2001) (in Chinese), 'Decision Making Process behind the Foreign Exchange Reform in an Earlier Period of Economic Development in Taiwan', presented in the Second National Conference on Empirical Economics, Kaohsiung, Taiwan: National Sun Yat-Sen University.
Lin, Chung-Hsiung (1995) (in Chinese), *One Hundred Years of Taiwan Economic Experience*, Taipei: San-Tong.
Mai, Chao-Cheng and Chien-Sheng Shih (eds) (2001), *Taiwan's Economic Success since 1980*, Cheltenham, UK: Edward Elgar.
Pang, Chien-Kuo (1992), *State and Economic Transformation: The Taiwan Case*, New York: Garland.
Stokes, Donald (1997), *Pasteur's Quadrant*, Washington, DC: Brookings Institution Press.
Tsai, C.-Y. and Yun-Peng Chu (2004), 'Building Bridges among Universities, Research Institutes, and Industry in Taiwan', presented at the Stanford - ITRI Seminar on Industrial Technology Development, Stanford: Stanford University.
Wade, Robert (1990), *Governing the Market: Economic Theory and the Role of Government in East Asian Industrialisation*, Princeton, NJ: Princeton University Press.
Wang, Chao-Ming (1995) (in Chinese), *Memoir*, Taipei: China Times Publishing Co.
Wong, Poh-Kam and C.-Y. Ng (eds) (2001), *Industrial Policy, Innovation and Economic Growth: The Experience of Japan and the Asian NIEs*, Singapore: Singapore University Press.
Woo-Cumings, Meredith (1999a), 'Introduction: Chalmers Johnson and the Politics of Nationalism and Development', in Woo-Cumings (ed.), pp. 1 - 31.
Woo-Cumings, Meredith (ed.) (1999b), *The Developmental State*, Ithaca, NY: Cornell University Press.
Wu, Rong-I and Shiu-Yin Lin (2003) (in Chinese), 'The International Competitiveness of Taiwan's Innovations in Industrial Technologies', presented at International Conference on Industrial Technology Innovations, Taipei: Taiwan Institute for Economic Research.
Wu, Yu-Shan (1994), *Comparative Economic Transformations: Mainland China, Hungary, the Soviet Union, and Taiwan*, Stanford, CA: Stanford University Press.
Yang, Ai-Li (1989) (in Chinese), *Sun Y.-C.: A Biography*, Taipei: Commonwealth.

Yang, Ai-Li (1998) (in Chinese), *Godfather of IC: The Legend of the Strategy of Morris Chang*, Taipei: Commonwealth Magazine.

Yauchihara, Tadao (1965) (Chinese translation by H.-W. Chou), *Taiwan under Japanese Imperialism*, Taipei: Research Department of Bank of Taiwan.

5. Moving towards High-Tech Industrialisation: The Case of Malaysia

Tham Siew-Yean and Haji Mat Zin Ragayah

1. INTRODUCTION

The pursuit of technology development in Malaysian manufacturing can be traced back to the first formal framework for industrial development in the country as encompassed in the First Industrial Master Plan (IMP1: 1986–95). In this plan, the development of indigenous technological capability was recognised as an indispensable component of the industrialisation process for attaining international competitiveness.

In 1990, the National Action Plan for Industrial Technology Development (NAPITD) was launched to develop a strong technological foundation for industrial development while science and technology (S&T) development was identified as a national priority. Subsequently in 1991 technology development was again declared a priority in Prime Minister Mahathir's Vision 2020. Out of the nine challenges outlined in Vision 2020, 'establishing a scientific and progressive society, a society that is innovative and forward-looking, one that is not only a consumer of technology but also a contributor to the scientific and technological civilisation of the future' highlighted the critical need for technology development in the country in order for Malaysia to be a fully developed nation by the year 2020.

The need for technology deepening within the manufacturing sector was further recognised in the Second Industrial Master Plan (IMP2: 1996–2005) that was launched in 1996. In this plan, an integrated approach using value chain-based manufacturing was proposed to rectify the dichotomous manufacturing structure that had evolved. IMP2 also shifted from the sector-based approach in IMP1 to a cluster-based development strategy. While IMP2 continued to focus on manufacturing activities, it also sowed the seeds for a move towards the service sector as the Manufacturing ++ orientation in the plan called for a strategic shift from the 'P-economy' or production-based economy to the 'K-economy' or knowledge-based economy. The adoption of information-intensive and knowledge-driven processes in manufacturing and related activities was deemed to enhance the technology content in the industrialisation efforts of the country. At the same time, knowledge-based activity is expected to foster new sources of growth that will enable the production possibility frontier of the country to shift outwards.

The launch of the K-economy Master Plan in 2002 further accentuated the shift towards the service sector in the technology goals of Malaysia as technology acquisition is clearly not confined to industrial development alone. Rather, the acquisition of technology is deemed to be indispensable for the purpose of knowledge stock-building and utilisation that will enable the country to shift to the K-economy.

It can therefore be seen that the development of technology has been an important policy goal of the country since the launch of IMP1 in 1986. What has been the impact of these policies on high-tech industrialisation in the country? The objective of this chapter is to review the policies that have been used to promote high-tech industrialisation in Malaysia and to assess the impact of some of these policies, focusing in particular on the impact of the technology policies and government support for SME development.

2. OVERVIEW OF INDUSTRIALISATION IN MALAYSIA

Before gaining Independence in 1957, there was no active programme of industrialisation and the manufacturing sector played a minimal role in the economy. Government interventions were biased towards the rural areas. However, after Independence, the government actively promoted industrial development in a move to diversify the economy from its over-dependence on the two primary commodities of rubber and tin and to encourage growth. Since then, the Malaysian industrial strategies have undergone four distinct stages (Jomo and Edwards, 1993: 18; Anuwar, 1994: 712; Rokiah, 1996: 32–49) as summarised in Table 5.1. Nevertheless, as pointed out by Rokiah (1996: 32), these four stages are not mutually exclusive and have overlaps with one another.

Phase 1: Import Substitution Industrialisation (ISI) Strategy, 1957–1970

Phase 1 spanned three development plans: the First Malaya Plan (1956–60), the Second Malaya Plan (1961–65), and the First Malaysia Plan

(1MP: 1966–70). According to both Chee (1987: 7) and Rokiah (1996: 32), the policies implemented during this phase were based on the International Bank for Reconstruction and Development's report of 1957. This report recommended an ISI strategy. Contrary to what was adopted by most developing countries, fiscal incentives for 'pioneer companies' rather than tariff protection were given in order to stimulate industrial development up to the mid-1960s. Tariff and non-tariff protection were increasingly used to promote the growth of new industries only after the mid-1960s.

To a certain extent, the ISI strategy did stimulate growth and help to diversify the economy. Tan (1981: 100), Bruton (1992: 263) and Rokiah (1996: 35) all showed that the manufacturing sector recorded high growth rates per annum in the 1960s. However, in a country with a small domestic market, the ISI strategy very quickly ran out. Furthermore, the heavy imports of capital and intermediate goods for the production of final consumer goods worsened the balance of payments problem while growing unemployment compounded the economic problems of the country. Growing inequality – between income groups as well as along intra-ethnic lines – and persistent poverty was said to be one of the proximate causes of the May 1969 'race riots' which culminated in the proclamation of the New Economic Policy 1971–90 (NEP) in 1970.

Phase 2: Export-oriented Industrialisation (EOI) Strategy, 1970–1980

This phase coincided with the Second Malaysia Plan (2MP: 1971–75) and Third Malaysia Plan (3MP: 1976–80). The NEP was launched with the objective of attaining national unity and fostering nation-building through poverty eradication and economic restructuring so as to eliminate the identification of race with economic functions. With the introduction of 2MP, the goals of development had shifted from growth and diversification to growth with equity. Subsequently, the industrial policies formulated and adopted by the government since the early 1970s were also for the realisation of the objectives of the NEP. The policies emphasised the growth of the manufacturing sector through export orientation and called for more active participation of Bumiputeras[1] in industrial activities. Foreign direct investment (FDI) was encouraged as it was expected to provide capital, management expertise, industrial technology and overseas market access.

Outward orientation actually started with the Investment Incentives Act (IIA) of 1968, which offered generous tax exemptions to export-oriented (EO) firms that emerged alongside the protectionist IS firms that marked the growth of a dual manufacturing sector with little structural link

Table 5.1 Historical Background: Phases of Industrialisation in Malaysia, 1957-Present

Periods	Background and General Conditions	Development Objectives/Strategies /Agenda	Specific Policy Measures	Sectoral Foci
Phase 1 1957-70 ISI Strategy	Av. GDP = 6% p.a. High unemployment, reaching 8% Manufacturing share of GDP = 13.9% in 1970	Market-led Development: *laissez-faire* policies for industrial development; extensive government intervention to promote agriculture & rural development - provide social and physical infrastructure	Pioneer Industries Ordinance, 1958	Agriculture & rural development Manufacturing: Simple consumer goods Domestic market-oriented
Phase 2 EOI Strategy 1970-80	Av. GDP 7.8% p.a. Unemployment fell slightly from 7.4% to 5.7% Manufacturing share of GDP = 19.6% in 1980	State-led Development, 1971-1985: Implemented New Economic Policy 1971-90 - goal of national unity to be attained by eradicating poverty and restructuring society	Investment Incentives Act, 1968 (IIA) Industrial Co-ordination Act 1975 (ICA)	Agricultural diversification & rural development Manufacturing increasingly emphasised: Export orientation Free trade zones Electronics and textiles for exports
Phase 3 ISI Strategy 1980-5	Av. GDP 5.1% per annum Unemployment increased from 5.7% to 6.9% Manufacturing share of GDP = 19.6% in 1980	State-led Development, 1971-85: Implemented New Economic Policy 1971-1990— goal of national unity to be attained by eradicating poverty and restructuring society. Heavy industries	Continued with IIA & ICA Heavy industries esp. national car project	Manufacturing sector emphasised: Consumer durables, intermediate & capital goods Domestic market-orientated

Phase 4 EOI Strategy 1980-present	Av. GDP 5.9% p.a., 1981-90; rose to 8.7% p.a., 1991-95, fell to 4.9% p.a., 1996-2000 & 3.0% p.a., 2001-03 Unemployment fell to 5.1% in 1990, 2.8% in 1995 Manufacturing share of GDP = 33.4% in 2000	Liberalisation and Globalisation, 1986-present: National Development Policy 1991-2000 (NDP) - aimed at balanced development in order to create a more united and just society. Vision 2020 - to be an industrialised and developed nation by 2020 National Vision Policy 2001-10 - to attain a united, progressive and prosperous Malaysian society	Industrial Master Plan 1986-95 (IMP1) Promotion of Investment Act 1986 - provided incentives for manufacturing, agriculture and tourism Industrial Master Plan 2 1996-2005 (IMP2)	Manufacturing: IMP1 identified 12 sectors including: Rubber products Palm oil products Food processing Wood based products Chemicals Electronics & electrical Transport equipment Machinery & engineering Iron & steel Textiles & apparel IMP2 emphasises: Manufacturing ++ Cluster approach: internationally linked resource based policy driven

Source: Compiled by authors.

between the two (Rasiah, 1995: 77). This dual structure was partly the result of the establishment of the free trade zones and licensed manufacturing warehouses that had operated as export enclaves since 1972.

Nevertheless, the switch in industrial strategy was successful as the share of manufacturing employment rose from 8.7 per cent in 1970 to 15.8 per cent in 1980 (Table 5.2) while manufactured export grew from about RM615 million to more than RM4.8 billion over the same period. By 1980, the share of manufacturing in the gross domestic product (GDP) had reached 22.4 per cent and, for the first time, the share of this and the construction sector together in the GDP exceeded that of the agriculture

Table 5.2 Manufacturing's Share of GDP, Employment, and Exports, 1970-2002

Year	Manufacturing value added as % of total GDP*	Manufacturing employment ('000)	Manufacturing employment as % of total employment	Manufacturing exports as % of total exports
1970	13.9	290	8.7	11.9
1975	17.4	398	10.1	21.9
1980	19.6	802	15.8	22.4
1985	19.1	836	15.1	32.8
1990	27.0	1,290	19.5	62.8
1995	33.1 (27.1)**	2,027	25.7	79.6
1996	34.2 (29.1)**	2,230	26.4	80.5
1997	35.7 (29.9)**	2,375	27.1	81.0
1998	34.4 (27.9)**	2,277	27.0	82.9
1999	(29.9)**	2,343	26.4	85.5
2000	(33.4)**	2,558	27.6	86.6
2001[e]	(32.8)	2,574 ***	27.3 ***	86.4
2002[f]	(33.2)	n.a.	n.a.	n.a.

Notes:
*: In 1978 constant prices.
**: In 1987 constant prices for numbers in parentheses.
***: January–August 2001.
e: Estimate.
f: Forecast by Ministry of Finance.
n.a: not available.

Source: Tham (2003: Table 6).

sector. Three major industries were responsible for the rapid growth in manufactured exports in the 1970s – electrical and electronics, textiles and clothing as well as wood products. These industries together accounted for 66 per cent of Malaysia's manufactured exports in 1980, up from 25 per cent in 1970. Moreover, the EO manufacturing sector also became an important vehicle to achieve the NEP objectives, namely, creating employment opportunities and eradicating poverty as well as restructuring Malaysian society to correct economic imbalances by ethnic groups and regions.

Phase 3: Import Substitution Stage 2, 1980–1985

By the end of the 1970s the Malaysian government realised that the export of manufactured goods was confined to a narrow range of products and there were minimal industrial linkages with the manufacturing sector. The development of the EO industries was based on the availability of cheap labour, but as the labour market tightened and wage rates rose, there was an increasing concern that Malaysia might lose its competitiveness.

Consequently, Malaysia had to widen and strengthen the foundation and structure of the manufacturing sector, although EO industries would still be encouraged. Furthermore, Malaysia needed to improve the capabilities of the workers and there was also a desire to produce industrial raw materials and intermediate inputs as well as machinery and capital goods so as not to be too import dependent. It was felt that these objectives could be achieved by going into heavy industries due to their potential spin-off effects. Thus, import substitution was reintroduced in the third phase of industrialisation when the government established the Heavy Industries Corporation of Malaysia (HICOM) in 1980 to promote heavy industries in Malaysia.

Projects under HICOM include PROTON – the national car project, steel mills, petrochemical plants and cement factories. Tariff protection was the main incentive in promoting heavy industries, plus further protection through price control, import restriction, duty exemption and other investment incentives under the pioneer status (Rokiah, 1996: 43). However, the performance of these heavy industries had generally been weak and cost asubstantial amount of resources to the economy. Higher interest rates and the recession of the mid-1980s, as well as the balance of payments, fiscal and debt problems soon got in the way of heavy industrialisation plans, forcing the government to slow down and trim projects already in progress, while cancelling or delaying others still at the planning stage (Jomo, 1990: 133). Hence, the vigorous promotion of this stage lasted only over the Fourth Malaysia Plan (4MP: 1981–85).

Phase 4: Export Orientation Stage 2, 1980–present

In order to get over the malaise of the worst recession Malaysia has experienced since independence, the government turned to the private sector as the engine of growth. The mid-1980s recession and the accompanying outflow of capital resulted in a depreciation of the ringgit that lowered production costs, the relaxation of the Investment Coordination Act and the granting of complete ownership under the Promotion of Investments Act (PIA) that was enacted in 1986 made Malaysia a very attractive destination for FDI. These factors, combined with favourable external market conditions and appreciating East Asian currencies, led to a resurgence of export-oriented manufacturing, largely under the auspices of foreign capital (Jomo and Edwards, 1993: 34).

In early 1986, the government launched the Industrial Master Plan (IMP1) that served as a guide for the development of the manufacturing sector during the 1986–95 period. IMP1 provided a long-term indicative plan for the development of specific subsectors, policy measures and areas of special emphasis. It had three broad objectives (Malaysia, 1996a: 7):

1. to ensure a continued rapid expansion of the economy through the accelerated growth of the manufacturing sector to meet the objectives of the NEP;
2. to promote the optimum and efficient utilisation of the nation's natural resources through value-added manufacturing activities; and
3. to lay the foundation for the development of indigenous technological capability.

The incentives provided under IMP1 include the removal of the 30 per cent foreign equity restriction subject to the fulfilment of export conditions, while generous incentives were provided and licensing procedures liberalised (Tham and Mahani, 1999: 57). However, the buoyant economy of the late 1980s led the government to review and set more stringent conditions for the approval of new applicants, for the first time stressing local sourcing as a crucial condition for the provision of the incentives (Rasiah, 1995: 82).

The Second Industrial Master Plan (IMP2) that was to chart the growth of the manufacturing sector for the 1996–2005 period was introduced at the end of 1996 to succeed IMP1. IMP2 would be implemented during the Seventh Malaysia Plan (7MP: 1996–2000) and the Eighth Malaysia Plan (8MP: 2001–05). IMP2 envisaged the transformation of the manufacturing sector to one with higher value added, technology- and knowledge-based activities that would require development of the manufacturing sector to integrate all the components of the value-added chain. The cluster industrial development approach was adopted so that related industries are grouped together to take advantage of common externalities (Ariff et al., 1996: 41) to develop a more interlinked structure of Malaysian manufacturing (Tham and Mahani, 1999: 60).

Three broad types of potential industrial clusters were identified, each requiring a different set of strategies in order to evolve into a mature and globally competitive cluster (Malaysia, 1996: 31–2). The first is the internationally linked clusters, which are primarily multinational corporation (MNC) driven, such as the electrical and electronics industry. Second is the resource-based cluster, which include wood, rubber, palm oil and petroleum-based industries that have high a degree of indigenous involvement and ownership and are supported by local research and development (R&D) capability. The final cluster is policy driven, involving strategic industries that are fundamentally technology driven, identified through government policy initiatives and are critical for the development of certain capabilities, industries and competencies within Malaysia.

Over the slightly three decades of industrial development, the manufacturing sector has grown to be an important contributor of growth in the country as shown in Table 5.2. Manufacturing value added as a percentage of GDP grew from 13.9 per cent in 1970 to 33.2 per cent in 2002 while its contribution to total employment increased from 8.7 per cent to 27.3 per cent over the same period. However, it is the growth in manufacturing exports that is truly astounding, transforming the country from a predominantly primary-commodity exporter to an exporter of manufactured goods. Certainly this transformation shows that Malaysia did succeed in embarking on low-tech industrialisation but is it able to sustain and upgrade its industrialisation efforts? To answer this question, the next section will examine the strategies that have been used to shift the country towards high-tech industrialisation.

3. MALAYSIAN STRATEGIES FOR HIGH-TECH INDUSTRIALISATION

Lall (1996: 29) defined technological capabilities (TCs) in manufacturing as the skills, technical knowledge and organisational coherence that are required to make industrial technologies function in an enterprise.

Based on this definition, the five factors that he identified as the main influences on TC acquisition are: (i) the incentive structure facing the firms, (ii) the availability of the right quantity and quality of skills, (iii) the availability of technical information and support services, (iv) the availability of finance, and (v) the technology policies of the country. The strategies pursued in Malaysia are discussed in the context of these five factors, starting with the technology policies of the country; the other four remaining factors are then discussed; under the enabling environment of the country.

Technology Policies

Malaysia pursued a hybrid policy for deepening technological development in its industrialisation process. This includes direct government support and intervention that are geared at enhancing technology transfer from FDI, inducing the multinationals that are operating in the country to upgrade their operations technologically as well as specific policies targeted at promoting investment in indigenous R&D.

Foreign direct investment (FDI)

As shown in the previous section, fiscal incentives have played an important role in the FDI policies of the country since its import-substitution days. With the launch of IMP1, the incentives as encompassed in the 1986 PIA, became linked with the industrial targets of the country. Consequently, the basis for granting the incentives was changed from investment to product and activity based. This was in line with the industrial targeting that was practised in IMP1 as FDI was sought to channel investment into the 12 targeted subsectors in the manufacturing sector. Hence, under the PIA, both pioneer status and investment tax allowance is awarded for only promoted products and activities.

As industrialisation progressed in the country, the government increasingly directed its incentives system towards technology deepening in the manufacturing sector and for meeting the performance requirements of the country. In the case of technology deepening, special incentives are given for 'high-technology' companies, 'high-technology' being defined as investing at least one per cent of gross sales in local R&D or at least 7 per cent of the workforce being science and technical graduates (MIDA, 2003: 13). Companies that satisfy these criteria and design, develop or manufacture of selected technologies are eligible for an investment tax allowance (ITA) of 60 per cent on qualifying expenditure within five years, or are granted pioneer status with full tax exemption for five years (Table 5.3). The government also provided special incentives for investment in R&D for companies conducting R&D in Malaysia for their own business and for other companies as well as contract research.

Incentives are also tied to performance requirements beginning with the export incentives that are offered under the PIA, 1986 to facilitate the shift towards export promotion in the country. Double deduction for the promotion of exports, single deduction for the promotion of exports, double deduction on export credit insurance premiums, and tax exemption on the value of increased exports illustrate some of the export incentives used to promote exports (MIDA, 2003: 16 and 41).

Over time, the performance requirements in Malaysia moved beyond export promotion and instead increasingly focused on technology transfer and the deepening of linkages between MNCs and the domestic economy. For example, incentives were given to foster the development of the small and medium-sized enterprises (SMEs) in the country and their linkages with the MNCs, as explained later.

Since technology transfer involves training as well as increasing the local content, incentives were also given to encourage the training of employees. As shown in Table 5.3, under the industrial linkage schemes, tax deductions are also provided for the training of employees while the matching grant under the Human Resource Development Fund (HRDF) launched in 1993 helps to defray or subsidise the training costs of the firms.

In line with the shift in emphasis towards the information, communications and technology (ICT) subsector and K-economy, incentives are also provided for firms that undertake ICT activities in the Multimedia Super Corridor (MSC) and for firms that are granted 'Strategic Knowledge-based Status' (MIDA, 2003: 34–5). The MSC is a special 15 × 50-kilometre zone that was launched in 1996 for firms that create, distribute and employ multimedia products and services. The main incentive for foreign companies is the pioneer status that permits tax exemption for a period of five years or an ITA of 100 per cent. However, incentives are not the only main attraction of the MSC. The MSC also provides other benefits such as duty-free importation of multimedia equipment, world-class physical and IT infrastructure, globally competitive telecommunication tariffs and services as well as no censorship of the internet. Similarly pioneer status or the ITA is also awarded to firms granted 'Strategic Knowledge-based Status', based on their potential to generate knowledge content, high value-added operations, usage of high technology, usage of a large number of

Table 5.3 Major Investment Incentives (manufacturing sector)

Tax Incentives	Tax Concessions
Pioneer Status	Exemption of income (for high-tech companies and companies in an approved industrial linkage scheme) for five years, thereafter a 30% corporate tax, an added incentive for Sabah, Sarawak, Labuan[1] and designated Eastern Corridor of Peninsula Malaysia,[2] a 15% corporate tax
Investment Tax Allowance	An allowance of 60% (80% for Sabah, Sarawak, Labuan and designated Eastern Corridor of Peninsula Malaysia) of qualifying capital expenditure incurred during the first five years. The allowance can be utilised to offset against the 70% (85% for Sabah, Sarawak, Labuan and designated Eastern Corridor of Peninsula Malaysia and 100% for high-tech companies) of the statutory income in the year of assessment. Any unutilised allowance can be carried forward to the following year until the amount has been used up Different incentives given to companies specialising in R&D activities, an allowance of 100% for R&D,[3] Contract R&D[4] and technical/vocational training companies (50% for in-house R&D[5] companies) in respect of qualifying capital expenditure incurred during the first ten years. The allowance can be utilised to offset against the 70% of the statutory income in the year of assessment. Any unutilised allowance can be carried forward to the following year until the amount has been used up
Reinvestment Allowance (RA)	An allowance of 60% of capital expenditure incurred by the companies. The allowance can be utilised to offset against the 70% (100% for Sabah, Sarawak, Labuan[1] and designated Eastern Corridor of Peninsula Malaysia[2] and companies which can improve significantly in

Tax Incentives	Tax Concessions
	productivity) for of the statutory income in the year of assessment. RA is given for a period of 5 years beginning from the year of first reinvestment Upon expiry of RA, companies producing promoted products/engaging in promoted activities are eligible for Accelerated Capital Allowance on capital expenditure where 40% of initial rate and 20% of annual rate will enable capital write-off within 3 years
Incentives for Industrial Adjustment	Incentive given to manufacturing sector for reorganisation, reconstruction, or amalgamation within the same sector, enhancing industrial self-sufficiency, improving industrial technology, increasing productivity and enhancing efficient use of manpower and resources
Incentives to Strengthen the Industrial Linkages Scheme Incentives for Large Companies Incentives for Vendor	Tax deductions for expenditure incurred for training of employees, product development Pioneer status or an ITA status for five years with 100% exemption on the statutory income
Incentives for Export	Double deduction for promotion of exports, double deduction on freight charges, double deduction of export credit insurance premiums, tax exemption on the value of increased exports, industrial building allowance and export credit refinancing scheme
Incentives for Promoting Malaysian Brand Names	Double deduction for expenditure on local advertisement Professional fees paid to companies promoting Malaysian brand names
Training Incentives Pre-Employment Training Double Deductions for	Single deduction on training expenses incurred prior to the commencement of business Double deductions on expenses incurred of employees trained at approved training institutions

Tax Incentives	Tax Concessions
Expenses Incurred for Approved Training Human Resources Development Fund	
Infrastructure Allowance	Companies which are engaged in the manufacturing or commercial sector in East Malaysia and the Eastern Corridor whereby expenses incur on qualifying capital infrastructure are given an infrastructure allowance of 100%. The allowance can be utilised to offset against the 85% of the statutory income in the year of assessment. Any unutilised allowance can be carried forward to the following year until the amount has been used up
Incentives for R&D and Contract R&D Company	Eligible for Pioneer Status with full income tax exemption at statutory level for five years, or an ITA of 100% on qualifying capital expenditure within 10 years. The ITA can be used to offset against 70% of the statutory income in the year of assessment Eligible to apply for ITA 100% on qualifying capital expenditure incurred within 10 years. The ITA can be utilised to offset the 70% of the statutory income in the year of assessment Eligible to apply for ITA of 50% on qualifying capital expenditure within 10 years

Notes:
1. Only applicable to the hotel and tourism industry
2. Kelantan, Terengganu, Pahang and the district of Mersing in Johor.
3. A company that provides R&D services in Malaysia to its related company or any other company.
4. A company that provides R&D services in Malaysia other than its related company.
5. R&D carried out in Malaysia within a company for the purpose of its own company.

Source: Malaysia, Economic Report 2002/2003 (Investors Guide).

knowledge workers and possession of a corporate knowledge-based master plan.

Promoting indigenous research and development

Public funding for developing indigenous R&D is mainly implemented in the form of various grants that are targeted at public R&D institutions, universities, statutory bodies or companies that have at least 51 per cent Malaysian ownership. The main R&D grants are: (i) Intensification of Research in Priority Areas Fund (IRPA), (ii) Demonstrated Application Grant Scheme (DAGS), (iii) Industrial Grant Scheme (IGS), and (iv) Multimedia Grant Scheme (Table 5.4). Grants are also given for the commercialisation and technology upgrading of Malaysian industry such as: (i) Commercialisation of R&D Fund (CRDF), (ii) Technology Acquisition Fund (TAF), and (iii) the Industrial Technical Assistance Fund (ITAF) (Table 5.5).

Strategies Directed at Providing the Enabling Environment

Incentive structure

Macroeconomic policies as well as both the external and internal competitive regimes are the three main factors that can affect the firm's investment in TC (Lall, 1996: 37). The overall protection of intellectual property also plays a critical role in technology development. Of these four factors, Malaysia has maintained a good record in macroeconomic management and this subsection will therefore focus on the competitive regime of the country. The overall legal framework of the country is also considered to be adequate while Malaysia is in the process of enacting new laws or amending existing ones in compliance with its obligations under the World Trade Organisation (WTO).

The trade regime in Malaysia is relatively open compared to other developing countries. Tariffs have been the main instrument used in Malaysian trade policies while quantitative restrictions and import licensing are relatively less utilised. The average unweighted nominal tariff rate (NRP: nominal rate of protection) increased from 10 per cent in 1963 to a peak of 22 per cent in 1979 before it was decreased to 17 per cent in 1982 in line with the government's decision to liberalise tariff protection (Mahani, 1998: 19). The trend to reduce tariffs was continued throughout the subsequent years except for the heavy industries during the second round of import substitution that was launched in 1980.

The effective rate of protection (ERP) showed a slightly different pattern with the ERP actually declining slightly from 40 per cent in 1973

Table 5.4 Summary of Research and Development Grant Schemes

R&D Grant Schemes	IRPA	MGS	IGS	DAGS
Purpose	To support R&D activities in the public sector on areas which address the need of Malaysian industry for the enhancement of the national socio-economic position	To help local or joint-venture companies to develop relevant multimedia technologies and applications which would contribute to overall MSC development	To encourage Malaysian companies to be more innovative in using and adopting existing technologies and creating new technologies, products and processes which will benefit the national economy	To encourage Malaysians to adapt and customise existing IT and multimedia technologies in applications compatible with local culture and to promote the development of local software and content industries for greater competitiveness in global market
Implementing Agency	MOSTE	MDC	MOSTE	MIMOS
Priority Areas	Priority research areas of 7MP	Multimedia	Key technology areas that support IMP2 clusters with prospects for early commercialisation	IT
Eligibility	Public R&D institutions (GRIs), universities government and statutory bodyies	MSC-Status Companies, which are at least 51per cent Malaysian owned	Malaysian majority-owned companies	Malaysian citizen or locally registered organisations with at least 51per cent Malaysian equity and/or government agency

R&D Grant Schemes	IRPA	MGS	IGS	DAGS
Collaboration	Inter-agencies and private sector	Not specified	Local GRIs/universities	Host/user community at one or more local agencies
Funding Duration	1–3 years or more in exceptional cases	2 years	3 years	1 year
Max. Grant Support	100 per cent direct cost	50 per cent of total project cost	70 per cent of approved cost	70 per cent of total project cost
Status	Under 7MP, 3,705 projects valued at RM698.3 million were approved, of which only 10.3 per cent were awarded to the manufacturing sector*	Since its launch in 1998, 19 projects worth RM38 million have been approved*	By the end of 7MP, 58 projects valued at RM138 million have been approved*	Since its establishment and up to July 1999, 61 applications have been received. Only 9 of them were finally approved with a total grant of RM7.8 million Out of the total 9 projects approved, 5 fall into the E-community priority areas, 3 in E-economy and one under E-learning

Source: Akademi Sains Malaysia (2001, Chapter II: 29). Except for * which was extracted from Malaysia (2001: 343–44).

Table 5.5 Summary of Commercialisation and Technology Acquisition Grant Scheme

Grant Schemes	CRDF	TAF	ITAF
Purpose	To enhance competitiveness and capability of the Malaysian industrial sector by promoting the commercialisation of indigenous technology; and to accelerate the commercialisation of R&D results undertaken by local universities and research institutions, companies and individual researchers or investors	To promote technology upgrading through the introduction and utilisation of modern and efficient technology in the manufacturing and physical development of existing and new products/ processes, and; to enhance the competitiveness level of firms to enable them to complete globally	Development of new product, new design or new processes Improvement and upgrading of existing process Improvement of existing products or designs
Implementing Agency	MTDC	MTDC	SMIDEC,SIRIM, MATRADE
Eligibility	Company is locally incorporated and has at least 51% Malaysian equity/ ownership. Joint-venture companies between local firms and universities/ research institutions (both local and foreign)	Company is locally incorporated and has at least 51% Malaysian equity/ ownership Company must produce evidence of the signed technology agreement, where applicable A consortium of firms can also apply	Company incorporated under the Companies Act 1965 A manufacturing company with an annual sales turnover not exceeding RM25 million *and* with full-time employees not exceeding 150 At least 70% of their equity held by Malaysians, of which not more than 25% held by large companies
Amount Supported	50 to 70% of the total cost	50 to 70% of the total cost	50% matching grant

Grant Schemes	CRDF	TAF	ITAF
Activities Funded	The eligible project costs are market survey and research; product/ process design and development; and standards and regulatory compliance and intellectual property protection	Purchase of high-tech equipment and machinery; technology licensing; acquisition of patent rights, prototypes and design; placement of Malaysians in foreign technology-based companies and foreign technology institutes; expert sourcing programme, and; information dissemination seminar/ workshops	Consultancy costs; cost of acquiring technology and skills through training; service cost for related schemes, testing and calibration; cost of testing materials and developmental equipment used in designing and prototypes; other costs as per contract
Total Fund	RM63 million	RM118 million	RM120 million
Status	As at July 1999, 123 applications have been received by MTDC for CRDF with 21 applicants approved, 40 rejected, 46 under evaluation and 16 have withdrawn. RM15.9 million has been approved to these companies with the electronics sector accounting for the highest number of approved applications (6) and amount (RM5.3 million). RM3.9 million has been disbursed to 11 companies and only one has received the full disbursement of the approved grant	As of 14 July 1999, 160 applications have been received for TAF with only 23 approved with a total grant of RM34.9 million; 47 applications have been rejected and the remaining are either under evaluation or have withdrawn However, only RM18.4 million has been disbursed: 15 companies have received full disbursement and 2 only partial disbursements. The rest are still waiting	As at December 2000, 2,679 projects have been approved worth RM53.6 million*

Source: Akademi Sains Malaysia (2001: Chapter II: 30). Except for * which was extracted from MITI (2002: 8).

to 38 per cent in 1978 when the NRP also increased slightly from 19 to 20 per cent. The ERP subsequently rose to 46 per cent in 1982 (Mahani, 1998: 19). According to Rokiah (1996: 174), the ERP in 1987 was 28 per cent, but the range in the ERP was rather wide with some sub-sectors such as wood products, furniture, non-metallic mineral products, basic metals and transport and equipment, receiving relatively high levels of protection.

Commitments under the WTO have hastened the process of tariff liberalisation in the country as the average tariff rate has fallen progressively. Nevertheless, there is still a wide dispersion in tariffs as well as tariff peaks. Tariff peak is highest in automobiles where the MFN (Most Favored Nation) applied rate can be as high as 300 per cent. These tariff rates have not yet been bound in the WTO. Tariff escalation is also observed as the applied MFN tariff for raw materials in manufacturing is only 3.0 per cent while those for semi-processed and fully-processed products are 7.7 and 13.6 per cent, respectively. Bound rates are also higher than the applied rates.

In terms of domestic competition, the main constraint in ownership faced by domestic enterprises is the equity requirement under the NEP. Moreover, there is no competition policy in the country and the regulation of competition has been virtually non-existent (Lee, 2003: 3). Nonetheless Nor Ghani et al. (2000: 18) have found that even in the absence of any kind of competition policy, the ten-year trends from 1985–94, using standard measures of market concentration such as the four-firm concentration ratio and the Herfindahl index, suggest that between 40 and 45 per cent of the Malaysian industrial sector has been evolving towards a more competitive environment. State-owned enterprises are in the process of being privatised, thereby increasing the role of market forces in their operations.

Skills and human resources

Although industrialisation started with import substitution in the 1960s, the education policies of the country were not directed at the industrial needs of the country. Instead, in the 1960s, the focus of government efforts in education was directed at consolidating the diverse school system into a cohesive national education system with a common national curriculum and examinations (Ruhizan and Ahmad Basri, 2003: 9). The main tasks of education at that time were to achieve national unity and universal education.

Between 1970 and 1990, education became a vehicle of the NEP for the reconstruction of society through correcting the economic imbalances between the races and bridging the gaps in racial harmony. The focus of education policies during this period was on positive discrimination to fulfil the redistributive goals of the NEP. At the same time, the Malay language replaced English as a medium of instruction in schools and public universities, although vernacular schools and private schools are allowed to operate.

It was the launch of Vision 2020 in 1991 that redirected the focus of education policies towards the development of a 'world class quality education system which will realise the full potential of the individual and fulfil the aspirations of the Malaysian nation' (Ministry of Education's Mission Statement as quoted in Ruhizan and Ahmad Basri, 2003: 10). Hence, while the share due to secondary and higher education remained relatively unchanged during the 3MP, 4MP and 5MP periods, higher education's share of actual development expenditure jumped to 40.2 per cent under 6MP in an attempt to meet the manpower requirements of a rapidly growing economy. Under 7MP and 8MP, however, the emphasis has shifted to industrial training in order to redress the shortage of skilled and semi-skilled manpower in the country (Lee, 1998: 189). Thus, the share allocated for industrial training can be seen to increase significantly from 4.9 per cent of actual spending during 6MP to 9.3 per cent under 7MP and further again to 16.6 per cent under 8MP.

The emergence of acute shortages in skilled and semi-skilled manpower as well as professional workers such as engineers and R&D scientists and technologists led to several significant changes in the education and training policies of the country. For example, the establishment of the HRDF in 1993 aimed at increasing private sector participation in skills development. Amendments to the Universities and University Colleges Act 1971 in 1996 as well as the enactment of the Private Higher Education Act 1996 paved the way for the corporatisation of public universities that will eventually reduce the fiscal burden of the country, while deregulation of the private sector's role is targeted at expanding the supply of educational services. Quality control over the supply of tertiary education is also addressed in the establishment of the National Council of Accreditation. At the same time, the ratio of enrolment of science to arts students is targeted at 60:40 by the year 2010 in order to increase the supply of S&T manpower in the country.

Technology infrastructure support

According to Lall (1996: 92), technology infrastructure support includes the metrology, standards, testing and quality (MSTQ) system, other forms of technical support by the government, including technical support and extension for SMEs and public R&D institutions.

Public technical support institutions

The Standards and Industrial Research Institute of Malaysia (SIRIM) is in charge of metrology and standards in Malaysia. SIRIM maintains the primary physical standards and provides the testing and calibration services used by industry. It also functions as a secretariat and the registration body for the registration of the Quality Consultant Scheme (RQCS) which aims to monitor the quality of consultants for ISO 9000 standards.

Apart from SIRIM, the National Productivity Corporation (NPC), under the Ministry of International Trade and Industry (MITI), monitors the productivity performance of the country. Related to this function, NPC also provides courses for improving the productivity of firms, especially the SMEs. At the same time, NPC also provides benchmarking services for the firms.

Public R&D institutions

Malaysia has several research institutions of international repute that are primarily engaged in upstream research in primary commodities such as the Rubber Research Institute of Malaysia (RRIM), the Palm Oil Research Institute (PORIM), the Forest Research Institute of Malaysia (FRIM) and the Malaysian Agricultural Research and Development Institute (MARDI). These institutes have established world-class infrastructure for research and they have been increasingly drawn into downstream research that can contribute positively to the technology development of resource-based industries.

However, the two public research institutes that conduct industrial R&D are SIRIM and the Malaysian Institute of Microelectronic Systems (MIMOS). SIRIM was corporatised in 1996 in order for it to pursue its commercial activities more aggressively. Nevertheless, it continues to function as a national development agency for improving product and service quality as well as developing and transferring advanced technologies to Malaysian companies in order to improve their productivity and efficiency. In R&D, SIRIM concentrates on contract R&D for the industrial sector. Its clients include domestic firms as well as MNCs from the automotive, ceramics, chemicals, electrical and electronics, machinery, engineering and plastics subsectors. Apart from contract research, SIRIM also funds joint research projects under its Joint Research Venture Program (JRVP), which also provides facilities and expertise to assist industry partners in turning innovative ideas into marketable products through equity, cost and task sharing. SIRIM also undertakes government-funded R&D programmes that aim at building the long-term capabilities and capacities in advanced technologies for the country such as the development of the Artificial Intelligence Development Laboratory.

MIMOS, established in 1985, has a narrower objective: it aims to support Malaysia's industrial technology in microelectronics systems. This objective is supported by four main types of activities in MIMOS:

1. conducting MIMOS-based R&D,
2. providing product and research services to industry,
3. disseminating information through courses and seminar, and
4. creating new high-technology local firms.

MIMOS officially launched Malaysia's first Internet Service Provider (ISP) in 1992, which provides JARING (Joint Advanced Research Integrated Networking) internet services to the public. In the field of semiconductors, MIMOS has set up two wafer fabrication plants: on 1 February 1997 and 2 February 2002. The first Malaysian 16-bit microprocessor chip, Pesona, was fabricated on 1 February 1997. Corporatised in 1996, MIMOS is currently in the process of restructuring to separate its three functions, that is R&D, policy work and business, in a bid to become profitable.

Support for SMEs

The Small and Medium Industries Development Corporations (SMIDEC) was established in 1996 as a specialised agency for the promotion and development of SMEs in Malaysia. SMIDEC provides advisory services, fiscal and financial assistance, infrastructural facilities, market access and other support programmes. There are several SMIDEC schemes that are geared towards stimulating investment and technological upgrading of the SMEs:

1. *Vendor Development Programmes* (VDP), launched in 1993 is essentially a matchmaking scheme that is aimed at increasing subcontracting from the MNCs. MNCs that signed formal agreements as anchor companies can choose from the designated vendors to supply a broad spectrum of components. In turn, MNCs and their affiliates which offered guaranteed purchasing contracts and technical support to local suppliers, received incentives. The local vendors or supplier would receive subsidised credit from local banks as well as technical support from government institutes.
2. *The Industrial Linkages Programme* (ILP) within SMIDEC was launched in 1996, focusing on technology transfer and industrial deepening. Leading manufacturing sectors such as the electrical and

electronics products, chemicals and petrochemicals, pharmaceuticals, textiles and apparel, automotive, motorcycles, marine and aerospace development, machinery and equipment and resource-based industries were targeted for this programme (UNCTAD: 2001: 207). The ILP has three main components, that is fiscal incentives, business matching and programmes, which support industrial linkages such as technology development, skills upgrading, export and market development as well as the provision of SME industrial sites.

3. In 1999, the government launched the *Global Supplier Programme* (GSP) aimed at strengthening the competitiveness of the Malaysian SMEs so that they become suppliers not just to the foreign affiliates of the MNCs but also to the world. In other words, Malaysian SMEs are encouraged to become global suppliers with the assistance of participating MNCs, funding from the state and federal government based on the HRDF as well as tax incentives. Through the GSP, Malaysia's supporting industries are being prepared to participate in the new supply chain management (SCM) system of the MNCs. One critical component of the GSP is that it requires significant commitment from the large corporations since they have to 'adopt' local SMEs and assist them in upgrading their leadership skills and technology. In this way, it is envisaged that the SMEs will become integrated into the supplier development programme of the MNCs.
4. *RosettaNet* is the latest government initiative to propel Malaysian SMEs towards global supply through the adoption of Rosetta Net e-business standards (Ernst, 2002: 53).
5. *Industrial Technical Assistance Fund (ITAF)* provides matching grants for SMEs to undertake consultancy and feasibility studies, product development and design, quality and productivity improvement, and market development activities.
6. *The Soft Loan Scheme for Modernisation and Automation* aimed at stimulating SMEs to seek and acquire technology that is embodied in new machinery and equipment.

Finance

Public funding for technology development in terms of fiscal incentives is under the jurisdiction of the Ministry of Finance (MOF). R&D grants, on the other hand, are managed by the Ministry of Science, Technology and Environment (MOSTE), while funds for the commercialisation and technology upgrading are under the management of MITI. MOSTE also manages the HRDF for S&T.

The private sector also contributes to a cess fund that is managed by the Ministry of Primary Industry (MPI) from the exports of primary commodities such as rubber, palm oil, cocoa, timber and tin. This fund is then used to fund R&D activities that are relevant to each commodity. Apart from the MPI, the Ministry of Energy and Multimedia also collect a cess from the energy industry that is used to fund R&D activities in the energy sector.

Of particular interest is the venture-capital industry in terms of the supply of risk capital that is necessary for funding innovative activities. Venture capital (VC) was first introduced into Malaysia in 1984 but the government became actively involved in the industry only with the setting up of the Malaysian Technology Development Corporation (MTDC) in 1992 to finance the commercialisation of technology from local universities and research institutes (Chin, 2001: 231). Chin further noted several measures taken by the government to promote the VC industry. These include the removal of the requirement that VC companies have to invest 100 per cent of their funds in venture companies and new technology projects involving high risks. Conditions such as VC companies not being allowed to invest more than 10 per cent of their funds in any one company and no more than 25 per cent in any one industry were also abolished. Efforts were also made to deepen the capital market to provide new avenues for VC companies to raise capital from the public to fund VC start-up operations or expansion programmes with no track records required if the companies are deemed by MTDC to be technology based.

In 1997, the Malaysian Exchange of Securities Dealing and Automated Quotation (MESDAQ) was launched to meet the financing needs of technology-based companies (Malaysia, 2001: 414). MESDAQ provided both an avenue for high-growth and technology companies to access public funds and an early exit route for venture capitalisation.

4. IMPACT OF POLICIES ON THE DEVELOPMENT OF HIGH-TECH INDUSTRIALISATION

Based on Hatzichronoglou's (1997) classification, the four main manufacturing activities in high-technology industries are (1) aerospace, (2) computer and office machinery, (3) communication and the electronics and the (4) pharmaceutical subsectors. Malaysia's association with high-tech industrialisation and high-tech exports is mainly in the electronics sub-sector. The industry has been and still is the leading sector in Malaysia's manufacturing, being the largest contributor to manufacturing output, employment and export. A brief overview of the performance of the electronics subsector is first presented before the impact of some of the Malaysian policies on the technology development in the country is analysed, focusing in particular on the impact of the technology policies and government support for SME development.

Performance of the Electronics Industry in Malaysia

The electronics industry started in 1970 with a handful of companies and fewer than 600 workers. However, by the year 2001, it had grown to employ more than 335,200 workers and more than 900 companies (Table 5.6). It is the only sector where the majority of investment comes from foreign investors; in 1999 and 2000, foreign investment accounted for 84.5 and 83.6 per cent, respectively, of approved capital investment in the electrical and electronic product industries.

Structurally the composition of output in this sector has shifted over the years. In the early 1970s, the electronics industry was dominated by the electronic components sector, accounting for 80–85 per cent of output. In fact up to 1984, electronic components still accounted for 85 per cent of total output while the consumer and industrial electronics subsectors contributed only 12 and 4 per cent, respectively. However, by 1990, the distribution among component electronics, consumer electronics and industrial electronics was 55, 24 and 21 per cent, respectively. This shift in the production structure was continued throughout the 1990s. By 2000, the output structure for the electronics industry was 39.1 per cent for the electronic components, 12.6 per cent for consumer electronics and 38.3 per cent for industrial electronics.

Table 5.6 Performance of Malaysia's Electronics Industry, 1995–2001

Year	Output		Employment		Exports		Imports	
	RM (bn)	Growth %	No.	Growth %	RM (bn)	Growth %	RM (bn)	Growth %
1995	73.1	26.3	313,000	12.6	85	28.0	63.8	29.2
1996	79.2	8.3	339,000	6.7	91.7	7.9	68.0	6.6
1997	85.6	8.1	343,300	1.2	107.7	17.4	76.2	12.1
1998	103.5	20.9	320,600	−6.6	146.0	35.6	96.7	26.9
1999	126.8	22.5	361,100	12.6	179.1	22.7	108.5	12.2
2000	167.1	31.8	401,700	11.2	230.9	28.9	166.1	53.1
2001	140.9	−15.7	335,217	−16.5	199.7	−13.5	141.2	−14.9

Source: MIDA (2002), ***Business Opportunities in Malaysia's Electronics Industry.***

The electronics industry is also the single largest manufacturing subsector that is contributing to total exports. In the year 2000, this sector accounted for 44.7 per cent of total exports in Malaysia, increasing from 36.5 per cent in 1997. Principal markets for Malaysia's exports of electronics are the United States, Singapore, Chinese Taipei, Japan and Hong Kong, China. The most important product that is exported from this subsector is semi-conductor devices. Malaysia is currently one of the world's leading host economy for semi-conductor assembly, testing and packaging.

Technology Development in Malaysia

While the previous section clearly shows the rapid growth of the electronics sector, it does not show whether this rapid growth is accompanied by technological change within the MNCs that dominate this sector and the development of linkages with the domestic economy. More importantly, it leaves as yet unanswered the contribution of technology policies and SME support towards the development of innovative activities and linkages as well as the constraints in technology development. These issues will be examined in turn in the following subsections.

R&D and innovation activities

While the public sector has historically led in the R&D efforts of the country, the participation of the private sector has increased over time. This can be observed from Table 5.7, where the contribution of government research institutes (GRIs) and institutes of higher learning (IHLs) was 46.0 and 9.2 per cent, respectively, in 1992 while the contribution of the private sector was 44.8 per cent. But by the year 2000, the contribution of the private sector has grown to 57.9 per cent while spending on R&D from the GRIs and IHLs constitute 25.0 and 17.1 per cent, respectively. R&D expenditure from the private sector thus grew from RM246.3 million in 1992 to RM967.9 million in 2000. The overall increase in R&D expenditure over time has contributed to the increase in R&D spending as a percentage of GDP from 0.37 per cent in 1992 to 0.5 per cent in 2000 (MASTIC, 2002a: 10).

The manufacturing sector continued to be the main recipient of the private sector's R&D expenditure. Sectorally, based on the 2000 National Survey of R&D, the electrical and electronics (e&e) subsector is the largest sector in terms of the number of private sector companies that are engaged in R&D activities. It is also the largest sector in terms of its contribution (25 per cent or RM275.6 million) to the total private sector expenditure on R&D for that year (MASTIC, 2002b: 135). Out of the 35 companies that reported R&D spending in this sector, 13 are foreign owned and these 13 companies spent 92 per cent of the reported R&D expenditure in this sector. Based on this survey, it would appear that the foreign firms in the e&e sub-sector contributed significantly to the R&D activities of the country.

Similarly, the National Survey of Innovation (1997–99) found the e&e subsector to be the largest, with 38 companies engaged in innovative activities or 17.5 per cent of the total 217 companies that reported innovative activities in this survey (MASTIC, 2001: 11). In terms of ownership, 49 per cent of the innovating firms are 100 per cent wholly owned, while 24 per cent are wholly foreign owned. The remaining 29 per cent are under joint foreign–local ownership. While the survey covered innovation in both products and processes, the economic downturn in 1998 may have contributed to the relatively large response of firms (67 per cent) having innovating projects-in-progress as opposed to the smaller number (33 per cent) that reported having new or improved products or processes. Nonetheless, 23 out of the 217 innovating companies reported being involved in product innovation while another 28 reported being involved in both product and process innovation.

Table 5.7 Overview of National R&D Expenditure, 1992–2000 (RM m)

	1992	1994	1996	1998	2000
Overall Expenditure	550.7 (100.0%)	611.2 (100.0%)	549.1 (100.0%)	1127.0 (100.0%)	1671.5 (100.0%)
Private Sector	246.3 (44.8%)	292.6 (47.8%)	400.1 (72.8%)	764.1 (66.0%)	967.9 (57.9%)
GRI	253.7 (46.0%)	164.9 (27.2%)	108.7 (19.8%)	247.3 (22.0%)	417.5 (25.0%)
IHL	50.7 (9.2%)	150.9 (24.6%)	40.4 (7.4%)	133.6 (12.0%)	286.1 (17.1%)
NPO	0	2.9 (0.4%)	0	0	0

Notes:
GRI: Government research institute;
IHL: Institute of higher learning;
NPO: Non-profit organisation;
GERD: Gross expenditure on R&D.

Source: MASTIC (2002a, p. 50).

While the national surveys on R&D and innovation do not provide much detail on the kind of innovation that has occurred, Hobday's study of 20 electronics firms in Malaysia based on in-depth interviews between 1994 and 1995 shed considerable light on this issue (Hobday, 1996: 74). His survey showed that although the firms were not engaged in long-term or basic research, nonetheless considerable innovative activity was carried out, focusing on improvements to products and processes as well as 'soft' innovation such as managerial or organisational improvements. Extensive improvements and modifications to capital equipment were carried out by most of the firms in his sample, some of which led to patents and own-brand sales abroad. Thus he concluded that although the innovation by the 12 MNCs in his sample was more incremental rather than radical in nature, it is nevertheless clear that they are no longer screwdriver operations. Based on his evidence, he contends that substantial innovation has occurred, albeit of the continuous improvement variety and it is this innovation that has contributed to the 'high-tech' export success of Malaysia.

SME and indigenous technology development

In 1996, of the estimated 20,200 manufacturing establishments operating in Malaysia, 93.8 per cent or 18,948 are SMEs (SMIDEC, 2002: 31). These SMEs contributed 27.3 per cent to manufacturing output, 25.8 per cent to value added, 27.6 per cent of fixed assets and 38.9 per cent of employment. SMEs as a whole exported 20.8 per cent of their output, accounting for only 10.8 per cent of total exports of manufactured products.

In the e&e subsector, there were 599 SMEs, constituting 3.2 per cent of the total number of SMEs in the country (SMIDEC, 2002: 34). The export contribution of SMEs in this group is small, comprising only 17.3 per cent of their gross output in 1996. However, in the all-important electronics subsector, 42 per cent of the firms are SMEs in 1996. They accounted for 1.9 per cent of gross output and 3.7 per cent of value added and 4.8 per cent of employment. Moreover, the majority of Malaysian-owned e&e firms are SMEs while the large ones are majority foreign owned.

Based on SMIDEC (2002: 19–20), as at 2000, 63 vendors or 11 per cent of the SMEs in the e&e sector have participated in the VDP. Sixty, or 10 per cent, of the SMEs in the e&e subsector have participated in the ILP, generating sales of RM115.6 million. Nine MNCs have adopted 10 SMES under the aegis of the GSP.

The relatively small number of SMEs that are involved in the VDP and ILP as well as the GSP is supported by the studies on technology transfer in the e&e subsector. While backward linkages have evolved over time, these involve a relatively small number of firms. Hobday (1996: 76) found that the number of suppliers to MNCs was viewed as too low for the needs of the MNCs. Moreover, most of the major suppliers to MNCs were other MNCs (for example, from Taiwan) and the majority of the locally owned firms were low value-added producers.

The 1998 National Survey of R&D revealed that only one-third of the companies carrying out R&D activities were SMEs, and they contributed 10.8 per cent of total R&D expenditure (Malaysia, 2001: 247). However, by the year 2000, about 60 per cent (117 firms) of private sector R&D was contributed by SMEs, or 23.5 per cent of the total private sector R&D expenditure (MASTIC, 2002a: 35). Thus, there is increasing participation of SMEs in R&D but again this is likely to be of the process improvement variety as found in Hobday's survey of the four linkage firms in the semi-conductor industry (Hobday, 1996: 73).

Given that external funding from Malaysian sources was identified as the major contributor (79 per cent) to the private sector's source of R&D funding (MASTIC, 2002a: 42), it would appear that the numerous grants that have been launched by the government, especially in the late 1990s, have contributed to the increase in SME research efforts. ITAF was identified as an important Malaysian external fund for financing private sector R&D (MASTIC, 2002a: 42). Nevertheless it is pertinent to note that although this fund was launched in 1990, as of December 2000 the value of the approved projects amounted to only RM53.6 million or 45 per cent of the amount that has been allocated for this fund (See Table 5.5). More importantly, of the four schemes launched under ITAF, ITAF2, which is earmarked for process and product development, has the lowest uptake, accounting for 12.4 per cent of all applications as at December 2000 or 24 per cent of the total value of approved projects under this fund (SMIDEC, 2002: 7).

Apart from ITAF, another important external Malaysian source of funding for private sector R&D is the CRDF (MASTIC, 2002a: 42). As shown in Tables 5.4 and 5.5, the CRDF and other grants launched by the government in the late 1990s have had numerous projects approved. Although this augurs well in terms of indigenous R&D, it should be cautioned that the number of firms and projects that are involved is still very small as compared to the vast numbers of Malaysian establishments in the manufacturing sector, many of whom are of the SME variety.

Appropriation of incentives and motivation for technology improvement

While the government has provided numerous incentives for R&D as explained in Section 3, the uptake of the incentives was mainly for double-deduction expenditure for R&D (Malaysia, 2001: 342). The number of projects approved increased from 247 in 1995 to 531 in 1999 with a corresponding value of RM42.2 million and RM279.2 million, respectively. For the year 2000, however, the number of projects approved fell to 162 with a value of RM51.0 million. Nevertheless, six companies in the electronics subsector with a total of 42 projects garnered RM26.7 million or 53 per cent of the total value that was granted for double deduction for R&D in the year 2000 (MASTIC, 2002b: 216).

In contrast to the large number of projects granted approval for double deduction for R&D, only 50 companies were granted Pioneer Status or Investment Tax Allowance (ITA) for in-house R&D activities as well as R&D for other companies for the whole of the Seventh Malaysia Plan period (1996–2000) (Malaysia, 2001: 342). These companies were mainly in the electrical and electronic products and transport equipment industries.

More importantly, only 56 (or 37 per cent) out of the 153 manufacturing firms that reported R&D activities in the year 2000 have benefited from the fiscal incentives provided by the government (MASTIC, 2002a: 44). The two reported incentives that have been enjoyed by the majority of these 56 companies are double deduction for R&D expenditure and exemption on import duty, excise duty and sales tax on machinery/equipment, materials, and components parts and samples used for R&D activities. Similarly, the National Survey on Innovation (MASTIC, 2001: 18) found that only 62 companies (or 28.6 per cent) of the 217 innovating companies in the manufacturing sector have benefited from the tax incentives while 38 companies or 22.8 per cent of these companies reported government support, assistance and incentives to be 'not relevant'.

In terms of the problems encountered by the companies when applying for incentives, the majority of the private sector companies in R&D activities reported that the procedures for application were not clear (MASTIC, 2002a: 194). While this is an improvement compared to the finding from the 1998 Survey where the majority reported that they were not even aware of the availability of incentives, it still indicates that the delivery system of the incentives may need to be reviewed.

Again as in the case of grants and other government programmes, while numerous incentives are offered by the government, it would appear that only a few have been able to access these funds. It is therefore not surprising to find that the main motivations for innovation activities were to improve product quality, open up new markets, increase market share and reduce cost (MASTIC, 2001: 15). In other words, firms were driven to innovate in order to compete. Similarly, competitive pressures have been found to be an important motivating factor for MNCs to transfer technology (Tham, 2003: 28). Since the 1980s, automation has increased in tandem with the rapid changes in technology and the demand for precision. With increasing automation, MNCs had to increase the capabilities of local plants, thereby leading to continuous process improvements in the Malaysian subsidiaries. By increasing local capacity, MNCs have also been able to reduce costs and increase productivity, thereby enhancing their competitiveness.

While incentives have also played a role in the transfer of technology, it is important to note that the more successful cases of technology transfer as found in the case of Penang have worked due to the strong proactive enabling environment provided by the state government (Tham, 2003: 23). It would therefore appear that incentives need complementary factors in terms of state support such as training and matchmaking of MNCs with the suitable SMEs for it to contribute to the formation of linkages with the domestic economy.

Given that parent companies are commercially motivated to expand plants and to increase exports in local subsidiaries, the main constraint in technology transfer and increasing R&D and innovation at the local subsidiaries depends in the main on the host country's absorptive capacity and competencies in the technology upgrading. Without these, it is extremely doubtful that effective technology transfer can be executed or that local subsidiaries can hope to be upgraded in the MNC value chain even if more incentives are provided.

Main Constraints in Technology Development

The shortage of R&D personnel was found to be the most important factor limiting R&D activities in the 2000 R&D Survey, followed by limited financial resources. Similarly in the Innovation Survey, the main factors that are hindering innovation are the high cost of innovation, the recent economic crisis (the survey was conducted over the period of the crisis) and the lack of skilled personnel. Although government funding may have contributed to the increase in SME participation in R&D, these firms still face severe constraints in terms of skilled personnel. For example, in the Inter-Firm Linkages and Technology Development Study carried out in 1997, it was found that the average number of scientific, engineering and technical (SET) personnel per 100 employees for small firms was 2.4, compared with 3.2 for the medium and 4.1 for the large ones (Malaysia, 2001: 247). For foreign firms, the average was 6.1 SET personnel per 100 employers.

Human resources

Government spending on education has contributed greatly to the increase in educational attainment of the labour force in Malaysia. Table 5.8 reveals a distinct improvement from 1967–2000. The percentage with no formal education has dropped from 41.0 to 6.1 per cent in 2000, while the percentage with secondary or tertiary education has increased from 8.6 to 67.7 per cent. Some 13.9 per cent of the labour force had attained tertiary education in 2000, but this is still significantly smaller than some other FDI-dependent economies such as Ireland, where the achievement is almost double as the proportion at work with tertiary education is reportedly close to 30 per cent.

However, output at the tertiary level still shows a greater concentration in arts and humanities (58.4 per cent by the end of 7MP) (Table 5.9). Technical, engineering, architecture, surveying and others continue to produce the smallest output, although this has increased over time. Similarly, the output of skilled and semi-skilled workers has also increased over time, at an average annual rate of 9.8 per cent during 7MP, of which 65 per cent was in the engineering trades (Malaysia, 2001: 108).

The increase in science and technical graduates has more than doubled the number of research personnel involved in R&D activities from 9,136 in 1996 to 23,262 in the year 2000 (Malaysia, 2001: 349 and MASTIC, 2002a: 13). Nevertheless, Malaysia's current estimation of 15.6 researchers per 10,000 labour force pales in comparison to the reported 60 (1998) for South Korea, 74 (1999) in the United States, 83.5 (2000) for Singapore and 136 (1999) for Japan.

Table 5.8 Percentage Distribution of Educational Attainment of Labour Force (selected years)

Educational Level	1967	1974	1985	2000
No Formal	41.0	20.8	14.1	6.1
Primary	45.0	51.5	39.7	26.1
Secondary	8.0	25.0	4.5	53.8
Tertiary	0.6	1.8	4.7	13.9

Source: Nagaraj and Lee (2002).

Table 5.9 Output of Degree Courses,[1] 1986–2000

Course	5MP Number (%)	6MP Number (%)	7MP Number (%)	8MP[2] Number (%)
Art & Humanities including Economics, Business & Law	27,780 (53.0)	49,018 (62.0)	87,882 (58.4)	161,102 (48.4)
Science including Medicine, Agricultural Science, Pure Sciences & Others	17,510 (33.0)	19,642 (25.0)	38,273 (25.4)	100,967 (30.4)
Technical, Engineering, Architecture, Surveying & Others	7,550 (14.0)	10,508 (13.0)	24,343 (16.2)	70,650 (21.2)
Total	52,840 (100.0)	79,168 (100.0)	150,498 (100.0)	332,719 (100.0)

Notes:
1. Output for First Degree, Masters, PhDs and post-degree Diplomas.
2. Estimates.

Source: Malaysia (2001, p. 350).

More importantly, the emergence of graduate unemployment in 2002 is indicative of some of the problems in tertiary education. Although it is possible that the employment market in Malaysia may have been adversely affected by the global slowdown, nevertheless it is pertinent to note the comments from the Ministry of Human Resources on this issue. As stated in the *New Sunday Times*, on 24 November 2002, 'only 600 of the 13,000 graduates who registered with the Ministry of Human Resources for jobs in the service and marketing sectors in the first 10 months of the year were employed, while the rest were rejected due to their poor command of English and lack of communications skills'. Apparently there were 3,300 job opportunities in these sectors, but only 600 successful applicants. To a large extent, the unemployment problem indicates the mismatch between the skills required by the private sector and the output of the university as well as the quality in the tertiary output.

Financial resources

While public funding has benefited some firms that have managed to access these funds, the number of VC companies (VCCs) reported have increased from 20 in 1995 to 30 in 1999, with a total of RM726 million invested in the manufacturing sector. However, the funds were used mainly for acquisition or buy-out financing, second-stage financing and bridging financing. Significantly, VCCs are reported to be still reluctant to provide seed capital financing due mainly to the risk-averse attitude towards investment (Malaysia, 2001: 413). Similarly, financing for start-ups was substantially reduced from RM81.1 million in 1997 to RM7.3 million in 1999. Although more data are needed before it can be ascertained whether the economic crisis has contributed to this sharp decline, nevertheless it can be seen that VCCs have not contributed substantially to overcoming the financing problems in technology development in the country.

MESDAQ is still relatively new. Since commencement of trading in April 1999, 20 technology-based companies have been listed (*The Star*, 26 June 2003: Stocks, p. 21). Another 14 technology-based companies have been listed in the Kuala Lumpur Stock Exchange (KLSE) main board while three are listed in the KLSE second board. Therefore altogether 37 technology-based companies have been able to access public funds.

It would appear that the current funding mechanisms, be they government or private, have met the financing needs of only a small number of private firms. Thus funding continues to be a hindrance to innovation and R&D activities for private firms. It is also likely that the general shortage of skilled personnel has also affected technical personnel, who are required by government and private institutions to effectively assess the risk and returns in the application for funds.

Coordination, implementation, monitoring and assessing problems

The various schemes for technology development clearly show that Malaysia is not short on vision or direction in terms of policy formulation. But, these policies involve a large number of government institutions. For example, in the case of human resources alone, it is not just the Ministry of Human Resources that is involved. Rather manpower projections require close coordination with MITI and MOSTE. Similarly, while the management of S&T in Malaysia is centralised, there are many ministries, agencies and departments involved. Given that there is to some extent inter-ministry, inter-agency and inter-department rivalry, it is important to have effective coordination.

Similarly, effective implementation requires constant monitoring and assessment. According to UNCTAD (2001: 188), in the case of technology transfer, there has been no systematic assessment of the effectiveness of policy instruments in fostering local input linkages and technology transfer in Malaysia. The lack of formal systematic *ex post* evaluation programmes is rather incongruous given the government's concern over these issues as reflected in the numerous policy initiatives that have been taken to enhance technology deepening in the country. Assessment efforts are constantly hindered by the availability of data in this country. This requires up-to-date data collection and analysis.

5. POLICY IMPLICATIONS

Human Resources

There are current isolated attempts at addressing the weakness in English proficiency in the country. Science and mathematics are now taught in English while students entering university are required to take the Malaysian University English Test (MUET), although passing it is not compulsory. Unfortunately, weak language and communication skills at the tertiary level are to a large extent inherited problems. Furthermore, the quality of tertiary education institutions and programmes is strongly determined by the quality of secondary school graduates and this is in turn linked to primary school programmes. Thus, there is a need for a *comprehensive* rather than a piecemeal approach towards improving the quality (and not just the English proficiency) of graduates and tertiary education in Malaysia. In other words, there has to be an objective evaluation of the entire educational system.

This assessment should not only consider improving the quality of the programmes at all levels, it should also consider improving the quality of the service providers in these programmes or the teachers and lecturers. Public universities should move towards more independent funding to reduce the fiscal burden of the government and to enable the public universities to be more responsive towards market needs in both teaching and research. It would also push the public universities and research institutes towards seeking private funding for R&D. The current level of collaboration in the private sector is still weak as out of the RM131.9 million R&D funds that were outsourced from the private sector, only 3.9 per cent went to IHLs and GRIs.

Financing Mechanism

A review of the funding mechanism covering both the supply and demand for funds will enable the government to ascertain the problems in funding, especially in terms of the demand for and accessing of the funds. Given the relative underdeveloped nature of private sector funding for technology development in the country, government provision of funds will certainly help to bridge some of the gaps in financing from the private sector. However, it is important to know where the blockages are in the financing mechanism before it can be fine-tuned to meet the financing needs of technology development. It is equally important to survey the firms to ensure that the supply of funds is meeting the demands of industry.

Coordinating, Implementing, Monitoring and Assessing the Effectiveness of Policies

A coordinating body needs to be set up to strengthen the coordination of S&T policies with the industrial and broader economic development policies of the country. This body can also be used to enhance the implementation of policies as well as to monitor and assess their effectiveness. The use of government funds that have been disbursed need to be closely monitored to prevent wastage of resources. In this regard, all firms receiving government funding should have performance indicators to assess their development and to ensure that their competitiveness is increasing.

6. CONCLUSION

Malaysia has pursued industrialisation for more than four decades. In this period, it has clearly succeeded in low-tech industrialisation as shown in the structural transformation of its economy from the production and export of primary commodities to that of manufacturing and in particular the export of e&e products.

Although the need for industrial deepening was identified as early as IMP1, it is possible that MITI as a whole was more concerned about the achievements of the NEP objectives during the 1970s and 1980s than in technology issues (Anuwar, 1992: 89). Thus, efforts to address technology deepening in industry began in earnest only with the promulgation of the NAPITD in 1990. However, development in the second half of the 1990s was interrupted by the emergence of the Asian financial crisis in 1997 and the domestic economic crisis in 1998. Although the economy did recover in 1999, effectively the crisis has derailed industrial development for at least two years. Consequently, the drive towards accelerating technology development and industrial deepening has only been in effect for approximately 10 years. In this short 10 years, the government has formulated a vast array of policy measures that are targeted at increasing technology transfer, encouraging the MNC subsidiaries to upgrade their operations, attracting FDI in high-tech industries, accelerating R&D and innovation activities in the country, promoting indigenous R&D development and strengthening the capabilities of the SMEs. What then has been the impact of this vast plethora of government support on technology development and industrial deepening?

Empirical evidence from the two most recent national surveys on R&D and innovation indicate that R&D and innovation efforts have increased. This concurs with the available literature that the rapid growth in the electronics subsector is not completely hollow as claimed by some. However, these R&D and innovation efforts are of the process and product improvement variety rather than frontier development. Furthermore, it would appear that while both foreign and local firms are involved in these efforts, R&D expenditure of the foreign firms far supersedes that of the local firms. More importantly, the total number of firms involved in these efforts is quite small so that Malaysia has yet to amass a critical mass of firms that are involved in R&D activities.

Since firms are reported to rely mainly on external rather then internal funding for R&D, it appears that the increase in government funding has stimulated an increase in private R&D expenditure and innovation activities. Again, the numbers reported are small. The two main fiscal incentives that have benefited the private sector are double deduction for R&D activities and duty-drawback for imported inputs while the rest of the incentives are less well utilised. It would appear that the delivery mechanism needs improvement, as the main deterrent encountered in firms' recourse to these funds lies not in ignorance of the availability of funds but in the application process as the procedures are not clear to the applicants. As in the case of R&D spending, the numbers that have been able to access these incentives are rather small.

Similarly, the number of SMEs within the e&e subsector which are involved in the various government programmes that have been formulated for the promotion of SMEs and linkage development such as the VDP, ILP and GSP are small, approximately 10 per cent. While backward linkages have evolved over time within this subsector, it is important to note that the number of local suppliers that have benefited from these linkages are even smaller as some of the suppliers to MNCs are not Malaysian but foreign SMEs that are operating in Malaysia. Certainly anecdotal evidence shows that several successful local firms have emerged in the e&e subsector, for example, Globetronix and Eng

Hardware, but it is troubling to note that these are isolated cases. Moreover, SMEs have underutilised the ITAF, which has been set up for them, especially ITAF2 which is geared towards their technology development.

Since only a small number of firms have managed to access public funding for R&D and innovation activities for technology development, it is unlikely that private funding can do better. The relatively risk-averse nature of VCCs and shallow capital market in the country have not helped in the development of private funding. Thus funding and the high cost of innovation activities continue to be cited as a hindrance to innovation activities from the private sector.

While the private sector is forced by increasing global competition to increase its R&D and innovation activities as well as to transfer technology, it is hindered by the local absorptive capacity, especially in terms of R&D personnel and technical and engineering support staff. As in the case of industrial policy, the education policy of the country was more concerned with nation-building and the implementation of the NEP until the 1990s. Thus manpower development was not geared towards the industrial needs of the country and mismatches between the supply of manpower and the demand from industry as well as the supply and demand for different types of skills have emerged over time. While there are increasing efforts to liberalise the education and training sector in order to improve the supply, these mismatches have yet to be resolved. Without substantial improvements in the matching of manpower supply and demand, more programmes and more funding or more incentives will merely go to waste. Building local innovative capacities must begin with building local human capacities. In this regard, the piecemeal approach towards improving supply, such as the current efforts to teach science and mathematics in English, may not be adequate. Rather, the whole delivery system needs to be reviewed, from the curriculum to the effectiveness of teaching and training.

It would certainly require better coordination among the numerous ministries, agencies and departments that are involved in technology development and industrial deepening. There has to be a greater emphasis on implementation, monitoring and assessment in order to increase the effectiveness of the policies in the country. New initiatives may be plagued by the same old implementation problems that have not been resolved due to insufficient attention to implementation issues. If coordination is hindering effective implementation, new initiatives will still suffer from coordination problems if left unresolved. Similarly, continuous monitoring and assessment will ensure that the nation reaps the full benefits of these policies.

Finally, given that policy focus on technology development is relatively new in the country, it is unrealistic to expect the country to make the technology leap in such a short space of time. Malaysia must first learn to crawl up the technology ladder.

NOTE

1. Literally means 'son of the soil'.

REFERENCES

Akademi Sains Malaysia (2001), 'A Review of the National Science and Technology Policy', Draft Report submitted to the Ministry of Science, Technology and the Environment (MOSTE).

Anuwar, Ali (1992), *Malaysia's Industrialization: The Quest for Technology*, Singapore: Oxford University Press.

Anuwar, Ali (1994), 'Industrial Development in Malaysia', Chapter 27 in National Institute of Public Administration, *Malaysian Development Experience: Changes and Challenges*. Kuala Lumpur: National Institute of Public Administration (INTAN), pp. 706–25.

Ariff, Mohammad, Mahani Zainal Abidin and Tan Eu Chye (1996), 'Trade and Investment Policies in Developing Countries: Malaysian Case Study', Kuala Lumpur: Faculty of Economics & Administration, University of Malaya.

Bruton, Henry J. (1992), *Sri Lanka and Malaysia: The Political Economy of Poverty, Equity, and Growth*, New York: Oxford University Press.

Chee Peng Lim (1987), *Industrial Development: An Introduction to the Malaysian Industrial Master Plan*, Petaling Jaya: Pelanduk Publications.

Chin Kok Fay (2001), 'Financing Manufacturing in Malaysia: Experience, Issues and Challenges', Chapter 8 in K.S. Jomo (ed.), *Southeast Asia's Industrialisation: Industrial Policy, Capabilities and Sustainability*, London: Palgrave.

Ernst, Dieter (2002), 'Global Production Networks in East Asia's Electronics Industry and Upgrading Perspectives in Malaysia', East-West Centre Working Papers, Economics Series, No. 44, March.

Hatzichronoglou, T. (1997), 'Revision of the High Technology Sector and Product Classification', STI Working Papers 1997/2, Paris: OECD.

Hobday, Michael (1996), 'Innovation in Southeast Asia: Lessons for Europe?', *Management Decision*, 34/9: 71–81.

International Bank for Reconstruction and Development (1957), *Report on the Economic Development of Malaya,* Baltimore, MD: Johns Hopkins Press.

Jomo, Kwame Sundaram (1990), *Growth and Structural Change in the Malaysian Economy*, Basingstoke: Macmillan.

Jomo, Kwame Sundaram and Chris Edwards (1993), 'Malaysian Industrialization in Historical Perspective', Chapter 1 in K.S. Jomo (ed.), *Industrialising Malaysia: Policy, Performance, Prospects*, London: Routledge, pp. 14–39.

Lall, Sanjaya (1996), *Learning from the Asian Tigers: Studies in Technology and Industrial Policy*, London: Macmillan.

Lee, Cassey (2003), 'Implementing Competition Policy in Malaysia', Paper presented at Workshop on Malaysia: Selected Economic Issues and the

Challenges Ahead, at the Institute of Southeast Asian Studies (ISEAS), 27 March, Singapore.

Lee, Kiong Hock (1998), 'Labour Market Issues: Skills, Training and Labour Productivity', Chapter 12 in Soon Lee Ying and Shyamala Nagaraj (eds.), *The Seventh Malaysia Plan: Productivity for Sustainable Development*, Kuala Lumpur: University of Malaya Press, pp. 181–201.

Mahani, Zainal Abidin A. (1998), 'Malaysian Industrial and Trade Policies under the New International Trading System', Report submitted to the Economic Commission for Latin America and the Caribbean, LC/R.1805.

Malaysia (1996), *The Second Industrial Master Plan 1996–2005*, Ministry of Trade and Industry.

Malaysia (2001), *Eighth Malaysia Plan, 2001–2005*, Kuala Lumpur: Malaysian National Printing Corporation.

Malaysia Industrial Development Authority (MIDA) (2002), *Business Opportunities in Malaysia's Electronics Industry*, Kuala Lumpur: MIDA.

Malaysia Industrial Development Authority (MIDA) (2003), *Malaysia: Investment in the Manufacturing Sector (Policies, Incentives and Facilities)*, Kuala Lumpur: MIDA.

Malaysian Science and Technology Information Centre (MASTIC) (2001), *National Survey of Innovation: 1997–99*, Kuala Lumpur: MASTIC.

Malaysian Science and Technology Information Centre (MASTIC) (2002a), *2000 National Survey of Research and Development*, Kuala Lumpur: MASTIC.

Malaysian Science and Technology Information Centre (MASTIC) (2002b), *2000 Malaysian Science and Technology Indicators Report*, Kuala Lumpur: MASTIC.

Ministry of International Trade and Industry (MITI) (2002), 'SMI Development Plan: 2001–2005', Kuala Lumpur: Malaysian National Printing Corporation.

Nagaraj, S. and K.H. Lee (2002), 'Human Resource Development and Social Re-engineering: Which Part of the Field are We Leveling?', Paper presented at the CEDER Conference on Sustaining Growth Enhancing Distribution: The NEP and NDP Revisited, 15–16 May, Auditorium Faculty of Built Environment, University of Malaya.

Nor Ghani Md. Nor Zulkifly Osman, Ahmad Z. Abdullah and Chin Yit Jun (2000), 'Trends in the Malaysian Industrial Market Structure', *Jurnal Ekonomi Malaysia*, 34: 3–20.

Rasiah, Rajah (1995), *Foreign Capital and Industrialization in Malaysia*, New York: St. Martin's Press.

Rokiah, Alavi (1996), *Industrialisation in Malaysia: Import Substitution and Infant Industry Performance*, London: Routledge.

Ruhizan, M. Yasin and Ahmad Basri (2003), 'Development in Education in Response to Globalisation: A Comparison between Japan and Malaysia', Paper presented at the 2nd Workshop on Japan's Response towards Globalisation: Lessons for Malaysia, 11 June, IKMAS, Bangi.

SMIDEC (2002), *SMI Development Plan (2001–2005)*, Kuala Lumpur: Malaysian National Printing Corporation.

Tan Tat Wai (1981), 'Industrial Development Strategy for Malaysia', *Jurnal Ekonomi Malaysia*, No. 3 & 4 99–118.

Tham Siew-Yean (2003), 'Foreign Direct Investment in Malaysia', First Draft of Report submitted to the Asian Development Bank, 1 May.

Tham Siew Yean and Mahani Zainal-Abidin (1999), 'Industrial Institutions: The

Case of Malaysia', Chapter 5 in Colin Barlow (ed.), *Institutions and Change in Southeast Asia*, Cheltenham: Edward Elgar, pp. 55–71.
UNCTAD (2001), *World Investment Report 2001: Promoting Linkages*, Geneva: UNCTAD.

6. High-Tech Industry Development in the Philippines: At a Dangerous Crossroad?

Gwendolyn R. Tecson

1. INTRODUCTION

The development of high-technology industries has become the major objective of many contemporary developing countries. However, a great divide seems to separate those who are able to develop such industries from those who are not able to. And the hard reality is that more countries seem to constitute the latter group. Why is this so?

Technology upgrading and deepening seem to stand at the heart of this great divide. Technology acquisition *per se* does not guarantee development of high-technology industries. While technology must be initially acquired, adopted and eventually mastered by the acquiring country, the final test of whether industrial technology will transform the acquiring country from a developing to a newly industrializing economy (NIE), or indeed into a developed country, is the latter's ability to *upgrade* and *deepen* such technology (Lall, 2003). Failure to do so leaves a country stagnating on the lowest rungs of the technological ladder.

Where does the Philippines stand, given its current position and its potentialities?

This chapter argues that the Philippines stands at a dangerous crossroad along the road to development. Whether it is able to get over that impasse will depend crucially on public policy reform and private sector initiative. How, in the first place, did it get to that crossroad? Are there signs that it will ever be able to cross over? Section 2 takes a brief look at industrialization in the Philippines, using the technology definition adopted in this chapter. Then it takes a long look into the evolution of the electronics industry in the Philippines, an industry that represents the 'high-technology' industrialization to which most developing countries aspire. In particular, it delves into the industry's modes of technology acquisition, using the classical division of import-substituting and export-oriented stages, as well as the industry characteristics that flow from such modes. Section 3 discusses the main policies and institutions that fostered these modes of technology acquisition, including the international developments that helped shape it. Section 4 discusses the weaknesses of the electronics industry, particularly that of its inability to cross the technological divide, making it an industry that stands dangerously at the crossroad of development. It looks into the economic, political and institutional factors that shaped its current state. Section 5 discusses the elements of a way that may lead it out of such an impasse. Finally, the concluding section looks soberly at the possibilities of the electronics industry and assesses the country's ability to cross over the technological divide.

Definition of 'High Technology'

High-technology industry has been defined in various ways and fora. In the present case, we adopt the OECD classification of *high-technology sectors* depending on their technology intensity, that is, taking into account both the level of technology specific to the sector (that is, measured by the ratio of R&D expenditures to value added) and the technology embodied in the purchase of intermediate and capital goods. Hatzichronoglou (1997) provides the derivation of 'high-tech' sectors and products which we shall by and large adopt in this study, depending on the availability of data[1].

Brief Glance at Philippine Industrialization

Table 6.1 presents the shares of industries broadly classified according to the technology classification given by Hatzichronoglou (1997).

Table 6.1 Shares of Philippine Industries by Technology Classification, 1979 and 1998 (%)

Technology Classification	Share of Total Value Added	
	1979	1998
High and medium-high technology	22.51	33.15
Medium-low technology	14.70	19.43
Low technology	62.78	47.09

Source: Philippine Statistical Yearbook, 1981 and 2003.

It can be seen that by the end of the 1970s, low-technology industries dominated the manufacturing sector, making up about 63 per cent of the sector's value added. About twenty years later in 1998, the share had gone down to a little less than half or 47 per cent share of value added of the sector. In contrast, high and medium-high technology industries made up a little over a quarter of value added of manufacturing in 1979; by 1998, such industries made up about a third of the sectoral value added. Most of the increase was due to the rise in the share of export-oriented, foreign-dominated electronics industry, as will be seen below.

The Philippines, like most developing countries, started off its industrialization process with import-substituting consumer goods industries: food, beverages and tobacco, textiles and clothing and so on. (Table 6.2). These were fostered behind high tariff and non-tariff walls, which gave the consumer goods industries the highest effective rates of protection, at the expense of export production. It proved, however, to be the 'easy' phase of industrialization until the limits of a thin domestic market were reached and the peso had to be floated. Amidst the export successes of its newly-industrializing neighbours as well as the acute need for foreign exchange, a policy shift took place towards a more outward-looking industrial development strategy with the Export Incentives Act of 1970 and the opening up of the domestic market to more foreign investments through the enactment of the Investment Incentives Act three years earlier. But it was a half-hearted shift to export orientation because the still highly protective tariff system proved to be the primary source of bias against exporting. It was only in the early 1980s that the Philippine government started to bite the bullet with the start of the tariff reform policy. There was a brief attempt to undertake secondary import substitution with 11 major industrial and heavy-capital-intensive industries underway (copper smelting, petrochemicals, fertilizer production and so on). However, all these had to be halted, given the severe foreign exchange crisis that translated itself into a debilitating economic crisis. It was only with the end of the Martial Law period and resumption into office of the new administration of Pres. Corazon Aquino in 1986 that trade liberalization was resumed and, together with it, a more credible export-oriented industrialization strategy. The liberalization period, however, when the economy embarked on a more credible export-oriented strategy, coincided with a lower economic growth and lower ratio of manufacturing employment than the import-substitution period that preceded it.

In the OECD classification, the industries that were considered 'high-technology' industries were: aerospace, computers & office machinery, electronics–communication, and pharmaceuticals. With trade

Table 6.2 Phases of Industrialization in the Philippines

Periods	Economic Performance	Specific Policy Measures
1957–1969 Import substitution	Real GDP growth: 4.5%; employment in manufacturing: 12%	High protection; tariff escalation
1970–1982 Shift to export orientation	Real GDP growth: 6.2%; employment in manufacturing: 11%	Investment Incentives Act of 1967; Export Incentives Act of 1970; creation of export-processing zones; de facto devaluation & subsequent floating of peso
1983–85 Severe economic crisis	Ave. real GDP growth: (4.2%); employment in manufacturing: 10%	Subsidized loans; tariff protection to 11 major industrial projects
1986 – Export orientation	GDP growth: 3.3%; employment in manufacturing: 10%	Trade liberalization pursued; Foreign Investments Act of 1991; creation of more export-processing zones

liberalization, most segments of the highly protected and foreign-dominated pharmaceutical industry migrated to other Asian countries, while no important aerospace industry exists in the country. Thus, in order to facilitate discussion of the issues, this chapter will limit the analysis to the electronics industry, both consumer (radio, TV and communication equipment) and industrial electronics (computers & office machines, telecommunications, electronic integrated circuits and microelectronics) which roughly constituted about 12 per cent of the country's manufacturing value added (MVA) in 2000. Since the OECD classification is based on ISIC Rev. 2 (International Standard Industrial Classification) it covers only manufacturing industry and thus excludes services, which is a new development in the country's electronics sector.

2. GENESIS AND GROWTH OF THE ELECTRONICS INDUSTRY IN THE PHILIPPINES

Industrial technology, for a developing country like the Philippines, is *acquired* technology. It is acquired from countries or firms that developed it, because a developing country, by definition, usually does not have the wherewithal to be its source. The traditional forms of technology acquisition are *internalization* mainly via foreign direct investment (FDI or investments by foreign multinationals in affiliate companies) or *externalization (*through sale of proprietary technology). However, the second mode has taken many forms, such as joint ventures, sale of equipment, licensing, technical assistance, turnkey projects and franchising (Lall, 2003). Moreover, straddling between the two forms are inter-firm arrangements that involve neither FDI nor outright technology sales, and these are modes of international subcontracting or original equipment manufacturing (OEM) arrangements. For instance, while the growth of South Korea's industrial technology development was believed to have eschewed the former in favour of the latter and its various forms – where indigenous technological effort was normally larger – recent research has shown the importance of technological learning through OEM (Cyhn, 2002). Whatever the form of technology acquisition a firm from a latecomer developing country ultimately takes, it will usually involve a multinational corporation (MNC) that developed or at least mastered the technology.

The production of consumer electronics in the Philippines dates back to the import-substituting period of the 1960s (at the time, the industry being still mainly part of the electrical industry). Just like the rest of manufacturing, the industry expanded mainly through the sheer growth of domestic demand, a large part of which had to be met by domestic production due to the presence of protective barriers on import substitutes. Table 6.3 shows the average tariff rates in the industry in 1978, as compared with that of average tariff for manufacturing (1974).

Because of protection, the industry started off with the assembly of imported CKDs (completely knocked down) and SKDs (semi-knocked down) parts, interspersed with component parts that were domestically assembled under a foreign or local brand. Production units were either affiliates of foreign firms – majority ownership *had* to be in the hands of Filipinos[2] (e.g., GE Appliances, Hitachi) – or were mainly Filipino-

Table 6.3 Average Tariff Rates in the Appliance Industry, 1978 (%)

Sector	Average Tariff Rate
Appliance	71
Subsectors:	
Audio-Video	89
Cooking, Heating	39
Miscellaneous	68
Refrigerators and air conditioners	100
Parts	
Radio, TV parts	59
Manufacturing Sector (1974)	44

Source: Lapid (1996, Table 5.1).

owned firms, either in joint venture with foreign MNCs (e.g., Delta Motors in joint venture with Sharp and Daikin) or as licensees of foreign MNCs (Conception Industries with Carrier, E. Lim Enterprises with Sony).

Just like most new industries, foreign technology was ubiquitous in the manufacturing sector. Foreign licensing, however, seemed to be the preferred route because the Americans were not too keen on investing in the country,[3] while the Japanese, generally, had not yet found the need to transplant their production units overseas. Thus, Conception Industries assembled Carrier air conditioners, Motorola TV sets and Whirlpool refrigerators; Radiola was a licensee of Toshiba, Refrigeration Industries, Frigidaire, and Union Industries of Hitachi. Given the penchant of Filipinos for foreign brands, it is hardly surprising that Radiowealth, an indigenous company, had to struggle with its own brands (Yoshihara, 1985) and was too small a producer to even be counted.

Being an import-substituting industry, production was mainly for the domestic market. It continued to grow under a combination of high protection – including a local content policy – and demand from a growing middle class. The industry nevertheless followed the boom–bust cycles of the Philippine economy. Although quality and supply availability were supposed to be comparable with those produced by other ASEAN (Association of South East Asian Nations) countries, price competitiveness eluded the industry. As Table 6.4 indicates, direct price comparisons of selected appliances produced in the Philippines relative to those produced abroad show considerable price differentials, especially for air conditioners and electric fans.

A measure of efficiency utilizing the domestic resource cost–shadow exchange rate ratio[4] (DRC–SER) would readily show that appliance production and its components (Table 6.5) tended to be very inefficient. The ratios, however, improved in 1988 when tariffs started going down,

Table 6.4 Direct Price Comparisons (Pd/Pb) for Selected Appliances, 1985–1990 (%)*

Product	1985	1986	1987	1988	1989	1990	1991
Air conditioner	298	247	253	262	264	250	247
TV set	180	116	124	114	106	122	80
Radio	57	88	105	102	96	88	90
Radio phono	19	15	23	131	234	184	131
Electric fan	174	226	253	352	371	361	325

Note: * Pd = domestic price; Pb = foreign price, using Hong Kong unit price values.

Source: De Dios (1994, Appendix 8).

particularly on inputs, and the appliances industry, as a whole, began to be classified as 'mildly inefficient'. There is an *a priori* reason why they could not readily compete in export markets. Considerable production rationalization had to be undertaken when the tariffs began to go down in earnest in the 1990s.

In contrast, non-consumer or industrial electronics grew in an environment when MNCs began to look for overseas locations whose labour abundance would allow them to preserve their competitiveness. They were thus mainly produced by subsidiaries of MNCs, often wholly owned, whose preferred mode was FDI. Semiconductor manufacturer Intel Philippines came in as early as 1974, Temic Telefunken Microelectronics (German) also in 1974, Motorola (Phil) Inc. in 1979, Texas Instruments Philippines and Fairchild Semiconductor Hong Kong (Holdings) Ltd. in 1979, and Philips Semiconductors Philippines in 1981. The first entrants were thus mainly American and European in origin. Indeed, by 1995, the electronics industry was the third 'most foreign' industry in the manufacturing sector, with 84.1 per cent of output produced by firms with foreign equity, surpassed only by petroleum refining (100 per cent) and non-ferrous metals (89.8 per cent) (Hill, 2003, p. 236).

The Japanese semiconductor firms, on the other hand, arrived later, that is, in the 1990s, with Fuji Electric Philippines, Inc. and Sanyo Semiconductor Mfg. (Phil.) Corp., both establishing themselves in 1995 (Tecson, 1999). In the late 1980s, after the huge *endaka* of the 1985 Plaza Agreement, Japanese firms started to invest all over ASEAN to

Table 6.5 DRC-SER Ratios in the Electrical and Electronics Industry, 1983 and 1988

Industry	1983	1988
Appliances	1.94	1.48
Subsector		
Audio-Video appliances	3.34	1.64
Cooking and heating	3.11	1.05
Refrigerators and air conditioners	2.89	1.56
Miscellaneous appliances	1.09	0.97
Parts		
Radio and TV parts	4.31	13.15

Notes: SER is the shadow exchange rate, which ws 13.19 in 1983 and 26.37 in 1988.

Source: Lapid (1996, Table 5.12).

transfer some of their plants that had become uncompetitive as a result of the large yen appreciation. Given the severe economic and political instability of the country during the end and the aftermath of the Marcos regime, the Philippines was not ready to receive part of the *tsunami* of Japanese investments that found their way to the more FDI-friendly host countries in Asia, that is, Singapore, Malaysia and Thailand. The entire decade of the 1980s, when the rest of ASEAN boomed partly as a result of FDI, was for the Philippines a lost decade as far as FDI was concerned. When the situation had normalized in the mid-1990s, however, Japanese investors came in a big way into the country's electronics industry. A wave of investments into storage products (hard disk drives or HDDs) was ushered in by the entry in close succession of all four of the Japanese majors, beginning with Hitachi Computer Products (Asia) Corp. in 1994, followed a year later in 1995 by the almost simultaneous establishment of Fujitsu Computer Products Corporation and Toshiba Information Equipment Inc. Less than a year later, NEC Components Philippines was set up to produce printed wiring boards (PWBs) mainly for the automotive industry but also partly for HDDs.

Like the American and European investors before them, since these Japanese firms were mainly interested in taking advantage of the semi-skilled, labour surplus available in the country, they merely transferred the labour-intensive portion of semiconductor production to the country. Components in the form of silicon wafers were being packaged and tested in the country and shipped back to the country of origin as exports or to regional headquarters for distribution in other developed countries. The industry was wholly export- oriented. Similarly the HDD majors were attracted by the large supply of semi-skilled labour in the country. Interestingly, the American HDD makers preferred mainland Asia to the Philippines, having located mainly in Singapore and spreading to Malaysia, Thailand and eventually China (McKendrick et al 2000).

FDI begets FDI so that around the time of their setting-up shop in the country, that is between 1994 and 1996, the Japanese MNCs' major and minor suppliers in Japan and elsewhere in Asia followed them into the country. Nidec Philippines Corporation (HDD-related spindle motors; entered in 1995 and began production in 1997), Luzon Electronics Technology Inc. (engaged in slider assembly and HGA or head gimbals assembly, a 100 per cent-owned subsidiary of Hitachi Metals, Inc.), Tsukuba Philippine Die-casting Corporation (aluminium base plates for 2.5 inch HDDs), San Technology (voice-coil motor, HGA and HSA or head slider assembly), Mette (base plates and covers), Sunpino (FPCs or flexible printed circuits), and Precision Technology or Pretech (spacer rings). Singaporean-owned CAM Mechatronics also began producing HDD components in 1995. Laguna Electronics Incorporated, a company belonging to the Tsukiden Group and established in 1993, set up a second plant dedicated to assembly of IBM-licensed high-capacity HDDs, for NEC HK, with commercial operation starting late 1998. Other investments by Japanese majors were expected before the 1997 crisis, including an investment by TDK in magnetic heads (which eventually came in 1996). Fujitsu was in the process of setting up its media factory when the Asian crisis struck in 1997. In contrast, the American MNCs seem to have preferred the Asian mainland, with Singapore as the centre of the industry's operations in Southeast Asia.

Indeed, the mid-1990s can be considered by the Philippines as the heyday of FDI in the electronics industry, as can be seen from the data on investments in the electronics industry in the 1990s (Table 6.6). One can also observe the dramatic increase in exports during the second half of the 1990s, which not even the Asian financial crisis could abate. However, investments in the industry showed a remarkable downturn after 2000, that is, after the Y2K scare had died down.

Table 6.6 Investments and Exports of the Philippine Electronics Industry, 1992 -2002 (US$m; %)

Year	Investments	Exports	Industry Exports/Total Exports (%)
1992	40	297	28
1993	220	378	33
1994	1,290	489	36
1995	2,160	755	43
1996	1,080	10,610	52
1997	1,470	14,980	58
1998	670	19,870	67
1999	790	25,340	72
2000	1,240	27,170	71
2001	720	21,900	68
2002	270	24,260	69

Source: SEIPI (2003).

Eventually, though, a number of Filipino-owned firms began to be set up as contract manufacturers and subcontractors of MNCs, some of which are documented in Table 6.7. Of the 800 electronic firms presently in operation, 28 per cent are Filipino (SEIPI; BOI and PEZA files). Foremost among them are Integrated Microelectronics Inc. (IMI) and Ionics Circuits Inc, which used to be mainly involved in the packaging of semiconductors, assembly of heads for the HDDs, and printed circuit board assembly. However, IMI, following the OEM route, has since evolved into an array of electronics manufacturing services (EMS) including product design and development. Ionics Circuits, on the other hand, has diversified from head assembly into production of CD-Rom and CD-RW, as well as telecommunications equipment, cellular phones and pagers. Their products, having been subcontracted by foreign firms, are destined for the export market. In 1995, 71 per cent of the total output of the industry was exported, making the industry among the most highly export oriented in the manufacturing sector (Hill, 2003, p. 236).

Technological dependence on foreign investors came naturally as a consequence of the need for technology through foreign licensing and/or FDI. But the degree of technological dependence can be expected to decrease as domestic firms are able to 'own' the technology through continuous technological learning. For instance, South Korea is believed to have relied less on the traditional flows of foreign technologies via FDI and licensing than on government intervention or market forces. However, Cyhn (2002, p. 11) argues that this phenomenon does not preclude the fact that 'MNEs were still a very valuable source of TL (technological learning) for Korean firms', because OEM, particularly in the electronics industry, was an 'indispensable ingredient behind its TL success'. In a powerful way, the importance of OEM in technological learning of South Korean *chaebols* is demonstrated through successful case studies. In the Philippine case, however, while there is some expansion in product lines over time and the accumulation of production experience in the process, little movement can be seen beyond OEM. ODM (original design manufacturing), let alone OBM (original brand manufacturing), remains an elusive dream.

3. POLICY REGIMES, INSTITUTIONS AND INDUSTRY DEVELOPMENT

It has been discussed above that some form of tie-up with MNCs was at the origin of the establishment of both the consumer and the industrial electronics industries. The former was mainly established during the country's import-substitution phase by domestic firms with foreign

*Table 6.7 Profile of Filipino Electronics Firms**

Company Name	Year Est.	Major Product Lines
Automated Technology (Phil.), Inc.	1987	Contract mfg for semiconductor products, assembly of multichip packages, plastic & ceramic IC, MCM & hybrid w/SMT, COB, OPTO & special assemblies
Cambridge Electronics Corporation	1993	Printed circuit boards (PCBs): single & double sided, multilayered
Dyna Growth Industries, Inc.	1992	3.5" floppy disk drives
Electronic Assemblies, Inc.	1988	Assembly, design/devt, tem cycling & burn-in, chip on board, in-circuit & functional testing, turnkey arrangement, captive & merchant line, failure analysis, rigid & flexible PC, CD-Rom drive, UPS-PCBA, cell phone PCBA, computers
Fastech Synergy Philippines, Inc.	1983	TO92, SOT89, UA device, SOT23/25/26/223, TO263, etc. RD91, PCA RFL, saw filters, metal cans, photodiodes, PDIP, dicing, special services
Gemphil Technologies Incorporated	1980	RF/IF transformers, coils
Hosokawa Assembly Incorporated	1995	Microvibration motors, headphones, electrolytic capacitors, FDD parts assembly
Integrated Mciroelectonics, Inc.	1980	Electronics mfg services (EMS) incl. product design and devt, prototyping, metal stamping, plastic injection, electronics assembly operations, supply-chain mgt to complete box-build solutions
Ionics Circuits, Inc.	1974	PCB assembly. Magnetic head assembly, motherboards for computers, finished product assembly, RF &GPS design, HDD, CD-Rom, cellular phones, pagers, telecom equipment, CD-RW
Labtech Mfg. Ind., Inc	1987	Silicon diode transistor
Mark Electronics Corp	1982	Single, double-side, multiplayer PCBs

Company Name	Year Est.	Major Product Lines
MX Technology Corporation	1995	PCB assembly, plastic injection mold, diode wires, complete assembly & testing
Oriental Toolmaster Corp	1984	Precision moulds & die components, jigs & fixtures
PSI Technologies Incorporated	1988	TO220, TO251, TO252, Powerflex TM, etc. PDIP, metal cans, flatpacks
Ramatek Philippines, Inc.	1992	Delay lines, filters, pulse transformers
Synertronix Incorporated	1996	Printed circuit boards (PCBs)
Team Pacific Corporation	1987	Hermetics, optoelectronics, power devices, plastic dips
Trident Electronics Corporation	1989	ODM of products & systems for computerized timekeeping & payroll, data capture and access control
Yosiko Industries Incorporated	1990	IC shipping tubes, end plugs, end pins, paper tapes

Notes: * These are electronic firms that are members of SEIPI (Semiconductor and Electronics Industry of the Philippines, Inc.).

Source: SEIPI Information Book, 2003.

licensing agreements with MNCs. On the other hand, the industrial electronics industry was set up directly by FDIs of subsidiaries of MNCs that specialized only in the labour-intensive part of the production process. Hence the industry was wholly export oriented, in contrast with consumer electronics which were mainly domestic-market oriented, at least at the start. The mode of foreign intervention – that is, through licensing or FDI – was influenced by the prevailing policy regime. Although the first semiconductor MNCs came during a time of high tariff protection, they were operating in specialized regimes that allowed freedom of importation of materials/components as well as exportation of the final products. In the following subsections, we shall discuss the evolution of policy that shaped the mode of foreign intervention in the industry and eventually its development history.

Local Content Policy and Import Substitution

Apart from protection through tariffs, a local content policy was initially instituted under the Electronic Local Content Program (ELCP) in the

1960s. Another local content policy governed electric fans and cooking appliances, requiring that 80 per cent of the number of parts and components be locally produced. Assemblers were required to achieve a certain percentage of local content in the parts and components they use. This, bureaucrats argued, would then coerce manufacturers to procure such parts and components from local suppliers. However, it was later noted that the policy had little effect on the sourcing decisions of assemblers because, given the difference in prices of parts and components, it was possible to achieve the requirement even with low levels of local content. Moreover, given the lack of immediate sources of components and raw materials, firms were driven to set up their own plants to supply the needed intermediate products, no matter at what level of efficiency. Higher prices of such products could anyway be passed on to consumers due to the relatively high protection given to final products. Nevertheless the local content policy coupled with protection on assembly did induce the setting up of plants for the production of parts, but because it was inefficient and uncompetitive, it eventually needed rationalization when, under the Aquino administration, local content requirement policy on electrical and electronic appliances was lifted.[5]

Import Liberalization Policy and Export Growth

The government began to pursue a unilateral policy of tariff reduction starting in 1981. It was only during the severe economic crisis of 1983 to 1985 that it was effectively aborted due to the extreme balance of payments situation that prevailed. In 1987, policy reform moved in the direction of removing import quotas on manufactured imports. By 1996, only the hard-core quotas remained and these represented 3 per cent of total tariff lines. Starting in 1991, the second phase of the Tariff Reform Policy (TRP II) brought down average tariffs to around 16 per cent (Table 6.8). With the third phase (TRP III) starting in 1996, tariffs were brought down even further to reach an average uniform tariff of 5 per cent in 2004. However, given the present government's pronouncements on the need to review World Trade Organization (WTO) tariff commitments as well as its readiness to seek an ASEAN Free Trade Area (AFTA) waiver on petrochemicals, it is doubtful whether the uniform tariff rate of 5 per cent will be attained by the deadline it had earlier set for itself.

Because imports were being liberalized, the penalties on exports were also being gradually removed, weaning the economy from a purely import-substituting to an export-oriented one. The basically inward orientation of the economy retarded the growth of manufactured exports

Table 6.8 Average Nominal Tariff Rate, 1975–2004 (%)

Sectors	1975	1980	1985	1990	1995	2000	2002	2004
Agriculture	67.16	61.95	35.86	34.61	27.99	14.60	12.79	9.33
Mining	19.89	18.38	15.72	14.24	10.79	3.27	2.84	2.84
Manufacturing	41.42	39.07	27.09	27.10	14.00	6.99	5.03	3.34
Overall	44.32	41.37	27.60	27.84	15.87	8.06	6.02	4.19

Source: Austria (2002, Tariff Commission: Table 2).

because of the penalties imposed on them by the regime. However, after the trade policy reform, manufactured exports growth outstripped that of total exports so that by 2000, they constituted about 90 per cent of total exports, as against 60 per cent before the effective start of the reform.[6] Electronic exports benefited from import liberalization, eventually becoming the country's chief source of export earnings. As Table 6.9 indicates, industrial, rather than consumer electronics, make up most of the country's electronic exports. Such growth, however, must be seen in tandem with liberalization in the capital account sector, because electronic exports were basically driven by FDI of multinationals.

Foreign Investments Act

With the decline in nominal and effective protection of supporting industries and the phase-out of local content policies, the government has upgraded its incentive packages to MNCs seeking to relocate overseas in order to make them competitive with the incentive packages of other ASEAN countries. A strong boost to foreign investments, especially those that were export- oriented, came with the passage of the Foreign Investments Act (FIA) in 1991. This act liberalized existing regulations by allowing foreign equity participation up to 100 per cent in all investment areas not included in the Foreign Investment Negative List (FINL), so long as the enterprise was exporting at least 60 per cent (instead of 70 per cent under the Omnibus Investments Code) of their output. According to the World Bank (1997) the FIA of 1991 made the Philippine foreign investment regime resemble those of its Asian neighbours in structure and compared favourably in terms of equity allowances (except for Hong Kong). Interviews of Japanese assemblers conducted at Calabarzon revealed that they evaluated the available incentives as being on a par with those available in the rest of ASEAN.

Table 6.9 Exports of Industrial and Consumer Electronics, 1998-2000 (%)

Exports Category	1998	1999	2000
Electronic exports (US$ bn)	19.9	25.4	27.2
Industrial electronics	96.0	98.8	98.2
Of which:			
Semiconductors & other components	78.8	82.2	77.2
Electronic data processing	13.7	16.2	18.1
Consumer electronics	1.5	1.0	1.3

Source: Foreign Trade Statistics, National Statistics Office, various issues.

Export-processing Zones

Another direct boost to foreign direct investments came with the creation of export-processing and special economic zones. Export-processing zones (EPZs) were first established in the 1970s with the creation of the Bataan Export Processing Zone and the Baguio Export Processing Zone. Texas Instruments established itself in the latter and enjoyed, among others, the free import–export regime at a time of widespread protectionism. As Warr (1989) has analysed, however, the first EPZs were largely unsuccessful attempts at promotion, and resulted in the lowest social rate of return among processing zones established in other countries in the region (Indonesia, Malaysia, South Korea). However, together with the package of import liberalization policies after 1986, the EPZs that were established later attracted a large number of foreign firms, and at a pace that continually necessitated area expansion to accommodate new entrants. This indicates that the creation of EPZs *per se* will eventually fail to attract export-oriented firms, so long as the rest of the political and economic policies run counter to export orientation and to FDI. It was only after 1987, when unilateral liberalization policy was adopted in earnest, that the country began to attract more foreign multinationals.

An important stimulus to FDI came when a number of these newer EPZs, or ecozones, were allowed to be partly leased and developed by Japanese trading firms. The country's ecozone is said to be unique among East Asian programmes in that incentives are provided to ecozone developers and other infrastructure providers (World Bank 1997, p. 79). These facilitated the entry of Japanese electronic firms into the country through their marketing efforts in Japan. The cash-strapped Philippine government allowed private firms, including foreign ones, to lease and develop zones, opening up space and facilities for investors while encouraging export growth and the advantages of efficient industry-related infrastructure. The emergence of such privately developed ecozones that appear to be more dynamic and cost-effective than older government-owned and developed export zones has enhanced the overall climate of liberalization for foreign investments in the country.

A recent World Bank comparison of government incentives provided by other EPZs in the region suggests that the country's package of benefits and other incentives is competitive[7] and can easily be considered 'the most generous and flexible set of incentives available anywhere'. Duty-free import privileges are said to be without parallel in its provision of full exceptions in perpetuity for export and free trade enterprises for almost all project-related inputs, whereas other countries restrict such privilege to production-related items or those not available locally, or it is made available only once. The range of promoted activities in the country's ecozones is very broad, allowing services, utilities, infrastructure development and tourism apart from manufacturing. Being under the purview of the Philippine Economic Zone Authority (PEZA), ecozones enjoy the benefits of a more efficient bureaucracy that simplifies procedures, coordinates the activities of national and local public agencies affecting investors (such as payment of national and local taxes) and reduces the costs of importation. Among the incentives are exemption from *Societe Generale de Surveillance* (SGS) inspection and simplified import–export procedures. Problems associated with government bureaucratic procedures, especially Customs, tend to present problems for firms located outside the zones. Given the net benefits provided by cozens to investors, the result has been a massing up of supporting industries in the zones and around them. These tend to create real economies for assemblers and other manufacturers located in

Table 6.10 FDIs Classified According to Technology Level, 1973-2000

Technology Classification	1973–88		1989–94		1995–2000	
	Value	%	Value	%	Value	%
Medium-high/high technology	579.4	42.1	747.5	40.4	880.1	35.9
Of which:						
Machinery, appliances & supplies	81.3	5.9	443.8	24.0	508.7	20.7
Medium-low technology	313.5	22.8	792.8	42.9	431.4	17.6
Low technology	483.7	35.1	308.0	16.7	1142.5	46.6
Total FDI in manufacturing	1376.6	100.0	1848.4	100.0	2454.0	100.0

Source: Bangko Sentral ng Pilipinas, *Selected Indicators* (various issues).

the zones. Reduction in the cost of transport of supplies, greater and closer interaction between suppliers and assemblers, and among different suppliers, easier application of just-in-time (JIT) systems and agglomeration economies resulting from greater proximity are just a few of the benefits created by the recent observable trend of supporting industries being drawn to locate in and around the ecozones.

The start of the liberalization of the trade and foreign investment regime therefore ushered in the entry of export-oriented foreign investors in both the electric and the electronics sectors. Data on FDI into specific industrial sectors as classified by technology are presented in Table 6.10.

It can be seen that FDI into the sector has increased from about 6 per cent in 1975–88 to about a fifth of total FDI in 1995–2000. And because the electronics sector exports about 70 per cent of its output and is largely foreign- owned (84 per cent vs. 57 per cent average for the manufacturing sector in 1995: Hill, 2003, p. 236), these have captured an increasing proportion of total exports of the Philippines (Figure 6.1). In the 1990s, electronics exports began to replace garments as the chief manufactured export of the country.

Liberalization of the Services Sector

An important policy turnaround that helped lift the market as well as

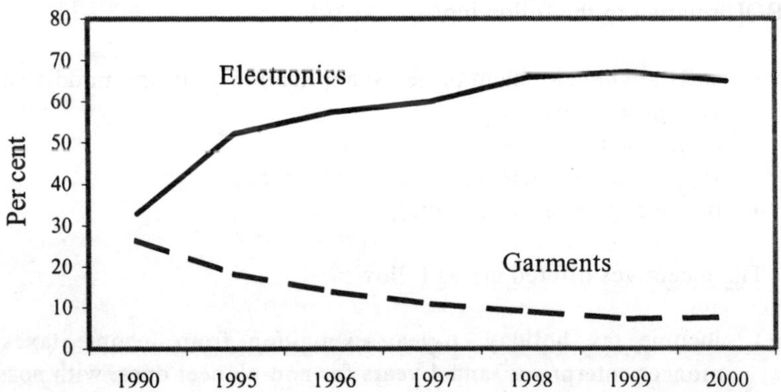

Source: NEDA Statistical Yearbook (2002).

Figure 6.1 Share of Electronics and Garments in Philippine Exports, 1990–2000

encourage more FDI into the industry is the liberalization of the services sector, particularly the telecommunications sector. Before 1993 the local telephone service was a virtual monopoly, thanks to the only national franchise held by a private telephone company, the Philippine Long Distance Telephone Company (PLDT). The company enjoyed 94 per cent of local exchanges, while the remaining ones were held by government and local private carriers (Abrenica and Llanto, 2003). In 1993, however, the government broke the monopoly by introducing competition in all market segments under the Universal Telephone Service Policy.[8] As a result, total access and use of telephones rose dramatically from 740,000 lines in 1992 to 6.9 million lines in 2000. This policy orientation came in tandem with technological change, as new competition set in from cellular phone operators. International carriers likewise came into the industry to challenge the incumbents. The cellular phone operators were awarded a licence provided that they installed 400,000 lines within three years. In the case of international carriers, the obligation was 300,000 lines. The rise in telephone density naturally provided a boost to ICT.

Incentives to Information Technology Services

In recognition of the growing niche in IT services, the Philippine government began to encourage investments in the area. Concretely, this came in the form of fiscal incentives (Executive Order 226, dated 24 November 2000) to activities registered under the Board of Investments (BOI) relating to the following:

1. software development projects: application software, middleware, system software;
2. IT-enabled services;
3. support and knowledge-based services, and
4. business process outsourcing.

The incentives offered are as follows:

1. income tax holiday: 6-year exemption from income taxes for pioneer enterprises[9] and 4 years for non-pioneer ones, with possible extension of 2 years but not exceeding 8 years.;
2. employment of foreign nationals;
3. additional deduction on labour expense, and
4. unrestricted use of consigned equipment.

Of the 708 firms listed as 'electronics companies in the Philippines' (SEIPI, 2003) 155 firms (or 22 per cent) are service-related firms, a large number of which are Filipino- owned. They offer all kinds of IT-related services, from call centres and internet providers to software development.

4. THE PHILIPPINE ELECTRONICS INDUSTRY AT A DANGEROUS CROSSROAD

By the usual standards of exports and even of FDI, the Philippines appears to be a developing country that has become visible on the technological map. McKendrick et al. (2000) have shown that electronic exports are increasing in absolute terms, even defying the global decline in terms of export growth. However, there are serious weaknesses in the industry, especially when seen in the context of the electronics industries of other countries in the region. It will be seen that many such weaknesses can be remedied only by medium to long-term policy actions, some of which may have to be specific to the industry. Following is a discussion of the industry's weaknesses, while the next subsection will suggest solutions that are the responsibility of both public and private sectors. Only with a concerted action by these two sectors – with the latter predominating – will a way out of the impasse be found.

Export Concentration

Commodity concentration is presumed by many LDC (less-developed country) policy makers to be positively related to export revenue instability. A study using evidence for several African countries, however, has shown that one cannot make such a sweeping statement. The final result depends on the variability of the export earnings of the major commodity export and its covariance with other exports (Tegegne, 1991). Nevertheless, the share of electronics exports to total Philippine exports has grown from 28 per cent in 1992 to 71 per cent in 2000. The global deceleration of demand (after the Y2K scare of 2000) reduced exports of the industry by US$5 billion or 87 per cent of the decline in total exports between 2000 and 2001. This shows that the degree of vulnerability of the country's total exports to variability of global demand for electronics.

And not only are the country's exports concentrated in electronics, making up about 70 per cent of total exports. Of 81 (5-digit) IT products exported by the country, only 11 of them made up about 93 per cent of the total.[10] Of these, semiconductor products (93102) imported on a consignment basis make up more than half of total IT exports

(Austria, 2000).

Moreover, Austria (2000) reports an alarming detail: 44 per cent of the country's exports are accounted for by products whose share is falling in world trade, while only 42 per cent are represented by products that are growing and dynamic. In contrast, the majority of exports of other ASEAN countries, the NIEs, China and Mexico originate in products whose share is rising in world trade.

Possible Loss of Comparative Advantage

Semiconductor manufacturing in the Philippines is currently confined to back-end assembly and test operations, which remains the labour-intensive portion of the value chain.[11] This implies further that industry investments being footloose, the country's vaunted comparative advantage in the (semi-skilled) labour-intensive portion can easily be overcome by other countries. The Philippines boasts of the 'Filipino worker' as the source of its comparative advantage because of his/her literacy, English-speaking ability, trainability and so on which presumably translate into steep learning curves of labour. Training of Filipino workers is said to be able to take place in 2 months, compared with 4 to 6 months of training in other countries. However, these are advantages that can easily be undermined, given the human capital investments taking place in neighbouring countries. Examples of possible erosion of the country's sources of comparative advantage are shown in Table 6.11, which gives comparable statistics for the Philippines and other labour-rich countries in the region.

The gap in the literacy rate has been eroded; China has the largest number of technical people per million inhabitants as well as the largest number of post-baccalaureate students in science and engineering; the hourly manufacturing cost in the Philippines is more than three times that in China and almost 13 times that in Indonesia. In the Philippines, minimum wages have increased by 93 per cent in 8 years as compared to 22 per cent for Thailand in the past 9 years (Deloitte Consulting, 2002), while there is no minimum wage legislation in Malaysia or in Singapore. Competitive advantage should therefore be created by moving up the value-adding chain.

And this is precisely where the obvious weakness of the country's electronics industry lies, namely its apparent inability to move fast enough up the value-adding chain. An important consequence is discussed in the next subsection.

Table 6.11 Indicators of Human Resource Qualifications

Country	Literacy rate (% of population) 1970	2000	Technical people per million inhabitants	No. of post-baccalaureate science & engineering students	Hourly manufacturing cost (US$, 2001)
China	82.4	97.7	200 (1995)	59,748	0.53
Indonesia	80.0	97.7	24 (1985)	–	0.14
Philippines	92.1	98.7	22 (1992)	5,520	1.78
Thailand	94.1	98.9	39 (1996)	4,928	–
Malaysia	81.3	97.6	32 (1996)	1,251	2.68

Source: UNESCO website; Deloitte Consulting (2002).

Lack of Technology Upgrading/Weak Backward Linkage

Semiconductor manufacturing dates back to the entry of American and European MNCs in the 1970s. While it started with back-end operations about thirty years ago, MNCs in the country have merely added simple test operations to their back-end activities, with very little movement away from labour-intensive processes.

Export-oriented investors in the industry locate in the EPZs whose main attraction is the duty-free importation of components and other materials for export processing. Hence, backward linkage with the rest of the economy tends to be weak in the zones. Very limited subcontracting activity is taking place due mainly to the absence of a local supply network that seems to have failed to develop in any significant way over the years. Despite the rather large international demand for semiconductors, no Filipino manufacturer of parts and components – such as lead frames or gold bonding wire – to semiconductor assembly has emerged. Instead, significant localization seems to have taken place only in outer packaging (e.g., carton boxes, plastic containers). The semiconductor assembly industry was, until recently, almost completely dependent on imported inputs. With the entry of Japanese semiconductor firms, some localization took place with the entry of Japanese component suppliers. And yet it is taken for granted that creation of a supplier base requires some amount of technology acquisition, through spillovers or interaction with technology sources (in this case, the MNCs) and through own effort at R&D. Spillovers occur over a long period if at all. Interaction with MNCs through a subcontracting or component supply relationship implies that local firms have already shown proof of a certain degree of technological competence or potential for further improvement for MNCs to be willing to invest in such interactions. This therefore makes R&D effort indispensable as a bottom line, and especially due to the proprietary nature of technology. But currently R&D conducted in Philippine electronics plants is mainly product related, or at best, focused on the production process of specific product(s).

It is generally recognized that the creation of a local silicon foundry or a fab would dramatically raise domestic value added. But whereas China has 31 fabs, Taiwan 28, Singapore 13, Malaysia 4, and even Vietnam has plans for 2,[12] the Philippines has none. Even the offer of Intel before the 1997 Asian financial crisis was allegedly turned down by the government due to the cost of the complementary infrastructure that was required.[13] The presence of a wafer fab will not only raise the proportion of domestic value added to the semiconductor industry but also trigger a wave of R&D and its requisite personnel of scientists and engineers. Malaysia is said to have set up its first fab in 1996 for the sole purpose of research and development in the field. MIMOS Semiconductors' (which owns the fab) CEO, Tengku Azzman Shariffadeen is said to have claimed: 'If you control the chip, you control the features and thereby the value of the product. This ability will impact our competitiveness'. The wafer fab which took the Malaysian government RM 174 million to build is thus supposed to 'endogenize' the Malaysian electronics industry. Now Malaysia is running four (4) fabs and has been able to roll out the first Malaysian-designed microprocessor called Pensona which is supposed to be the culmination of 10 years of work and RM 1 billion in government investment (Deloitte Consulting, 2002). Instead of encouraging the development of wafer fabrication, representatives of the Philippine government have just gone to ASEAN to ask for a waiver of tariff reduction on petrochemicals, in order to encourage the setting up of a naphtha cracker plant.

5. ELEMENTS OF A WAY OUT OF THE IMPASSE

The Philippine electronics industry appears to follow a trajectory of the Singaporean type, that is,. dependent on FDI, export and increasingly becoming service oriented.[14] It does not have the deep pockets that used to be characteristic of the *chaebols* of South Korea, so those knowledgeable in the industry do not harbour much hope for any local firm to develop its own brand (e.g., Samsung and LG Collins from Korea). On the other hand, the Singaporean model is based on a strong state, so that what appear to be private initiatives are at the very least a partnership between government and the private sector. Nevertheless, the Singaporean model remains efficient because it is constantly subjected to the test of the market. To minimize the inefficiency effects of capture by vested interests – especially in a soft state like the Philippines – markets have likewise to be kept open, both internationally and nationally so that the test of the market will prevail even with some government targeting.

Because the Philippines seems more akin to the Singaporean rather than the South Korean model, many of the examples of how the country can get out of an impasse will have the Singaporean experience in mind.

Public Policy is an Essential Ingredient

Government must be convinced that the industry needs assistance, both in terms of generic and a focused/targeted policy assistance. From a neoclassical standpoint, there are a number of market failures that can justify government intervention. These will require policy, both of a generic, macroeconomic policy environment and an industry-targeted type.

Macroeconomic policy environment

Continued trade liberalization and facilitation

It has been seen that the development of the electronics industry in the Philippines has taken the FDI route. MNCs, which preferred to internalize their proprietary technology through direct investment rather than joint ventures or licensing, used the Philippines as an export base due its low-cost labour. To favour MNCs the country created export-processing zones whose main feature was free importation of imported components and raw materials. To encourage further development, the Philippines must therefore ensure a liberalized environment on both the export and import sides. The development of the country's IT-related service industry requires open trading regimes (Mowery and Nelson, 1999). However, trade liberalization is not enough. Trade facilitation must go hand in hand so that hidden barriers, especially in customs administration, will not hinder the smooth flow of trade. Together, trade liberalization and facilitation will encourage the inflow of technology-bearing FDI.

Education policy

But while greater openness will encourage more technology flows, the capacity to absorb such technologies must exist. The Philippines has traditionally been high on the list of developing countries when it comes to education. In terms of gross enrolment ratio and education index for 1999 it scored higher even than Singapore and was second only to South Korea (Table 6.12). But other countries in the region have been investing a good deal in human capital formation: as a share of GNP, the Philippine expenditure in education was only higher than the lowest in the group of Asian countries, Indonesia, while during the 1995–97 period, other countries in its near-NIE cohort – Malaysia and Thailand – invested more. The poor performance of students in science and maths international exams, the high degree of unemployment of tertiary graduates, and the proliferation of non-engineering/technical students among those enrolled in tertiary levels, have all been pointed out as weaknesses of an educational system that does not seem responsive to the needs of a knowledge-based economy. The curricula of engineering and technical schools typically do not address the technological needs of industry and receive little input from the industrial sector.

Some state and technical colleges have tried to address this problem by requiring their students to undergo hands-on training or apprenticeships in electrical and electronics firms to become eligible for graduation. However, the poor quality of engineering and technical education cannot be adequately compensated for by on-the-job training (OJT). The implication is that the private manufacturing sector, probably even more than it already does, will have to invest quite substantially in worker

Table 6.12 Indicators of Education

Country	Public expenditures on education as % of GNP		Combined primary, secondary and tertiary gross enrolment ratio (%)	Education Index 1999
	1985-89	1995-97	1999	
Singapore	3.9	3.0	75	0.87
Rep. of Korea	3.8	3.7	90	0.95
Malaysia	6.9	4.9	66	0.80
Thailand	3.4	4.8	60	0.84
Philippines	2.1	3.4	82	0.91
China	2.3	2.3	73	0.80
Vietnam	–	3.0	67	0.84
Indonesia	0.9	1.4	65	0.79

Source: UNDP, Human Development Indicators, 2001.

training. Yet, unlike in other countries such as Singapore and even Malaysia, the Philippines does not make available any meaningful incentives to encourage more of it to be undertaken.

Investment in worker training had been the hallmark of Singapore's approach. Early in the game the government set up a large number of training institutes for precision engineering skills; later, with the need for skill-upgrading, worker training programmes were opened especially at polytechnics and formal educational institutions. Although the skills that were taught were of the generic engineering type, specific industries, such as the hard disk drive industry, eventually benefited from this reservoir of trained technicians (McKendrick, et al,, 2000). A number of subsidized skills programs were likewise made available to corporations, including MNCs. Thus a hard disk drive major used a facility to send engineers for training in the United States while other majors can take advantage of the specialized training offered on storage technology.

S&T/R&D policy
A number of theoretical and empirical studies on FDI indicate the crucial role of absorptive capacity of the host country if it is to obtain significant benefits from FDI.. Spillovers from FDI may not be feasible in the absence of adequate human capital or investments in R&D (Saggi, 2000).

Generally speaking, the Philippines is characterized by a low level of R&D effort (Table 6.13). Cororaton (2001) *cites* a 1996 UNESCO World Science Report that ranks the Philippines as 73rd among 91 countries in terms of R&D effort. The low number of researchers per million inhabitants in Table 6.13 compared to other countries in Asia is in stark contrast with that in the more advanced economies like South Korea and Singapore. The 0.03 per cent increase in the share of the country's GNP between 1981 and 1992 (0.19 and 0.22 per cent respectively) pales before the increases in South Korea and Singapore and even India during similar time periods.

Not only is there underinvestment in R&D in the Philippines which shows up in weak manpower training and lack of R&D facilities. There is also a weak linkage between technological generation, adaptation and use (Cororaton, 2001). The country's private sector spends only about 25 per cent of R&D expenditures, in contrast to Japan's 83 per cent, Korea's 82 per cent, Singapore's 62 per cent and Thailand's 40 per cent. As Table 6.14 indicates, it is the Philippine government that accounts for more than half of the country's R&D expenditure. Such government expenditure does not translate into industrial innovation either. Indeed, the criticism for the country's Department of Science and Technology, which was set up in 1996 to spearhead the country's S&T effort, was that

Table 6.13 Indicators of R&D in Selected Asian Countries

Country	All R&D personnel		Researchers per million inhabitants		Expenditure on R&D as % of GNP	
China	665,100 ('94)	787,000 ('96)	349 ('88)	154 ('96)	0.67 ('88)	0.66 ('97)
India	218,995 ('82)	336,589 ('94)	131 ('82)	149 ('94)	0.56 ('80)	0.73 ('94)
Indonesia	28,484 ('82)	36,185 ('84)	110 ('80)	182 ('88)	0.35 ('80)	0.07 ('94)
Rep. of Korea	58,970 ('82)	135,703 ('96)	725 ('82)	2,193 ('96)	0.57 ('80)	2.82 ('96)
Malaysia	4,563 ('92)	4,436 ('96)	182 ('83)	93 ('96)	0.40 ('92)	0.24 ('96)
Philippines	14,578 ('92)	-	112 ('80)	157 ('92)	0. 19 ('81)	0.22 ('92)
Singapore	2,741 ('81)	9,497 ('95)	485 ('81)	2,318 ('95)	0.30 ('81)	1.13 ('95)
Thailand	10,621 ('87)	12,802 ('95)	105 ('87)	118 ('95)	0.39 ('80)	0.13 ('96)

Note: Figure in parentheses refers to year when data are available usually beginning and end years.

Source. UNESCO website.

its attempt at commercialization was merely in support of its own institutes, with very little connection with real market demand (Abrenica and Tecson, 2003).

Government expenditures would then generally be absorbed by the salaries of 36 per cent of R&D personnel that are found in government, not including the salaries of those in higher education. The latter consists of state universities and colleges, the most prominent of which is the University of the Philippines which accounts for 46 per cent of existing R&D personnel in the country. In contrast, the private sector commands only 11 per cent of such human capital at its disposal.

Singapore promoted R&D both broadly and narrowly, that is, through broad macroeconomic policy and narrowly focused, industry-specific policies. It initiated a number of R&D schemes that subsidized private investments in R&D through the National Science and Technology Board (NSTB). Later in the 1990s, investment in research infrastructure by the Singaporean government in disk drives was undertaken in order to keep the industry in Singapore. Through the Magnetics Technology Centre (established in 1992 and later upgraded into the Data Storage Institute or DSI), for instance, the government worked hand in hand with hard disk drive companies towards collaboration in R&D. Furthermore, corporate

Table 6.14 R&D Expenditure by Major Sector, 1989 – 1996 (in current prices, million PhP)

Sector	1989	1991	1993	1995	1996
Higher education	210.8 (12.9)	290.0 (14.7)	380.0 (17.9)	457.1 (15.2)	532.0 (15.5)
Government	903.5 (55.1)	1019.6 (51.8)	1036.3 (48.9)	1433.2 (50.7)	1742.5 (50.9)
Non-government	130.9 (8.0)	135.7 (6.9)	155.6 (7.3)	207.7 (7.3)	249.9 (7.3)
Private industry	393.5 (24.0)	523.3 (26.6)	547.5 (25.8)	730.7 (25.8)	879.2 (25.7)
Total*	1638.7 (100.0)	1968.7 (100.0)	2119.4 (100.0)	2828.6 (100.0)	3423.6 (100.0)

Notes: * Figures in parentheses are in per cent; totals may not add up due to rounding.

Source: Patalinghug (2003, Table 1).

R&D received direct financial support from the government. Mistakes were committed in the process of subsidizing, but the important thing is that the Singaporean government knew when to walk away from such mistakes.

Fiscal incentives

Although the country is said to have 'levelled the playing field' when it comes to fiscal incentives in the EPZs, the country's ecozones are still badly rated, relative to those in other ASEAN countries. Major drawbacks are poor infrastructure and high power rates, especially of electricity. Electricity cost is highest in the region, particularly due to generation charges that are raised by the excess capacity in electric plants. Thus, the operating budget of Intel is said to be 41 per cent for electricity while it is 25 per cent for Texas Instruments. This is very high, considering that the electric cost of capital-intensive manufacturing is said to be less than 10 per cent for Malaysia (Deloitte Consulting, 2002). Moreover, the same report speaks of the voltage fluctuations (+/- 10 per cent) in the country being higher than in say Malaysia (+3 per cent/-5 per cent). Fiscal expenditures to improve the infrastructure and lower the power rates are therefore key to more investments in the industry, both by multinationals and by domestic entrepreneurs.

Industry-focused policy

A reading of the country histories of the development of a strong electronics industry – whether hardware or software – has required some form of government intervention, usually deliberate (see, for instance, Shin, 1996; Mowery and Nelson, 1999; Cyhn, 2002). Moreover, the policy actions of countries in the region and elsewhere show approaches that are targeted or at least focused on definite segments of the industry (McKendrick et al., 2000).

Parts, if not all, of the electronics industry have always found their way into lists of preferred industries drawn up over the years by Philippine development plans as well as the Board of Investments (BOI) because of their growth potential. For instance, the Department of Science and Technology (DOST) aligned its choice of industries for promotion[15] with the 'export champions' of the Department of Trade and Industry (DTI), because only the latter were entitled to fiscal incentives doled out by the BOI. Then again, electronics have been identified among other industries as the 'leading-edge sectors'[16] which were supposed to be the target for S&T support by DOST. But the choice of industries was so wide as to indicate that government could not have been serious in its fiscal support.[17] And the private sector must have given little meaning to the offer of government support, as the incentives given by the BOI to encourage private R&D were taken up by only 11 firms over a period of 6 years (Patalinghug, 2003).

Industry-focused incentives will be difficult to justify, let alone achieve, in the present situation of the industry. The irony is that the government would not even bat an eyelid in providing hidden subsidies through protection via higher tariffs or non-tariff barriers to certain other 'preferred' industries. However, given the highly politicized context of industrial development, industry-focused policy will have to consist of creative ways by which government can assist the industry to move out of the impasse. Policy assistance must be in the direction of how to prod the private sector to upgrade and deepen its acquired technology. This will require at the outset an in-depth study of the industry's needs and the problems associated with technology 'owning'. For instance, a more focused segment of engineering schools or public – private sector partnership in the building of needed human capital may be pursued.

On the supply side, the barriers to entry, especially in terms of capital requirements, have to be lowered. In particular, the development of a wafer fabrication plant which impacts not only on the industry value added but also on the creation of human capital deserves serious study. On the demand side, industry development, but more importantly, all-

Table 6.15 The Global IT Market by Technology (US$ million)

	1998	2003 forecast	Compound annual growth rate (%)
Hardware	296,613	394,996	6.0
Packaged software	135,548	271,718	14.9
Services	291,255	473,633	10.2
Data communications	43,422	67,491	9.2
Total IT	766,837	1,207,838	9.5

Source: Kraemer and Dedrick (2000, Table 1).

round increases in productivity, can be triggered by a concerted effort of government to adopt computerization of its services by its many departments and bureaus. This will foster the creation of a technological culture, especially among the up-and-coming generations like that which characterizes the youth of today's NIEs.

But the Private Sector Must Remain the Prime Mover

In fact, the private sector remains the prime mover of the industry and must continue to remain so. In particular, the private sector must find its own niches in the dynamic world of electronics. One area that has been identified as worthy of targeting by the SEIPI member firms is that of Electronic Manufacturing Services (EMS). This consists of contract manufacturing services to produce a variety of electronics products for high-technology electronics OEM, such as Compaq, Philips and so on. Although traditionally in printed circuit board (PCB) assembly, EMS firms have branched out into supply management, mechanical assembly, cabling, system integration, logistics, and even design activities. Of the ten firms that account for 60 per cent of an estimated market size of US$106 billion in 2001, none has invested in the Philippines, although they are present in other countries in the region (Deloitte Consulting, 2002).

However, unless domestic firms set their goals beyond assembly and simple testing and venture into the higher-value-adding activities, including design, the electronics industry in the Philippines will remain consigned to labour-intensive back-end activities that are footloose in character. This has been the experience in other segments of manufacturing with the closing of plants and their simultaneous resurrection in other countries in the region.

Perhaps in recognition of the difficulties of creating value added through manufacturing, a new development is taking place in the field of IT services. Liberalization in the services sector, particularly in the telecommunications industry, has been partly responsible for this. In addition, whereas the financial requirement for technological catch-up and upgrading in the manufacturing sector sets up a formidable entry barrier, such a barrier hardly exists in the services sector. The government has responded to such growth by an offer of incentives (EO 226) to firms that would offer IT-related projects. Moreover, the services sector seems to be among the most dynamic in the IT industry as shown in Table 6.15.

6. CONCLUDING REMARKS

The Philippine electronics industry represents the country's drive towards high-technology development. However, the industry presently stands at a dangerous crossroad. It is characterized by a concentration of its exports, particularly in semiconductor packaging which makes up more than half of all its electronic exports. For almost three decades of development since the first MNC came to its shores to take advantage of its labour abundance, Filipino-owned firms remain subcontractors, concentrating on back-end assembly and testing. Although the number of firms has increased and their product line has widened to include new lines, no firm has gone beyond OEM. While it is true that trade liberalization has encouraged FDI and hence technology flows, the injection of more competition into the economy falls short of addressing the problem of technological upgrading. Unless it is able to 'own' – that is, upgrade and deepen – the technology it has acquired, the country will eventually have to be prepared to lose its precarious comparative advantage to other countries in the region who are investing even more in human capital formation.

This chapter has identified the essential elements required for the country to get out of the impasse. The bottom line is the creation of a national technological infrastructure that will ensure the ability to absorb technology and thus prevent the erosion of its base of comparative advantage. While continued policy of trade liberalization and trade facilitation will encourage technology inflows through the FDI of MNCs, the capacity for technology absorption can only be guaranteed by more and better – quality investments in education and not only education of the generic kind, but one that breeds the critical mass of engineers and scientists that are required for the deepening of technology acquisition. Side by side with technological deepening, on the side of demand, a 'technological culture' will have to be fostered, so that increasingly domestic demand can be relied upon to keep the industry wheels moving. But the final burden of development rests squarely on the private sector which must find its own niches in the world market.

Lately, there has been a shift of government policy pronouncements to the development of the IT-related services sector, typified by the granting of incentives to the sector under EO 226. While such pronouncements mirror the growth of the services sector, it remains to be seen whether that small glimmer in the electronic firmament will ever grow into a star.

NOTES

1. However, an important caveat has to be stated. Flexible manufacturing has allowed the distribution of different production stages of a given high-tech product, exploiting the differences in comparative advantages of countries. Thus, the Philippines is a large exporter of 'high-tech' products, but it is involved only in the back-end operations involving the generally labour-intensive, rather than the R&D-intensive stage of production. To what extent exports of such products by the Philippines can be considered 'high-tech' is a question that is often glossed over in discussions and inter-country comparisons.
2. The Philippine Constitution at the time required that foreigners could not own a majority share of equity of a company.
3. GE Appliance (General Electric) and Philippine Appliance Corp. (Westinghouse) were not initially too keen on investing. However, General Electric was forced to take over a Filipino corporation (Del Rosario Brothers, or DRB Inc.) when the latter had overextended its credit. In the case of Westinghouse, it started as a licenser without equity of a Filipino corporation established in the mid-1950s – Philippine Appliance Corporation or Philacor – but ended up becoming a majority holder in 1970 when the Filipino group owners could not raise sufficient capital for expansion (Yoshihara, 1985).
4. Domestic resource cost analysis applies cost–benefit analysis to tradable goods production. It indicates the social cost of domestic resources used per unit of net foreign exchange saved by the activity through import substitution or earned through export production.
5. However, the local content policy for motorcycles and cars was retained, under refurbished programmes called the Motorcycle Development Programme and the Car Development Programme.
6. The trade policy reform started in 1981 with the reform of the tariff system, bringing down the average tariff from 41 per cent in 1980 to 28 per cent in 1982 and effective protective rate (EPR book rates) from 70.3 per cent in 1980 to 31 per cent in 1985 (Bautista, 1982). However, the policy reform had to be aborted because of the deep economic crisis triggered by the balance of payments crisis from 1983 to 1986.
7. This does not, however, preclude the fact that in some areas, the country's ecozone regime is less competitive than those in say Thailand or Malaysia. Such areas include: the lack of clarity of the ecozone concept, especially as it applies to non-manufacturing activities; the uncompetitive impact of the gross income taxation approach; unclear policies governing sales to the local market; and an excessively generous package of fiscal incentives in certain instances (World Bank, 1997, p. 79).
8. See Abrenica and Llanto (2003) for a discussion of the remaining barriers in the country's telecommunications industry, notwithstanding such liberalization of the sector.
9. Pioneer status is obtained if the project complies with any of the following criteria:

(1) it must utilize new or untried technology; (2) it must have substantial venture capital; (3) it will introduce a major innovation in software development; (4) the project cost should be at least PhP100 million. In addition, if registered projects are located in a less- developed area as identified in the Investment Priorities Plan (IPP) the firm shall be granted pioneer incentives, that is, an income tax holiday for 6 years.

10. These 11 5-digit SITC products made up between 91.5 per cent in 1991 and 94.9 per cent in 1997 of total IT exports of the country.
11. It is said that on a scale of 1–10, (with design as no. 1 and as 10 after-sales service), the Philippines' semiconductor industry is in no. 7, or 6 notches away from the skill and capital -intensive portion of the value chain.
12. The plan to have a wafer fab in Thailand was scrapped after the Asian financial crisis. However, there is nothing to prevent Thailand from reviving this plan.
13. Soon after the newspapers announced the scrapping of the plan to establish a wafer fabrication plant, the government announced the entry of FDI into the cigarette industry!
14. However it might be conceded that the resemblance stops there. The differences in technological orientation and policy between the two economies explain much of the difference in growth performance.
15. These are: electronics, garments, computer software, fashion accessories, gifts, toys and houseware, fresh and processed fruits, marine products, furniture, metal products, marble products, ornamental flowers, professional services and the construction industry.
16. These sectors are as follows: agriculture, aquaculture and marine fisheries, forestry and natural resources, metals and engineering, the extile industry, mining and minerals, the process industry, the food and feed industry, energy, transportation, the construction industry, information technology, electronics, instrumentation and control, emerging technologies, and pharmaceuticals.
17. One has only to view the list of industries in the current IPP to conclude that the government is using a 'shot-gun' approach to targeting!

REFERENCES

Abrenica, J.V. and G. Llanto (2003), 'Services', in A. Balisacan and H. Hill (eds), *The Philippine Economy: Development, Policies, and Challenges,* Ateneo de Manila University Press and Oxford University Press. pp. 254–80.

Abrenica, J.V. and G.R. Tecson (2003), 'Can the Philippines Ever Catch Up?', in S.Lall and S. Urata (eds), *Competitiveness, FDI and Technological Activity in East Asia,* Chelternham: Edward Elgar. pp. 288–304.

Austria, M. (2000), 'Assessing the Competitiveness of the Philippine IT Industry', PIDS Discussion Paper Series 2000-03, Makati: Philippine Institute of Development Studies.

Austria, M. (2002), 'Philippines in a Global Trading Environment: Looking Back and the Road Ahead', PIDS Discussion Paper Series 2002-15, Makati: Philippine Institute of Development Studies.

Bangko Sentral ng Pilipinas, *Selected Indicators,* Various issues.

Bautista, R.M. (1982), 'The 1981–1985 Changes and Effective Protection of Manufacturing Industries' UPSE Discussion Paper 82–13, Quezon City: University of the Philippines School of Economics.

Cororaton, C. (2001), 'Technological Innovations in Japan and S&T Experiences in the Philippines: Drawing Policy Lessons for the Philippines', PIDS Working Paper Series 2001-08, Makati: Philippine Institute of Development Studies.

Cyhn, J.W. (2002), *Technology Transfer and International Production: The*

Development of the Electronics Industry in Korea. Cheltenham: Edward Elgar.
De Dios, L. (1994), 'A Review of Remaining Regulated Commodities', Research Paper Series no. 94-08, Makati: Philippine Institute of Development Studies.
Deloitte Consulting (2002), *Competitive Landscape for the Philippines Semiconductor and Electronics Industry: Roadmap.* Manila: SEIPI.
Hatzichronoglou, T. (1997), 'Revision of the High-Technology Sector and Product Classification', Science, Technology and Industry Working Papers, OECD.
Hill, H. (2003), 'Industry', in A. Balisacan and H. Hill (eds), *The Philippine Economy: Development, Policies, and Challenges.* Quezon City: Ateneo de Manila University Press and Oxford: Oxford University Press. pp. 219–53.
Human Development Report (2001), New York: United Nations Development Program.
Kraemer, K.L. and J. Dedrick (2000), 'The Information Technology Sector and International Competitiveness: A Background Paper', Paper prepared for the Asian Development Bank Study on International Competitiveness of Asian Economies, Irvine, CA: Centre for Research on Information Technology and Organizations, University of California.
Lall, S. and S. Urata (eds.) (2003), *Competitiveness, FDI and Technological Activity in East Asia,* Cheltenham: Edward Elgar.
Lapid, Dennis (1996), 'Appliance Industry: Impact of Trade Policy Reforms on Performance, Competitiveness and Structure', in G. R. Tecson, M.S. Austria, V.S. Pineda, L.C. De Dios, D.D. Lapid, M.C. Medilo, C.P. Banzon, F.M.C. Trabajo, E.G.Q. Mendoza, *Catching up with Asia's Tigers,* vol II, Makati: Philippine Institute for Development Studies. pp. 233–78.
McKendrick, D., R. Doner, and S. Haggard (2000), *From Silicon Valley to Singapore: Location and Competitive Advantage in the Hard Disk Drive Industry,* Stanford, Ca: Stanford University Press.
Mowery, D.C. and R.R. Nelson (eds) (1999), *The Sources of Industrial Leadership,* Cambridge: Cambridge University Press.
NEDA Statistical Yearbook, National Economic and Development Authority, Various issues.
Patalinghug, E.E. (2003), 'The Philippine National Innovation System: Structure and Characteristics', PIDS Discussion Paper Series no. 2003-04, Makati: Philippine Institute of Development Studies.
Saggi, K. (2000), 'Trade, Foreign Direct Investment, and International Technology Transfer: a Survey', World Bank Policy Research Working Paper, no. 2349, May.
Shin, J. (1996), *The Economics of the Latecomers,* London and New York: Routledge.
SEIPI Information Book 2003, Manila: Semiconductor and Electronics Industry of the Philippines, Inc..
Tecson, G.R. (1999), 'The Hard Disk Drive Industry in the Philippines and Japanese Direct Investments', *The Philippine Review of Economics and Statistics.* 36:2, pp. 205–56.
Tegegne, A. (1991), 'Commodity Concentration and Export Earnings Instability: Evidence from Africa', Upper Montclair, NJ: Centre for Economic Research on Africa, School of Business, Montclair University.
Warr, Peter (1989), 'Export Processing Zones: The Economics of Enclave Manufacturing', *World Bank Research Observer*, 4:1, pp. 65–88.
World Bank (1997), *Philippines: Managing Global Integration,* Washington DC:

World Bank.
Yoshihara, K. (1985), *Philippine Industrialization: Foreign and Domestic Capital*. Quezon City: Ateneo de Manila University Press.
UNESCO website, www.unesco.org.

7. Thailand as a High-Tech Industrial Economy: An Impossible Dream?

Medhi Krongkaew, with Teeraya Krongkaew*

1. INTRODUCTION

Traditionally, Thailand's industries have been closely linked with agriculture. The major processing facilities from the post-Second World War years up to the late 1950s were rice mills, saw mills, sugar mills, ice factories, textile factories, and other household industries, some of which were owned by the state. Rice and rubber were Thailand's predominant exports, with manufacturing playing only a minor role in the country's economic activity. In the early 1960s, Thailand began its industrial development when the government took steps to adjust from a policy of direct state participation towards encouraging private sector participation (Sumeth, 1992). Under the new policy the state plays a supportive role in providing the manpower and infrastructure necessary for industrial development.

The nation's first National Economic and Social Development Plan, for the 1961–66 period, outlined the government's strategy of promoting private investment, particularly based on domestic natural resources and import substitution. Since then, there have been eight subsequent national development plans. The first four plans (1961–81) did not contain any explicit science and technology (S&T) capability upgrading policies. Specific S&T plans were first introduced in the Fifth Plan (1982–86), nevertheless the goals and the policies were rather unclear. The Sixth Plan (1987–91) provided an improvement as it aimed to strengthen the country's planning and coordinating capabilities, and to accelerate the use of technical capability in national development. Under the Ninth National Plan, the goal of industrial development strategies are 'to enhance the well-being of the Thai people by increasing

* Revised version of the paper presented at the Workshop on High-Tech Industries in East Asia, organised by the Jenn-Hwa Institute of Technology, Taipei, 24–26 August, 2003.

productivity in all industrial sectors employing technological efficiency that follows sustainable development path and economic stabilisation under risks as well as internal and external volatility'.

Thailand historically employed an import-substitution strategy based on high tariffs and requirements on domestic content. Since the late 1970s, Thailand along with most other ASEAN (Association of South East Asian Nations) countries has experienced significant economic growth, partly as a result of foreign direct investment (FDI). It has attracted a higher percentage of FDI to GDP relative to the world average and the average of developed countries. The growth was particularly impressive in the late 1980s to mid-1990s, when Thailand achieved double-digit GDP growth for a few years. However, since the early 1990s, there has been a major policy shift towards the establishment of manufacturing export industries, which were given many tax privileges. For instance, they could import machinery and other inputs tariff free, in addition to other tax incentives.

However, the lack of concrete and clear S&T policies in the past has contributed to a bottleneck in Thailand's industrialisation efforts and economic growth. The shortage in the technical human capital base has reduced the ability to create or receive effective technology transfer from foreign industrialised countries, playing down the nation's strong suits such as reliable transportation, telecommunications and electrical power networks. Without an adequate technical human capital base, effective absorption and transfer of know–how and technologies were not fully possible as locals were being trained to handle merely routine operations. In cases where foreign technologies were imported in full, the country has to incur substantial costs without any significant technology transfer or an acceptable return on investment.

The move towards export-oriented policies was quite successful, as Thailand is now one of the largest production bases of industrial electronics, automobiles, and automotive parts in Southeast Asia for export to the world market. As the level of competition intensifies, improvement in technological capability has become mandatory rather than optional. The government is aware of the issue and has initiated many programmes to support domestic technological capability, enabling it to better compete in international arenas.

Porter (2003) suggested that Thailand move to a new model of competitiveness as the country's labour productivity lags far behind leading Asian economies, and productivity growth is lagging badly. He also suggested that Thailand activates cluster development and applies strategies that promote higher coordination between government institution and industry associations, and clearer strategies within each industry. Private sector-led cluster competitiveness councils should be created and provided with extra funding and support, for clusters throughout the economy.

In the next section of this chapter, we shall explore the current state of high-tech development in Thailand, including the current efforts by the state and the private sector, the ICT (Information and Communication Technology) Master Plan, FDI in high-tech industries in Thailand, and recent experience in high-tech manufacturing industries. We shall then proceed to discuss the existing strategies for high-tech industries, and suggest some measures that the government should consider in order to promote growth in local high-tech industries.

2. CURRENT STATE OF HIGH-TECH INDUSTRY DEVELOPMENT IN THAILAND

The strong growth since the early 1990s slowed down markedly in 1997 following the financial crisis, which has been a big blow to Thailand's aspirations of becoming a serious high-tech player (see Table 7.1). Many projects were deferred or cancelled altogether. For example, entrepreneur Charn Uswachoke, CEO of Alphatec Group, had to temporarily stall the US$ 1.2 billion chip-fabrication plant that he originally planned to build with Texas Instruments Inc., and also backed out of a joint venture with the state to build a research and development (R&D) facility.

The country has traditionally been a low-cost assembly and export base for multinational companies, and not much value-added work is done domestically. However, Thailand now aims to create a supply chain for information technology (IT) products and to move beyond low-margin piecework, as the government has realised that the nation cannot achieve sustainable growth simply by making PC cases and packaging chips for multinational corporations (MNCs) such as Flextronics, Sony, and so on. It is studying and considering Singapore and Malaysia's method of subsidising the construction of chip-making facilities – which resulted in the establishment of regional manufacturing hubs for many well-known companies including Hewlett-Packard, Dell and Gateway.

There are several indicators available in determining Thailand's state of high-tech development. One of which is the High Tech Indicators or 'HTI' developed by the researchers at Georgia Tech's Technology Policy and Assessment Centre (TPAC) since 1987. The periodic indicator is collected and reported every three years, and results have been published in several recent volumes of the US National Science Board's *Science and Engineering Indicators*. High Tech Indicators 2003 compares the

Table 7.1 Industrial Capacity Utilisation by Product Group, 1995-2004

	1995	1996	1997	1998	1999	2000	2001	2002	2003	2004
Total	78.0	74.9	69.7	58.5	65.4	65.8	61.7	61.5	66.9	68.7
Total (exclude liquor)	78.0	74.9	69.6	58.4	65.2	66.0	61.9	61.6	67.0	68.8
Food	39.6	35.7	34.7	31.2	39.8	39.1	37.0	39.8	47.5	46.1
Canned seafood	38.4	34.8	37.2	41.8	48.3	36.1	36.8	37.3	44.9	43.1
Canned pineapple	45.4	41.1	33.1	26.8	52.0	46.3	46.7	42.4	53.2	53.4
Milk products	64.6	56.6	44.3	42.3	48.8	56.3	54.7	64.9	69.2	70.7
Sugar	31.8	28.8	30.1	19.9	26.2	30.0	22.6	27.7	36.1	33.0
Beverages	67.0	71.2	67.8	61.5	65.8	57.6	53.8	49.1	48.5	50.7
Liquors	79.5	75.4	89.7	79.4	107.9	8.5	15.8	40.3	42.2	48.3
Beer	69.8	77.7	69.8	65.6	69.9	64.7	58.3	50.6	48.8	50.1
Tobacco	75.2	84.2	75.8	60.4	54.4	53.7	52.1	53.8	55.8	60.7
Construction materials	96.7	79.0	72.9	44.6	49.8	53.1	59.5	66.8	62.9	67.4
Cement	96.7	79.0	72.9	44.6	49.8	50.1	52.3	56.8	57.4	62.8
Iron & steel products	56.5	58.7	44.8	30.5	42.3	49.1	50.4	64.1	69.1	68.5
Steel bar & shape steel	60.2	63.6	51.3	32.1	32.6	37.3	42.0	53.5	56.9	67.7
Galvanized iron sheet	65.8	68.9	68.9	43.2	49.3	60.7	71.5	85.7	76.1	71.8
P.C. wire & stand wire	58.4	59.8	50.2	28.7	27.0	28.0	31.4	39.6	46.4	53.6
Wire rods	42.5	48.6	52.6	27.6	35.2	39.6	50.2	59.1	54.2	57.7
Steel pipes	84.4	90.3	65.0	40.0	44.8	61.7	62.9	61.5	66.4	66.5
Rolled sheet	47.4	48.0	31.1	24.7	43.5	48.6	46.0	65.4	74.8	70.3
Tinplates	57.2	55.8	54.5	43.2	55.4	47.4	48.5	55.8	62.4	74.8
Vehicles & equipment	80.9	68.5	50.3	27.2	39.6	43.9	49.5	56.7	69.3	77.7
Passenger cars	61.3	54.9	39.4	13.0	23.6	27.7	45.0	49.3	69.1	72.2
Commercial cars	99.2	78.2	52.2	22.1	37.5	40.6	38.3	52.2	61.5	78.8
Motorcycles	90.4	65.6	48.4	26.8	33.9	40.4	45.5	52.2	66.9	80.0
Tyres	90.8	89.4	76.3	71.7	86.4	92.8	85.0	86.5	87.8	88.4
Batteries	86.9	74.1	65.0	57.1	70.5	75.8	69.6	73.8	75.7	83.3
Petroleum products	92.7	84.5	90.2	83.5	85.2	83.3	74.3	75.6	78.9	85.5
Electronic products	74.8	79.3	95.5	73.7	90.5	86.3	74.9	64.4	78.0	71.3

	1995	1996	1997	1998	1999	2000	2001	2002	2003	2004
Electrical appliances	86.8	87.5	68.5	57.7	61.1	55.2	50.7	53.3	56.4	61.4
Other	86.3	79.5	68.1	63.6	73.5	79.7	78.7	79.1	80.1	86.0
Tin metal	28.1	36.6	41.0	51.7	59.4	58.1	77.0	60.6	48.4	63.8
Zinc metal	81.4	99.2	79.2	86.4	92.6	95.1	90.7	91.8	100.0	96.7
Block rubber	64.6	64.1	51.0	40.6	54.4	67.0	72.2	86.7	85.6	90.7
Compressors	89.1	83.9	69.0	54.8	69.9	77.8	77.8	70.3	76.5	75.7
Glass sheet	97.6	78.4	74.5	54.0	46.9	63.0	59.6	57.5	67.7	76.0
Electric motors	86.3	79.5	68.1	63.6	73.5	79.7	78.7	79.1	80.1	86.0
Pulp & paper products	72.3	82.9	95.8	87.8	94.4	87.0	87.0	84.6	88.3	85.5
Synthetic fibres	90.1	84.7	69.3	79.9	101.2	100.3	97.2	101.5	95.1	102.3
Chemical products	89.5	93.4	76.9	78.7	85.3	89.2	87.6	92.3	95.8	95.2

Source: Bank of Thailand website.

technological competitiveness of 33 nations, including a large number of industrialising economies. TPAC has developed a conceptual model of the processes by which industrialising nations gain access to external technology and technical information, absorb that technology/information effectively, and institutionalise a science-based development and manufacturing capability. Throughout the 1990s, HTI pointed towards dramatically broadening, global high-tech competition. For the first time, the 2003 HTI raises a 'red flag'. The leading input indicators remain strong for the OECD, Asian Tigers and China. They present more of a mixed picture for other industrialising nations. Also, the general diminishing of expert optimism regarding future high-tech prospects demands careful assessment (TPAC website).

As shown in Table 7.2, the HTI for Thailand is quite low, signalling the need to improve productive capacity, and technological and socio-economic infrastructure as Thailand is lagging behind Malaysia, China and India in all of these factors. Moreover, the National Orientation (NO) indicates that, in the survey participants' views, Thailand's commitment to technology-based development along a number of dimensions (government policy, political stability, entrepreneurial spirit and acceptance of the idea that development should be technology based) is ranked at about the same level as Indonesia (47.7 vs 45.1), lagging far behind Singapore, Taiwan, South Korea, Malaysia, China, India and the Philippines (see Table 7.2).

Table 7.2 Georgia High-tech Indicator Values 2003

Country	Expert opinion current hi-tech capability	Expert opinion future hi-tech capability	TS	PC	TI	SE	NO
USA	42.7	44.4	93.9	82.8	92.7	86.4	79.8
Japan	47.0	46.7	81.6	80.3	73.8	67.0	76.4
Germany	45.4	43.1	61.4	65.0	65.5	77.5	75.6
UK	42.0	42.5	54.1	54.5	62.0	84.0	76.5
France	37.8	42.9	44.0	59.2	55.9	76.2	67.9
Netherlands	42.7	37.1	45.3	55.5	55.2	80.6	82.4
Italy	38.0	31.3	31.4	52.3	48.4	69.4	69.4
Switzerland	40.0	41.4	31.9	53.1	54.6	71.2	69.2
Sweden	43.3	38.8	34.6	53.0	53.2	78.9	75.5
Spain	30.0	28.0	20.3	35.1	33.7	68.8	55.1
Ireland	36.4	38.0	34.3	53.0	44.2	80.7	84.1
Canada	38.2	41.7	32.6	53.1	51.5	82.4	81.8
Australia	32.3	32.0	19.6	45.2	46.9	74.3	70.5
South Africa	24.6	26.7	12.5	30.4	35.9	49.1	53.7
NZ	35.0	30.0	21.0	44.7	49.8	75.9	75.1
Russia	30.0	27.1	17.5	46.5	51.2	60.4	48.6
Poland	27.0	31.3	14.9	42.5	36.6	68.3	63.7
Hungary	30.7	32.5	20.3	42.2	40.0	67.9	66.7
Czech Rep.	31.0	28.6	18.5	44.6	34.7	64.4	64.6
Singapore	40.0	36.7	52.4	57.7	46.7	84.2	83.8
Sth Korea	34.4	30.0	40.1	52.3	45.2	81.1	80.4
Taiwan	37.0	31.3	39.3	59.6	45.3	84.4	83.0
Malaysia	28.2	27.5	32.8	39.1	28.8	64.9	73.0
China	31.4	27.0	49.3	49.6	55.2	55.0	63.0
Thailand	**25.5**	**22.9**	**20.0**	**30.9**	**23.9**	**54.2**	**47.7**
Indonesia	25.0	25.0	14.0	27.7	20.7	39.1	45.1
Philippines	26.0	22.5	19.6	45.0	24.0	55.0	59.3
India	30.8	34.3	17.9	47.8	37.0	49.3	62.5
Mexico	25.0	21.1	27.3	35.4	28.7	49.0	55.2
Brazil	26.9	30.0	15.4	35.8	35.4	50.5	52.5
Argentina	25.0	23.3	12.7	32.5	31.0	52.4	44.9
Venezuela	18.9	19.1	7.5	20.3	19.5	46.0	37.2
Israel	38.8	42.5	26.7	49.4	52.3	85.0	79.5

Notes: The model included four 'input' or leading indicators of a nation's future capacity (roughly 15-year time horizon) to compete in international markets in high technology products. It derived three 'output' indicators of a nation's current international

competitiveness.
National Orientation (NO) indicates a country's commitment to technology-based development along a number of dimensions: government policy, political stability, entrepreneurial spirit, and acceptance of the idea that development should be technology based.
Socioeconomic Infrastructure (SE) indicates the strength of each nation's educational system, mobility of capital, and encouragement of foreign investment.
Technological Infrastructure (TI) captures the strength and contributions of a nation's scientific and engineering manpower, its electronic data processing (EDP) purchases, the relationship of its R&D to industrial application, and its ability to make effective use of technical knowledge. The composition of this indicator includes four expert opinion items plus a measure of purchases of EDP equipment, and number of scientists and engineers in R&D.
Productive Capacity (PC) concerns capabilities to manufacture technology-intensive products. It combines the value of electronics production with three survey items related to manufacturing and managerial capabilities to measure the amount and efficiency of resources available.
Expert Opinion Data were obtained from a survey of the International Technology Indicators Panel during summer and autumn, 2002. Surveys were largely completed through a website interface at http://tpac.gatech.edu/, augmented by e-mail, fax, and mail contacts. The resulting group of 371 experts (up considerably from 207 in 1996 and 303 in 1999) collectively provided 412 responses (up from 265 in 1996 and 336 in 1999). The average number of responses per country was 12.5, with a minimum of 10 per country.
High Tech Standing (TS) (the 'output') measures current high-tech production and export standing. TS incorporates three components: an expert opinion item (rating technology-intensive production), overall high-tech exports, and the value of electronics exports. As noted for the input indicators, the skewed distributions on statistical components exert a strong influence on the resulting indicator, even though each component is scaled separately for the 33-country set (S-scores). This effect appears for TS, with the US the benchmark (score of 100) for both overall high-tech exports and electronics exports.

Another good indicator is the Network Readiness Index which is a product of a joint study by Harvard University and the World Economic Forum.[1] The index indicates the strength in several areas such as:

1. Thailand's road infrastructure ranked 14th in the study;
2. Thailand is the world's second largest HDD manufacturer;
3. electronics and electrical appliance exports ranked 14th (1.73 per cent of the world market);
4. electronic equipment and parts exports ranked 12th in the study; and
5. the number of days to start a new firm ranks 19th (30 days).

Thailand has continued to upgrade its IT infrastructure and has employed many strategies to reduce the digital divide problem such as the expansion of fixed-line telephone service to all areas (the number of fixed lines is 12.6 lines per 100 persons), and the plan to provide internet access for over 4,550 schools and 9,000 for local government offices in 2003.

In assessing international competitiveness, the Institute for Management Development (IMD) has ranked the competitiveness of 49 countries worldwide and Thailand ranked 34th in 2002 (an improvement from 38th in 2001) indicating many factors that impede Thailand's competitiveness level such as the workforce education level, IT skills, IT infrastructure, and the low level of youth interest in science and technology. Despite these negative factors, Thailand's open society and close-knit family ties enable Thais to adjust quickly to changes. However, Thailand also has low unemployment, a low cost of living, and continued growth in exports.

According to the National Economic and Social Development Board (NESDB), Thailand has a very low R&D expenditure in the S&T field – only 0.1 per cent of GDP – of which 60 per cent is in R&D in IT and communications. Previously the government has allocated only 0.9 per cent – 1 per cent of the total budget for R&D expenditure (approximately 0.14 per cent of GDP). Comparing with other countries in the IMD study, Thailand has very low R&D expenditure, ranking 46th out of 49 countries surveyed in 2001. This has resulted in a slow development in ICT in the country due to the weak upstream industry and modest further developments in ICT products. As reflected in the export and import statistics, the ICT industry in Thailand still relies heavily on foreign imports and the government has not adequately encouraged the use of domestic content and increase in domestic consumption as an import-substitution measure (see Table 7.3).

On the global front, Thailand has joined many international forums such as World Trade Organisation (WTO), The ASEAN Free Trade Agreement (AFTA), and the Asia-Pacific Economic Conference (APEC). These economic cooperation groups generally call for trade liberalisation and facilitation through tariff reduction, increased international standards in the production process, and more stringent intellectual property rights protection. This 'more open' market has led to lower import duties and a squeeze in profit margins due to higher competition in the Thai economy resulting from an influx of cheaper imports. So far, trade liberalisation has played a significant role in encouraging some private firms to increase their technological capabilities and activities in order to maintain their competitiveness.

On the technology front, APEC believes that 'the New Economy' based on knowledge, associated with non-inflationary sustained growth, and investment in information and communications technology, is central to economic growth and development. APEC agreed to implement its policies on trade and the digital economy to help member economies benefit from the New Economy. Most APEC economies have agreed to

Table 7.3 Importance of ICT in Thailand's External Trade

Computer/equipment/parts	2000	2001	units
Total Exports	344,049	346,589	THB m.
Growth	12.8	0.7	per cent
Percentage of total exports	12.4	12.0	per cent
Total Imports	147,612	167,704	THB m.
Growth	59.7	13.6	per cent
Percentage of total imports	5.9	6.1	per cent
Ratio of imports to exports	42.9	48.4	per cent

Source: Data from the Ministry of Commerce, cited in ICT Master Plan, p. 13.

liberalise trade in ICT goods and services, produce an effective environment for services delivered through digital means, and protect intellectual property rights in cyberspace.

As technology transfer/creation can occur through several means, we have grouped the four main channels of occurrence below:

1. Knowledge spillages through trade and foreign direct investment (also called horizontal flows or flows due to demonstration effects). This occurs when local firms adopt technologies introduced by multinational firms through imitation or reverse engineering. In fact, it is usually an externality as most firms will attempt to minimise technology leakage to competitors; hence it is better classified as a 'spillover', not a 'flow' *per se*. In the case of FDIs, even if local firms are able to observe the technology brought along with FDI, they usually lack the absorptive capacity to adopt it due to the wide gap in human capital and product development capabilities between multinational and local firms. In an international study by Saggi (2002), it was found that at the aggregate level, evidence supports the view that FDI has a positive effect on economic growth in the host country.
2. Strategic transfer through supply chains (vertical flows). Multinationals may voluntarily transfer technology to firms that are current or potential suppliers of intermediate goods or buyers of their own products, with the aim of reducing the price and raising the quality and compatibility of inputs.
3. Labour turnover. This occurs when workers previously employed or trained by multinationals transfer important information to local firms, either by switching employers or starting their own firms and further

contributing to technology diffusion.

4. Internal research and development. This is perhaps the most costly and uncertain path, as the success of each research project usually varies significantly. At present, technology creation in Thailand through this channel is rather limited.

Industrialising countries gradually acquire rather than develop their own technology, often with the capital and equipment and product technology from industrialised countries. In Thailand, MNC subsidiaries are now playing an important role in technology transfer/development, which is generating significant spillover to the rest of the economy. However, technology acquisitions through horizontal or vertical flows are not perfect substitutes for internal R&D. All companies should have the long-term goal of uplifting their innovative capabilities, and they may start by strengthening their capability to assess, select, adapt and improve technology. Once they are able to acquire and adapt to new technology, it will serve as a starting point for future innovations.

In sum, the existing situations regarding high-tech industry development in Thailand appear to be perplexing, even paradoxical. On the one hand, as a result of its market-oriented development approach, the overall economic development is high by international standards, with active participation from many high-tech MNCs, generating good performance in high-tech manufacturing products such as computers and other electronic parts and equipment, and automobiles. Yet, on the other hand, Thailand is lagging behind many other countries in East and Southeast Asia in terms of its technological infrastructures and public spending on basic skills and technological R&D. These current situations will have significant repercussions on the future development of high-tech industries in Thailand as will become clear later.

Three specific issues will be discussed in the remainder of this section: (1) current efforts by the state and the private sector; (2) the ICT Master Plan as an example of a high-tech development plan; and (3) FDI in high-tech industries in Thailand.

Current Efforts by the State and the Private Sector

Establishment of public sector technological institutions

There are a number of industry-related technological institutions, some of which have been established since 1950 these have evolved and undergone various changes:

1. The National Science and Technology Development Agency (NSTDA) (a specialised agency of the Ministry of Science, Technology, and Environment).[2]
2. The Thailand Institute of Scientific and Technological Research (TISTR).
3. The Department of Science, Ministry of Science, Technology, and Environment.
4. The National Research Council of Thailand (NRCT) was set up in the 1960s under the Office of the Prime Minister to establish a national science policy and provide inputs into five-year plans, as well as funding in some R&D in universities.
5. The Thailand Research Fund (TRF) was set up in 1993 under the Office of the Prime Minister to fund R&D mostly in universities and public sector institutes, and occasionally in the private sector.
6. The Thai Productivity Institute (TPI) was established in 1994 as a non-profit independent foundation overseen by the Ministry of Industry. The two main objectives of the TPI are (a) to become the national body in charge of promoting general awareness and understanding of productivity through public lectures, publications, activities, exhibitions, and through information exchange with organisations in other countries; and (b) to provide direct consultancy and training services in quality assurance and productivity to all companies, both in the public and the private sectors.
7. The National Food Institute (NFI), which began operations in 1997, is overseen by the Ministry of Industry. Its aim of improving the quality and safety of food products is implemented through fee-paying laboratory services (chemical, microbiological and physical testing), consulting services, training seminars and workshops, contract research for the government, and publications on food safety and quality. The Food Industry Club strongly lobbied the government to encourage the NFI to provide a research facility to address non-tariff barriers erected by importing countries to protect their own food industries, and not focus only on fee-generating laboratory testing services. However, the Food Industry Club does not have much influence over NFI operations, since the NFI is currently supervised by the Ministry of Industry. The government expects the NFI to be financially self-supporting within five years of operation, hence it cannot act as a non-profit research institute and simply focus its efforts on research, non-tariff barriers, or redirect its laboratory services to small and medium-sized enterprises (SMEs) as the Food Industry Club would like (UNIDO, 2002).

Education

Thailand's education system has failed to produce a workforce that can compete in high-tech industries with countries such as Korea or Taiwan (see Table 7.4). Surveys by the Human Resource Institute, Thammasat University, and the then Ministry of Science, Technology, and Environment in 1984 showed that the ratio of engineering and science graduates to total employment is approximately 1:100 and 1:200, respectively. Language and communications issues are also of concern to some investors as Thai workers are not particularly competent with the foreign languages used in the parent companies – English, Japanese, and so on. Additional foreign language skills will assist Thai workers to better understand the related issues associated with imported technologies/machinery.

In a research study by the Human Resource Research Institution, Thammasat University, which examined the ability of Thai staff working in MNCs to obtain and retain knowledge in the science and technology fields (Wai, 2001), it is found that under the current situation the inadequacy of qualified staff in the S&T field has prevented the full transfer of knowledge from MNCs. It should be noted that 'technology transfer' in this study includes all kinds of transfer methods such as training, academic text/user's manual, and domestic technology diffusion. Approximately 75 companies in five industries were included, namely, (1) electrical and electronics, (2) food and beverages, (3) automobiles and parts, (4) materials, and (5) pharmaceuticals and chemicals.

The study found that most of the technology attainments occur through transfer from the parent company and through technology acquisition (purchase) process. The main method of transfer is staff training, which still has some obstacles such as communication problems (language barrier), high training expenses, difficulties in performance monitoring, and the parent company's policy which prevents full transfer in some cases. Quality control and process improvement and international work standards are the issues that trainers usually focus upon with the aim of cost reduction. MNC managers feel that it is necessary to develop staff competency and improve the Thai education system.

Wai (2001) emphasised that in order to facilitate the efficient transfer of technology, it is crucial that capable teams of trained workers exist and are ready to absorb the technology transfer. The gap between the technology transferor and the technology receiver could be reduced by encouraging employees to be more 'technology literate'. Apart from this, government policy also plays a major role in supporting and inducing MNCs to invest and establish production or even R&D centres in Thailand.

Table 7.4 Employed Persons by Level of Education Attainment for Whole Kingdom, 2003 to first quarter 2005

Level of education attainment	2003				2004				2005
	1Qtr (Jan.– Mar.)	2Qtr (Apr.– Jun.)	3Qtr (Jul.– Sep.)	4Qtr (Oct.– Dec.)	1Qtr (Jan.– Mar.)	2Qtr (Apr.– Jun.)	3Qtr (Jul.– Sep.)	4Qtr (Oct.– Dec.)	1Qtr (Jan.– Mar.)
Total	32,762	33,361	34,676	34,565	33,424	34,189	35,711	35,591	34,050
None	1,144	1,151	1,204	1,166	1,168	1,225	1,266	1,248	1,189
Less than elementary	12,354	12,415	12,960	12,812	12,092	12,071	12,692	12,391	11,566
Elementary	7,375	7,387	7,938	7,741	7,470	7,630	8,034	8,013	7,505
Lower secondary	4,331	4,606	4,749	4,753	4,636	4,890	5,029	5,074	4,923
Upper secondary	3,446	3,688	3,603	3,704	3,680	3,888	4,041	4,022	3,952
General	2,333	2,557	2,511	2,590	2,550	2,736	2,848	2,886	2,806
Vocational	1,098	1,116	1,073	1,098	1,115	1,138	1,178	1,120	1,130
Educational	15	15	19	15	15	14	16	16	16
Higher Level	4,005	4,018	4,122	4,283	4,261	4,337	4,499	4,675	4,717
General	2,016	1,961	2,016	2,145	2,187	2,212	2,313	2,425	2,493
Vocational	1,246	1,284	1,337	1,343	1,291	1,323	1,408	1,446	1,407
Educational	744	773	769	795	783	803	778	804	817
Other	17	33	22	8	17	19	23	25	23
Unknown	91	62	80	98	100	130	129	144	175

Source: Labour Force Survey, National Statistical Office (www.nso.go.th).

The ICT Master Plan as an Example of a High-tech Development Plan

The National Information Technology Committee (NITC) – a high-level policy body chaired by the prime minister himself, with members comprising executives from relevant public and private sectors – was established in 1992. Its main objectives are to develop ICT policies and plans on: (1) creating the digital opportunity, (2) ensuring the proper modernisation (IT laws), (3) forging the knowledge-based society, (4) developing infrastructure for e-government, and (5) developing infrastructure for the national economy.

In September 2001, the government approved the National ICT Master Plan 2001–06, which will operate under the main umbrella of the IT 2010 Policy (2001–10). To achieve the goals, IT 2010 identified five main areas that will be developed over the next decade as follows (see Table 7.5):

1. e-Society – aims to reduce the 'digital divide', improve the quality of life, enrich culture, increase knowledge in health and enhance public participation through social networks.
2. e-Education – aims to promote life-long learning, computer literacy, human resource development and virtual education.
3. e-Government – aims to improve public service via electronic service delivery, employment and legal infrastructure. The internal development was to be completed by 2004, and improvement in public service should be completed by 2010.
4. e-Commerce – aims to improve competitiveness through e-services (in tourism, international trade, finance, personal consumption, etc.) and provide an adequate underlying legal structure.
5. e-Industry – aims to promote e-manufacturing and IT-related industries (which should benefit the whole manufacturing sector), and international standardisation within 2010.

The IT 2010 policy has outlined the linkage between each of the five areas as follows:

1. The development of human resources will include skills enhancement for public employees (part of e-Government), development of students in the educational system (part of e-Education), life-long learning (part of e-Society), and development of knowledge workers (part of e-Commerce and e-Industry).
2. Innovation promotion will affect the management system in the public sector which is part of e-Government. It will also enhance the creative capabilities of Thai youth in the education system (part of e-Education), the transfer of local lore (part of e-Society), and R&D and technology transfer in the economy (part of e-Communication and e-Industry).
3. IT infrastructure and IT industry development will include development of (a) network and IT management of government enterprises (part of e-Government), (b) network for education and learning (part of e-Education), (c) IT infrastructure in the community

(part of e-Society), (d) e-Commerce and software industry (part of e-Commerce), and (e) supply chain management system and electronics industry development (part of e-Industry).

Put simply, the ICT Master Plan provides a major roadmap to develop and expand technology-related industries and services. The new technologies which include ICT, biotechnologies and genetic engineering, have initiated many new economic and social activities which are very different from the past. A knowledge-based learning economy will be the theme of the new economy in the 21st century, and hence Thailand must develop adequate qualified human resources in anticipation of this new paradigm shift. It also contains many goals/targets for public institutions as well in order to meet the expected needs from the private sector.

The government has established several institutions to support SMEs, but these are not sufficient for SMEs to receive full benefits from ICT. In a 2002 survey[3] conducted on SMEs in three industries – electrical and electronics, automotive and food – preliminary results suggest that small enterprises have not focused on utilising ICT in their business operations, mostly because management does not fully understand and appreciate the benefits of ICT usage. Therefore the government intends to help facilitate SMEs by ensuring higher utilisation of ICT through (i) establishment of an SME portal which contains a database on government regulations and subsidies, (ii) establishment of pooled retail sales, (iii) a good back office system for the OTOP Programme (One Tambon One Product),[4] and (iv) provide support to the supply chain management system.

FDI in High-tech Industries in Thailand: Recent Experiences

High-tech manufacturing industries

A good example of a high-tech manufacturing industry which has been significantly influenced by FDI is the automotive industry. The industry came into existence during the late 1960s and early 1970s, with several joint-venture assembly vehicles from completely knocked down (CKD) kits. Investment in auto parts and components began to expand in Thailand during the mid-1970s, however, the domestic market was not sufficient to provide the needed economies of scale that would make these local manufacturers cost competitive. It was then necessary for the government to take action to protect local assemblers and local parts and component makers, since a market for Thai-manufactured vehicles would not be able to grow as the existing tariff structure did not

Table 7.5 Summary of IT 2010 Policy Strategies and Measures

Promote exports Promote service industries Promote consumption from domestic producers	Upgrade production efficiency & expand market base through IT Utilize IT in agricultural production Focus on developing a reputable IT industry locally	Enhance both front and back office efficiency Promote good governance in the government process	Reduce the 'digital divide' Improve Thai quality of life Promote learning society	Create value added on existing equipment Reduce inequalities through social equity Plan 'quantum leap' in the long run
Strategies according to IT 2010 Policy				
e-Commerce	e-Industry	e-Government	e-Society	e-Education
Policy Measures				
E-Commerce law Secure electronic funds transfer system Promote awareness and understanding Promote SMEs Develop market for private sector through govt's e-procurement HR development Establish appropriate IT infrastructure and promote Thai IT industry	Establish Thailand Exchange Provide market information Promote R&D in the private sector Promote HRD in manufacturing sector & encourage knowledge transfer among the workers Promote IT industry to provide import substitution and export opportunities Encourage more IT usage in agricultural sector	Devise a Master Plan and arrange monitoring entities Improve work and database system, both for central and local govt Upgrade skills of public workers Adjust law and regulations Create IT infrastructure and promote Thai IT industry Promote awareness and confidence	Provide opportunity to access IT and knowledge Promote communities and life-long learning Develop workers' skills in accessing and using technology for educational purposes Promote IT usage to improve quality of life and create cultural & societal bonds Encourage appropriate IT infrastructure and support Thai IT industry	Upgrade teachers' skills through training Content development Establish good educational networking system Promote IT usage to upgrade S&T capabilities Provide IT infrastructure and support Thai IT industry

adequately protect assemblers from completely built-up (CBU) imports.

Therefore, the government instituted a ban on CBU imports and increased the tariff on CKD kits. Throughout the 1980s, the government protected local parts manufacturers by increasing the local-content requirements to over 50 per cent. This policy turned out to be very beneficial to the local manufacturers since Thailand's ability to provide cost-effective local inputs – raw materials, labour and infrastructure – began to attract a large number of industrial projects, many of which are Japanese investments. The local-content requirement was eventually scrapped on 1 January 2002 to pave the way for trade liberalisation. At present, manufacturing of auto parts involves a wide range of parts including engine, suspension components, brakes, clutches, steering, power transmission, chassis, electrical equipment, decorative accessories, tyres and auto glass.

Currently, Thailand is home to many regional production facilities of large international auto manufacturers, namely, Toyota, Honda, Mitsubishi, Nissan, General Motors, Ford, DaimlerChrysler, BMW, Hyundai and so on. Despite a much lower level of automation compared to other countries, the low labour cost enables assembly plants in Thailand to be cost competitive. In other words, Thailand's automotive sector's ability to compete internationally depends very much on low labour costs, which compensates for low levels of productivity. A 2003 Study by Sasin Graduate School of Business found that assembly plants in Thailand produce fewer cars per employee than the international benchmark countries (11 vs 45 per year). In addition, the capabilities of production engineers and technicians graduating from local educational institutions are often insufficient, and hence companies have to provide extensive in-house training to prepare new employees for their tasks. The Sasin study suggests that Thailand's auto industry – positioned by policymakers to be the 'Detroit of Asia' – may be marginalised and overtaken by China's in the future if changes and improvements are not made (Wichit, 2003).

On a brighter side, Thailand is the second-largest producer of pick-up trucks worldwide, trailing only behind the United States. The domestic demand for pick-up trucks for both commercial and private use is high, resulting in a large variety of models being produced to meet consumer needs. This is the main segment in the automotive industry in which Thailand has a clear comparative advantage. In April 2003, Thailand's Board of Investment (BOI) agreed to provide tax and promotional privileges to Toyota Motor Co. to help the auto-maker move its global pick-up truck production base from Japan to Thailand. It is expected that Toyota and its components and parts suppliers plan to invest THB

42.8 billion (approximately US$ 993 million) over the next two years in the project, which will create jobs for 3,000 Thais. The project has been launched in 2004, with an initial maximum production capacity of 200,000 light pick-up trucks a year (of which about 60 per cent would be geared for export). Moreover, Toyota is planning to establish a THB 2.7 billion R&D hub in Thailand soon.

Also attracted by the low-wage, low-tax, and good operating environment, the world's most important hard disk drive (HDD) maker, Seagate, has moved some of its less demanding operations from Singapore to Thailand. However, a large part of its operation still remains in Singapore due to the fact that skills underlying high quality, high speed, and high volume have accumulated there. In addition, HDD assemblers and their suppliers are geographically close – which drives learning and imitation among HDD assemblers and parts suppliers – and the government is providing the right incentives to attract high-value-adding technological activities (Ho, 2001). At present, Thailand's HDD production ranks 2nd in the world market, but the country does not have an R&D centre for HDD yet, even if this would enable further development of IT products for exports.[5] Therefore, in order to prevent the moves in production base to countries with lower wage costs such as China or Vietnam, it is crucial for Thailand to support R&D in the manufacturing sector, upgrade human resource capabilities, and enhance the value added of Thai products.

In October 2002, a group of investors pledged to inject up to US$ 1.1 billion into B-Square, formerly known as Submicron, the debt-laden wafer-fabrication project that has been sitting idle since 1997. The investment will come from a German group of companies that includes DRAM maker Infineon Technologies AG, and clean room manufacturer Meissner & Wurst, and several other financial institutions (Wilson, 2002). The Thai Asset Management Corporation (TAMC), which is the main government agency set up to clear the non-performing loans (NPLs) problem, will convert much of Submicron's THB 12 billion debt into equity, resulting in TAMC owning roughly US$ 100 million of B-Square's shares. This indirect investment of the Thai government shows that they support the semiconductor industry. Nevertheless, there is still much work to do before Thailand's first fabrication plant is successfully commercialised. A developed system which can ensure steady and pure supplies of power, water, and fuels is needed. In addition, an adequate number of qualified electrical engineers is also a crucial factor.

Mitsubishi Heavy Industries Limited has moved its production of Mitsubishi Heavy Duty residential air-conditioners from Japan to Thailand. Plans are also underway to produce commercial air-conditioners in Thailand, a move that the company believe will raise the competitiveness of the products in the world market (Sukanya, 2003). Similarly, Sumitomo Electric Group plans to invest at least THB 3 billion in Thailand in the future, focusing on automotive and other sectors such as magnet wires, ultra thin coils, power cords, ceramic coatings and flexible printed circuits. The group now has 23 local companies in manufacturing, engineering and sales/services, with more than 5,000 local employees, and a total investment of THB 10 billion.

In the case of Thailand, technology transfer through FDI in high-tech industries is not so prevalent. Since one of the criteria in determining investment privileges is the number of jobs that will be created, some foreign investors have opted to use older technology in order to create more jobs (become more labour intensive). For the manufacturing sector, a very large percentage of technologies being used in the production process are imported from overseas – mainly from Europe, Japan and Taiwan. Local Thai workers are trained to operate the machines, but in case of breakdowns, the overseas machine producers are still responsible for the maintenance function most of the time. There are hardly any Thai firms which uses locally made technology in large-scale high-tech production. To the local bankers, the use of locally made technology in large-scale production is often seen as an 'unfavourable factor' in the loan approval process – making it more difficult to obtain the necessary funding for the project.

Government policies

Thailand has lagged in the transition to an enterprise-based technological development system, partly resulting from the financial crisis that started in 1997. There were few effective measures to support firm-level technological activities – both to strengthen firm's own efforts, and to build the system to support technological development. The government has started to roll out several policies in an attempt to elevate the high-tech capabilities of Thai companies, many of which will be discussed later.

The first National IT Policy 'IT 2000' was approved by the Cabinet in February 1996. Seven years later, it was replaced by the second National IT Policy Framework 'IT 2010' policy – which has the mission to turn the country into a knowledge-based society and economy where science, technology and innovation are encouraged and valued, and where citizens are empowered with life-long learning opportunities and capabilities. The government wishes to move Thailand out of the 'dynamic adopters' group of the UNDP's technology achievement list and become a 'potential leader', to raise the proportion of knowledge workers from 12 per cent in 2001 to 30 per cent of all workers by 2010, and to increase the share of knowledge-based industries in the country to 50 per cent of GDP by 2010. The Ministry of Information and Communications Technology was established in late 2002. In the future, this new ministry could be responsible for co-ordinating the functions according to ICT Master Plan and reports to the government. It is expected that the Master Plan can be adjusted on a yearly basis if deemed necessary to ensure dynamic planning and effectively result in a rolling plan well into 2007–11.

Biotechnology

In a traditionally agricultural-based country like Thailand, biotechnology could play a major role in lowering the cost of production, and providing higher yields and better-quality crops. There have been attempts to apply biotechnology to improve crop production since biotechnology can offer a sustainable agricultural process by reducing the use of chemicals in agricultural production (Lim et al, 2001, p. 223). Thailand has successfully commercialised tissue culture such as micro-propagation of orchids and some ornamental species, but the country is not doing so well in other areas of the biotechnology race. The main factors restricting successful technology transfer and commercialisation of new methods of production are lack of coordination, inadequate funding, poor human resources and limited access to information. Technologies developed in educational institutions are not particularly oriented towards commercialisation and hence are not suitable for commercial production. There is a threat of substitution of Thailand's agricultural exports through genetically modified organisms (GMOs) – which can effectively 'create' similar products overseas, occasionally at less cost, although there are some concerns over its safety for human consumption in the long run. 'Bio control' products[6] are becoming more popular both domestically and globally, and the country may explore further in this area to make it more cost competitive and of higher quality.

In the Master Plan for Thailand's agricultural development, which was rolled out by the cabinet in early 1998, R&D was emphasised with the aim of raising production and reducing costs by introducing new technology. The government has recognised the strategic implications of the ongoing growth of the agricultural industry and has established the National Centre of Genetic Engineering and Biotechnology (BIOTEC), also under the NSTDA. This centre has been supporting and transferring technology for the development of industry, agriculture, natural resources, the environment and the socio-economy. Support on technology development and adoption has been given to both public and private institutions through R&D projects, human resource development, management and technical services, technology investment and information services. The Ministry of Agriculture and Agricultural Cooperatives has also utilised the internet for disseminating useful information such as agriculture best practices and industry news.

According to Morakot (2002), Thailand, mainly through BIOTEC and the Ministry of Agriculture and Agricultural Cooperatives, has begun to put more emphasis on research on agricultural biotechnology, exploiting its established competitive advantage on food production. A few successes could be cited. As one of the world's largest farmed shrimp exporters, Thailand has been a leader in the development of DNA probes for the rapid detection of major shrimp diseases which helped reduce substantially losses from the shrimp virus outbreak during 1995–97. With regard to rice, another major food product, Thai researchers have been able to breed high-value rice with higher resistance to typical diseases using molecular genetic methods such as DNA fingerprinting to a typical disease. Other types of progress include the use of biotechnology to control the waste in the cassava starch industry, and the establishment of bio safety guidelines on the release of GMOs into the environment. The establishment of the Biotechnology Centre at Mahidol University on 30 October 2003, is another milestone for the development of advanced biotechnology in Thailand.

Biotechnology can also play a major role in the pharmaceutical industry. A researcher at BIOTEC has recently succeeded in producing a human growth hormone which can be used to treat children with growth retardation syndrome. Further refinements are still needed for this unique achievement to reach its full potential, and when the project is fully completed it will reduce the government's budget for medication. According to the researcher, Thailand spends over THB 100 million annually importing human growth hormones to treat young patients (Heamakarn, 2003). Moreover, if economies of scale in production can be found, Thailand may even start to export some of these successful products of scientific innovation in the near future.

In December 2003, the Thai Cabinet approved a strategic plan aimed at guiding the nation towards the new goals of nano and bio-technology. The government believes that the two technologies are the global trend and can provide good business opportunities which Thailand can rely on in the long term. In addition, the Cabinet has approved a policy framework for the development of biotechnology, which will eventually be integrated into various national policies – such as making Thailand the 'kitchen of the world' and an Asian health hub. In order to ensure the policy's success, six sub-committees have been established to oversee each policy. For instance, the secretary general of the BOI chaired the sub-committee on developing biotech business, including promoting genetically modified products. The government intends to create the proper atmosphere for private investment in biotech development and hopes to facilitate the development of over 100 new biotech companies in an industry worth THB 30 billion. Moreover, the government expects the exports of food and agricultural products to expand to THB 1.2 trillion (Hassaya, 2003).

Software development

The global software industry still has high potential and is experiencing high growth. The OECD has estimated that software expenditure in 2001 totalled US$ 196 billion globally, mostly from developed countries. This provides ample opportunity for Thailand to increase its market share in the world market. Thailand's software development market currently accounts for less than US$ 1 million annually, but the government intends to increase that figure in the near future. The ICT industry in Thailand still lags far behind those in the US or Europe. Compared to other countries in ASEAN (excluding Singapore), the potential of Thailand's ICT industry appears more favourable, as the level of development is not significantly different, and Thailand's central geographical position may be beneficial. There is still a strong demand for Thai language and other customised software for the local market. However, as shown in Table 7.6, Thailand continues to lag behind in terms of the number of graduates entering the IT marketplace, and hence many Western companies seek to outsource software development projects to India, Malaysia, Singapore and China instead (Hall, 2002).

Table 7.6 Human Resource Requirements in Software Industries

Software Industry Human Resource Targets in 2003		
Title	2003	Ratio
Architects	100	2 per cent
Analysts/DB Designers	700	14 per cent
Developers	2,000	40 per cent
Project Managers	200	4 per cent
Test Engineers	700	14 per cent
System Engineers	700	14 per cent
System Administrators	700	14 per cent
Total	5,000	100 per cent

Source: Association of Thai Software Industry, www.atsi.or.th.

Thailand's 'software park' was set up under the umbrella of the NSTDA in 1999. Its main objective is to promote the development of Thailand's software industry through building a cluster of software development activities in the Software Park Complex in Bangkok. The Software Park supports the industry through four main channels: (1) technology transfer service, (2) business development service, (3) 'Incubator Centre' service, and (4) underlying technological infrastructure service. The goals are to upgrade the quality of software production, promote technology transfer, attract foreign investment, create an adequate pool of qualified software professionals in Thailand, and lower the piracy rate. In 2004, the Software Park housed close to 50 companies, most of which are small- and medium-sized start-up software developers. Many training courses are also offered here, ranging from new technology in software development, project management, software costing, product marketing and productivity improvement. Some believe that the Software Park has not utilised its full potential yet and there is still some room for improvement. The number of projects that it has attracted is still small on a national basis.

There is also a plan to create a one billion baht IT Knowledge Park in northern Thailand (Waltham, 2002, p. 1), with funding over a five-year period from the NESDB. The plan was announced during a three-day Asia and Oceania Computing Industry Organisation (ASOCIO) general assembly and symposium, attended by over 600 regional industry and government IT leaders. It was announced that the government would assist the Thai software industry by enhancing software human resource capability, by creating investment incentives, and by providing monetary resources. Two organisations, the Software Industry Promotion Board, and the Software Industry Promotion Agency, whose members are drawn from both the public and private sectors, may be set up in the near future.

In March 2003, the Asia Open Source Software Symposium 2003 was held in Phuket, Thailand. Many software developers from around Asia gathered to discuss many issues concerning open source software (OSS) and shared knowledge. The group came up with a very broad Phuket statement outlining that they recognise the importance of community in order to promote collaborative development and information sharing of OSS, and also recognise the necessity of harmonisation of the communities emerging in Asia. They share the understanding that OSS is crucial for skills development of professionals concerning software technology and hence will promote information exchange and collaboration on OSS in Asia.

The government will need to address several issues in order to promote the software industry: human resource development, standards for software process development, software marketing, the development of Thai software for international markets, cooperation between small software companies, and infrastructure development.

Science Park

The crisis which stalled Thailand's stellar economic growth in the last decade has signalled that there is a need to adjust from simply relying on low wages and utilising the country's abundant natural resources to a more technology-oriented approach. Thailand's competitive advantage is slowly eroding, with rising wages and depleting natural resources. It is necessary to improve the manufacturing process through science and technology to make higher-quality and more value-added products, while using natural resources more efficiently and effectively.

An initiative of the NSTDA, the 200-rai (approximately 32 hectares) Science Park, in which construction was delayed for a while due to the Asian crisis, is now developed to promote and facilitate R&D. The park will be equipped with techno-business incubators and will provide a base camp for NSTDA operations and activities. There will be a national genetic engineering and biotech centre, metals and material centres, the National Electronics and Computer Technology Centre (NECTEC) and the Innovation Development Fund (IDF). The objectives of the park are to establish research networks for universities and develop sustainable and competitive industry clusters. This is not only a local effort within the Science Park and the surrounding areas, but also a nationwide effort. Tax incentives will be offered to private companies who invest in R&D facilities at the park. These incentives include import tax exemption for machines, equipment and materials for R&D, exemption of corporate income tax for eight years, 200 per cent deduction of research expenses, accelerated depreciation of machinery and equipment, working permits and visas for foreign researchers and experts, and allowance for foreign ownership. The opening ceremony was held in June 2003. The success of this project remains to be seen.

3. STRATEGIES FOR HIGH-TECH INDUSTRIES

Against the existing background of technological advancements in Thailand referred to above, two questions may be pertinent:

1. What preparations do we need – manpower and skills or skill levels of the workforce; capital investment and technical know-how; local support and local adaptability; development of indigenous enterprises?

2. Which industrial sectors should be developed and which are high-tech – electronics; biotechnology; molecular technology; software designs; pharmaceuticals; communications technology; environmental technology?

There are many public agencies that can answer the above questions. But here we shall discuss the role of one of the most important public agencies: the Board of Investment (BOI)

The Office of the BOI was restructured in October 2002 with the aim of enhancing the agency's effectiveness in implementing the government's economic policy, and in strengthening its relationship with customers. A new strategy has been devised to transform the BOI into a genuine facilitation and investment services provider.

These strategies include:[7]

1. the transfer of the line of command of the BOI Office from the Office of the Prime Minister to the Ministry of Industry – in order to streamline the bureaucracy into a smaller and more efficient body;
2. internal change in the structure of the organisation, with three new divisions established, namely, Marketing, Management Information Systems, and Investment Facilitation (in charge of regional development and industrial linkage development); and
3. a new strategy to transform itself from a tax privileges provider into a facilitation role as an investment services provider.

The BOI has played a crucial role in promoting foreign direct investment in Thailand by offering tax privileges and some investment facilitation services. The country's investment policy remains one of the most liberal in the region. The two kinds of benefits that the BOI offer are: (1) tax-based incentives, such as tax holidays or tariff exemptions, and (2) non-tax privileges including guarantees, protections, permissions, and services. The non-tax privileges are available to all BOI-promoted projects regardless of location, industry or condition; while tax-based incentives depend on several factors, mainly the location of the project, and the industries that it is going to operate in. The BOI announced the 'Policies and Criteria for Investment Promotion' in April 1993, creating three investment promotion zones, in line with the government policy of decentralisation from the Bangkok Metropolitan Area. Promoted projects locating in the zones which are considered to be least developed receive additional incentives, including further tax and duty reductions, as well as allowances for infrastructure investment. The BOI offers:

1. exemption of import duty on machinery;
2. 8 years of corporate income tax exemption, and up to 50 per cent reduction for a further 5 years (for the 40 provinces which are least developed in Zone 3 only); and
3. exemption of import duty exemption on raw or essential material used in the manufacturing of export products for 5 years.

However, the zoning regulations were relaxed in 2003 to promote cluster development and hence improve industrial competitiveness. The BOI believed that clusters have historically led to new product development, and development of niche markets. Therefore, by lifting restrictions on where projects could locate, it is expected that industry-specific clusters would develop according to a variety of factors, such as manpower, utilities and infrastructure, which will vary between different industries. Most of the projects previously required to locate in Zone 2 or Zone 3 are now free to expand wherever they deem appropriate (except projects which may create harmful substances during the production process). Since 1985, the BOI has approved over 10,700 projects, with total investment of more than THB 2 trillion (US$ 50 billion), and creation of some 2.1 million jobs for Thai workers. In 2004, 1,053 projects were approved with a total investment of THB 371.3 billion, and expected 216,125 new employment positions (see Table 7.7).

In the years to come, the BOI intends to continue to assist investors wishing to establish their businesses in Thailand by providing information on investment opportunities, prospective joint-venture partners, and foster the development of subcontracting networks through its subsidiary, the BOI's unit for Industrial Linkage Development (BUILD). The BOI will increasingly focus on a new set of target industries, five of which have been identified: agro-industry, automotive, fashion, ICT, including electronics, and high-value-added services. The BOI also aims to conduct research into identifying promising investment opportunities in Thailand, promote technology transfer and support SMEs and community enterprises. Five incentive management divisions were set up to handle all aspects related to projects in the industries under their responsibility, including project appraisal, review of machinery and raw materials submitted for tax exemption, and project monitoring.

The five divisions are:

1. agro-industry division
2. metals, metal products, machinery and transport equipment division
3. electronics and electrical industries

4. chemical, paper, plastic, and light industries
5. services and public utilities

It is expected that investors will find this process more convenient and time-saving as all correspondence with the BOI can be conducted through only one division.

Last year, the BOI's flagship plan to entice foreign firms to establish regional operation headquarters (ROH) in Thailand was rolled out. ROHs are companies incorporated by Thai law providing management and technical services to their branches and associated companies from the Thai headquarters. Benefits for companies establishing regional operating headquarters have included a reduction in or exemption from

Table 7.7 Projects Receiving Promotion Certificates: 2002 to 2004, and 2005 (Jan–Aug) Investment Value in units of THB thousand million

	2002		2003		2004		2005 (Jan–Aug)	
	No. of project	Val.invt	No. of project	Val.invt	No. of project	Val.invt	No. of project	Val.invt
TOTAL	797	221.9	717	228.7	1,053	371.3	725	424.4
By Sector								
Agricultural Products	173	49.2	121	21.9	206	44.7	126	27.8
Minerals & Ceramics	11	3.4	12	2.3	26	18.0	23	24.8
Light Industry	78	18.8	54	5.7	78	13.9	67	9.2
Metal Processing	179	35.7	169	58.1	252	70.2	159	115.7
Electronics& Electrical	160	31.2	126	40.5	172	87.4	140	36.8
Chemicals & Paper	91	40.3	100	35.6	131	54.8	87	46
Services & Infrastructure	105	43.2	135	64.6	188	82.3	123	164.1
By Zone								
Zone 1	164	43.6	139	77.3	253	78.8	206	56.9
Zone 2	368	93.3	349	104.7	467	185.0	318	221.9
Zone 3	265	85.0	229	46.8	333	107.5	201	145.6

Notes: Projects receiving promotion certificates must have already registered a company with a percentage of paid up registered capital.

Source: BOI Investment Review and BOI website.

corporate income tax, a special depreciation allowance rate and special breaks for the personal income tax rates of expatriates working for the ROHs. To date the plan has not been very successful, due to a failure to promote its benefits as well as more competitive offers from neighbouring countries. Since cooperation between the BOI and the Revenue Department in designing the ideal corporate tax structures is lacking, only two companies, Exxon Mobil and Diethelm Travel, have taken advantage of the new regulation – which has been in effect since August 2002.

4. WHAT ELSE NEEDS TO BE DONE?

IT

In April 2003, experts from six IT associations staged a panel discussion to assess the ministry's initial performance and concluded that the ICT ministry has done a good job in training and improving IT skills in its first six months; however, it needs to focus more on e-law implementation and set up an e-transaction committee. They commented that the minister's 'open door policy' was helpful in allowing the private sector to communicate with administrators, although there was a concern that the ministry may not be able to prioritise matters due to its limited resources. In addition, the Association of Thai Computer Industry (ATCI) honorary president, Mr. Manoo Ordeedolchest, also commented that the ministry's initiative to create software clusters in three provinces – Chiang Mai, Khon Kaen and Phuket – would help diffuse software production capability throughout the country as it would be easier for people in these provinces and nearby areas to enter the software industry. He was also positive about the ministry's efforts over the past six months to cope with the skills shortage among ICT workers (Sasiwimon, 2003). In the 2003 study by the Sasin Graduate School of Business, IT executives report that nearly 50 per cent of graduates need to be significantly retrained before they are functional.

Recently, the number of students opting to study ICT as a major has increased, primarily due to the expansion in the IT products market, and hence IT-related employment. However, a survey by NECTEC (2002), found that even though the number of graduates has increased, Thailand still lacks quality IT professionals including system analysts, IT project managers and IT policy planners. This is because graduates are not qualified enough, and the courses offered at educational institutions do not meet industry needs. This field is very active and fast-paced, hence some universities and vocational colleges which do not adjust their curriculum on a regular basis have ended up producing substandard graduates who cannot directly enter into the IT workforce. Nevertheless, the increase in the number of IT graduates from vocational institutions is beneficial to the development in the ICT industry even if they do not work as IT professionals, as they can become competent IT users and this will support the IT-enabled professions in general.

Legal Infrastructure

Thailand has planned to legislate six new laws to prepare for the IT age: the Electronic Transactions Act, the Electronic Signature Act, the Universal Access Act, the Computer Act, the Electronic Fund Transfer Act and the Data Protection Act. NECTEC was appointed by the Thai Cabinet to coordinate and manage the drafting process. When all the laws are passed, the country will be equipped with the legal infrastructure required to enhance the domestic IT industry development.

The Electronics Transactions Act and the Electronic Signature Act were passed in December 2001. The Electronics Transactions Act defines the legal status of electronic records to be equivalent to paper documents. It also defines the scope of legal recognition of the transmission and reception processes for electronic data records. The Electronic Signature Act defines the digital signature as proof of identity of the signing party and that the signer approves the authenticity of the content that is being signed. The delayed endorsement of the other ICT-related laws such as the National Information Infrastructure Law, the Computer Crime Law, the Data Protection Law and the Electronic Funds Transfer Law were due to official process. The Computer Crime Law will be tabled to the Cabinet soon, while the universal access law is being considered by the Office of the State Juridical Council.

In addition, the country is enforcing the measures introduced in the Intellectual Property (Protection) Law on a stricter basis as Trade-Related Protection of Intellectual Property Rights (TRIPS) is part of the WTO's Agreements. In TRIPS, most processes and many products resulting from technologically-oriented research can be patented. This should reduce the negative externalities to the researcher caused by illegal usage of patented products, and in turn should increase the incentives to invest more in R&D so as to reap the benefits of new discoveries.

Research and Development

According to an estimate of R&D expenses for the Thai manufacturing sector (OIDF, 2002b, p. 5), the total R&D expenses which medium and large manufacturing firms spent in 1995, 1996 and 1997 were THB 602, 1,243 and 466 million, respectively. The number of full-time R&D staff employed by these companies totalled 5,291 persons, with 2,725 researchers and 2,566 supporting staff.

The amount of R&D funding support by the Thai government has been minimal at 0.14 per cent of GDP, compared to 0.5 per cent of GDP in some other countries. Comparatively, Thailand's R&D spending ranked 46th out of the 49 countries surveyed by the IMD. Moreover, the R&D outcomes are usually not concrete enough to be further implemented by the industry to make it more commercially feasible. The NSTDA has recognised this problem and intends to provide more support to the private sector to upgrade Thailand's R&D capacity. From now on the agency will forge a partnership with entrepreneurs to work out their needs, and conduct research accordingly. It is expected that the NSTDA will serve in both a consulting and a research role to all firms ranging from large-scale businesses to SMEs, and will absorb most of the R&D costs. The NSTDA would use Taiwan's Industrial Technology Research Institute and Canada's National Research Council which were successful in enabling the transfer of knowledge to SMEs and strengthening SME capacity as role models. R&D activity in Thailand had been neglected by many businesses, as they still consider R&D costs as 'expenses' rather than 'investments'. The tax incentives in place since 1996 for firms with R&D projects are not very enticing and have had limited results so far.

Current R&D incentives include tax allowances and soft loans. There are two main channels for providing tax allowances for firms undertaking R&D, (i) by the BOI through tax allowances and non-tax incentives, and (ii) by the Ministry of Finance through accelerated depreciation for machinery and equipment purchased for R&D purposes, and combined 200 per cent tax credit for R&D expenditures (UNIDO, 2002). Nevertheless, the tax schemes have had a limited impact on the technological activities of firms since they have burdensome operational and administrative requirements and in part are subject to the tax auditors' ability to draw a consistent distinction between R&D and other expenditures, and the NSTDA's final approval. The soft loan grants through the Research and Technology Development Revolving Fund (operated by the Ministry of Science and Technology or MOST) and the NSTDA soft loan facility are minimal at this stage due to limited budgets (see Table 7.8).

There are a number of firm-based technology development initiatives in Thailand (see Table 7.9). The NSTDA and the Department of Industrial Promotion (DIP) under the Ministry of Industry (MOI) initiated a series of pilot projects offering consultancy services in the early 1990s, in addition to some financial assistance. Many projects were operative by 2001, six under NSTDA, two in DIP and six in autonomous institutes under MOI. Nonetheless, most of them have remained relatively small, with a total number of delivery staff of just over 300. In general, the projects were inadequately recognised by policy makers and hence received a limited budget, which is also subject to change from one year to the next. Their field advisers were mostly inexperienced in industrial matters, and their services were usually limited to the Bangkok area.

According to a review completed in 2000, Thailand's industrial technology development system has relied almost exclusively on public institutions as suppliers of technology on behalf of industrial firms for over 40 years (Arnold, Bessant and Brimble, 2000) In contrast, other more industrialised countries in the region such as the Republic of Korea, Taiwan, Singapore, and to a lesser extent Malaysia, have developed a dual structure which has promoted technology creation and skills/capabilities development among private firms, as well as in public institutions. Private firms do not account for the larger proportion of industrial technology development in these countries.

Small and Medium Enterprise Development Bank of Thailand (SME Bank)

The SME Bank was officially set up in November 2002 to be the principal financial arm to promote the development of SMEs with the aim of focusing more on economic and social returns rather than on monetary profits. The bank provides financial support to SMEs operating in the sectors targeted by the government. It aims to strengthen the financial standing of new and existing SMEs and help them to achieve sustainable growth.

Traditionally, those who apply for a loan need to have some kinds of collateral such as property or other tangible assets, but for approximately 700 software developers (95 per cent of which are SMEs), asset guarantees are often not available. Therefore, most banks hesitate to approve loans to these software developers as it is seen as a high-risk business. To overcome this inherent problem, the SME Bank intends to work with the Association of Thai Software industry (ATSI) to help software houses get loans based on their intellectual property. ATSI is in the process of drafting regulations for measuring its capital value. This is a new attempt to support the growth of local software houses which, by nature, requires high investment. However, the success of this new idea remains to be seen, as it is obvious that in practice it would

Table 7.8 R&D Budget and Expenditure in the Public Sector, 1997–2000

Organisation	1997		1998		1999		2000	
	A	B	A	B	A	B	A	B
Office of the Prime Minister	300	300	100	100	400	400	850	850
Ministry of Defence	86	84	27	26	13	12	18	17
Ministry of Agriculture and Cooperatives	3,175	2,643	2,689	2,348	2,913	2,604	2,886	2,751
Ministry of Education	309	220	7	7	88	64	130	116
Ministry of Public Health	159	129	62	46	231	162	255	267
Ministry of Industry	106	81	35	20	76	61	70	67
Ministry of Science Technology and Environment	3,150	2,487	3,066	2,262	3,642	2,003	3,140	2,759
Ministry of University Affairs	2,026	1,279	1,475	1,298	970	939	1,146	1,119
State Enterprise	144	117	136	117	96	94	140	139
Total	9,455	7,340	7,598	6,224	8,429	6,339	8,635	8,085

Notes:
A: Budget.
B: Expenditure.

Source: www.nstda.or.th (Budgeting data from the Comptroller General's Department).

Table 7.9 Firm-based Technology Development in Thailand, 2001

Programme	Acronym/ Year launched	Firm-based services	Other activities	Core delivery staff	Firms/ year (2001)	Firms/ staff
NSTDA				26	146	
1. Industrial Technology Assistance Programme (w/ IRAP of Canada)	ITAP/2001 (ICS 92)	Diagnosis, identify. of specialists		15 ITAs	75	5
2. Standards, Testing and Quality Control	STQC/ 1993	ISO 9000 consultancy	TFQS Research	7 Analysts	57	8
3. Company Directed Technology Development	CD/1992	Low interest R&D loans		4 Analysts	14	4
4. National Centre for Genetic Engineering & Biotechnology	BIOTEC		R&D			
5. National Metal & Materials Technology Centre	MTEC		R&D			
6. National Electronics & Computer Technology Centre	NTEC		R&D			
Dept. of Promotion, MOI				85	175	
7. Consultancy Fund Project (previously in Northeast only)	CFP/1997	Consultancy		85	175	2
8. Invigorating Thai Business (with Thai-Japanese Assoc)	ITB/2002	Consultancy			(2,600)	
Autonomous, under MOI				193	240	
9. Thai Productivity Institute	TPI/1995	ISO 9000/ 14000 trg	Training	61	120	
10. Electrical/Electronics Institute	EEI/1998		Training, testing			
11. National Food Institute	NFI/1997	HACCP certification	Lab & testing services	42	40	3
12. Thai Textile Institute	TTI		Testing, training	(30)	80	
13. Thai Automotive Institute	TAI	Group & individual consult	Testing, training	(30)		
14. Thai-German Institute	TGI/1998	Company visits	Training	(30)		
Total				304	561	

Note: Figures in brackets are estimates.

Source: NSTDA (2002) Building Technological Capability in Thailand's SMEs, interviews.

be rather difficult to actually 'value' each software project plan in order to provide reasonable funding. Hence it is necessary for the SME Bank to educate its staff to understand and be able to implement the new policy effectively and fairly (Karnjana, 2003).

To facilitate the loan approval process, a joint committee comprising representatives of the SME Bank, Small Industry Credit Guarantee Corporation, the Information and Communications Technology Ministry and the IP Department, will be set up. The committee will screen the business opportunity and the quality of the software house before passing them on to the SME Bank for a loan consideration. Four criteria will be assessed at the outset (1) the estimated value in the development process (including technology, R&D and training), (2) the value of the business opportunity, (3) the level of certification and the number of customer reference sites, and (4) the estimated market value of the software (assessed from its potential impacts on other industries). In addition, the Small Industry Credit Guarantee Corporation will also work with the SME Bank on the project and may help by issuing a loan guarantee of up to 50 per cent of the loan, in the case that the SME Bank provides a lower amount than the software house expects.

In a 2002 study by the NSTDA/World Bank on 'Analysis of Financial Incentives for Enhancing Thailand's Industrial Technological Capabilities' (University of Wollongong/ The Brooker Group Plc. 2002), several interesting recommendations were presented. The authors recommended that the Thai government should:

1. recognise the critical role of firm-level technological development;
2. focus on building capacity in industrial firms by:
   - implementing a grant-based incentive structure for technological development at the firm level and
   - implementing a training support programme under the umbrella of the Skill Development Fund; and
3. place greater emphasis on the role of industry groups and associations and related mechanisms that allow firms to be involved in S&T policy decisions and implementation.

Several issues will be discussed in the final section on what else needs be done to improve the possibility of Thailand being a high-tech industrial economy.

Technological incubator

Thailand has many knowledge hubs including universities, colleges and governmental research institutions such as the NSTDA, the TRF, and the

TISTR.[8] Innovation in the knowledge-based economy is the link between knowledge and productivity which leads to improved competitiveness of the underlying economy.

The TISTR, a state enterprise under MOST, has conducted R&D activities, transferred technologies and provided scientific and technological services with the objective of encouraging the government's policy guideline, to support effective production processes of SMEs, and to optimise the quality of life of people in the rural areas.

Innovation Development Fund (IDF)

There are obvious links between manufacturing production and innovation. The level of Thailand's innovation as measured by US patenting is very low compared to other more developed countries. The IDF was established in 1998 during the Chuan government period. The IDF aims to accelerate the development of innovation in the country in order to help with the restructuring of Thai industries and agriculture in crisis, and to help increase their competitiveness. An IDF Board was appointed in November 1998. In March 2000, the first director was recruited, and an executive committee of the board was also appointed to oversee and to support the functioning of the office.

The IDF provides two types of support: (i) financial support in the form of grants, loans and investment for execution of innovation activities, and (ii) technical support is provided for technical needs as well as business needs for the project owners. This support is available from the project development phase, the project execution phase throughout the post-project phase. The IDF intends to support projects which are clearly innovative, and are feasible technologically, financially and commercially. All proposed innovation action projects will be screened for feasibility (both technological as well as financial) and potential benefits (social or public benefits such as pollution reduction or environmental improvement are also considered here).

The framework of projects qualified for support must be one of the following:

1. the study of innovation development policies and feasibility studies;
2. implementation of new technologies in government and private sectors;
3. joint investment between public and private sectors in initiating new businesses in Thailand;
4. selection of experts to consult in R&D, design, manufacturing, management and marketing;
5. training at all levels of personnel;

The IDF Office has received over 130 project proposals, mainly from SMEs, in addition to hopeful start-up entrepreneurs/technologists/ inventors. By 1994, 10 projects have been approved, with THB 35.6 million funding from the IDF (the total project value is THB 158.8 million). There are also 35 projects which are in the development phrase. The IDF currently has a Five-Year Plan consisting of three groups of strategic action plans, namely, (i) direct upgrading of innovative capability, (ii) strengthening the innovation culture, and (iii) building up an innovation administrative system.

However, since the IDF has only THB 100 million of seed capital, approximately one-third of which has already been disbursed, it cannot allocate large grants to each project due to limited funding. The maximum amount of funding for each approved project is usually capped at THB 5 million, with a fixed repayment schedule between 3 and 5 years depending on the nature of the project. The IDF does not charge interest on these loans, but may request royalty fees and/or other types of repayment as deemed appropriate. It is seeking additional funding from other sources, and now plays more of an intermediary role in matching investors with potential innovators/business owners. Common channels of funding for new innovators include the SME Bank, the Krung Thai Bank, B-Net and the THB 1.2 billion (with THB 1 billion contribution from the government) SME's Venture Capital Fund (SMEVC) managed by One Asset Management Limited. The common objectives are to provide financial support either in the form of lower-cost debt financing or pure equity injection, together with any applicable advice/ guidance on business operations and development, to these SMEs.

There are five manufacturing sectors which are the IDF's strategic development sectors: (i) food and herbs, (ii) rubber and rubber products, (iii) software and mechatronics, (iv) vehicles and parts, and (v) engineering and manufacturing design.

Business incubator

With regard to the software industry, we mentioned earlier that the 'Software Park' under the supervision of the NSTDA provides a 'Software Incubator Centre' with free support to new entrepreneurs/ graduates wishing to start their business in this industry – with an accepted business plan – but do not have the necessary means to do so. The centre provides a small workstation equipped with a personal computer, office equipment, intranet/internet and some other office support and is open for operation 24 hours a day. However, in practice the centre has not been fully utilised yet – as fewer new entrepreneurs/

graduates have been accepted on the programme than originally expected.

Industrial R&D

Many industrial R&D projects have been initiated through FDIs in Thailand. In June 2003, Toyota Motor Corporation announced the plan to build a THB 2.7 billion R&D facility in Samut Prakarn, the Toyota Technical Centre Asia Pacific Thailand (TTCAP-TH) (*Bangkok Post*, 13 June 2003). The facility, occupying 320,000 square metres of space, has been in operation since late 2004, and initially employed over 200 staff. This is Toyota's first R&D centre in Asia outside Japan. Other overseas facilities are in the United States, Europe and Australia. This establishment is intended to help industries to achieve world-class quality standards and ultimately to allow local suppliers to achieve first-tier status, instead of the second- and third-tier status in the current supply chain scenario. The TTCAP-TH is expected to incorporate a component and system development centre which includes chassis and body systems development, electronics and materials engineering.

5. SUMMARY AND CONCLUDING REMARKS

In this chapter we have shown that the environment surrounding the development of high-tech industries in Thailand is quite puzzling. On the one hand, the market-oriented, outward looking, development policy of the Thai government has resulted in the country enjoying rapid economic growth in the last four decades, with active participation from foreign direct investments from various countries. But, on the other hand, this rapid economic growth has not been matched by a commensurate increase in technological capability of the Thai entrepreneur and the labour force. Until recently the overall attainment of secondary level of education of the population has remained low, and so has the public support on industrial R&D. Thailand lags behind other more progressive economies in East Asia such as Korea, Taiwan, Singapore and Malaysia in terms of its success in high-tech industrial development.

Yet, the situation seems to have improved in the last few years. There have been numerous concerted efforts both in the public and private sectors to improve or upgrade the technological capability of the Thai people and society. Thailand now has its second National Master Plan for ICT, whose aim is to utilise ICT to achieve electronic advancements in industry, commerce, government administration, education and social development in general. Another National Master Plan on Biotechnology is in the making and will be announced soon. As a result of these public initiatives on technology upgrading, Thailand has started to operate its first Science Park, and has launched many high-tech-oriented projects such as the promotion of software industries. The general economic and investment promotion through such government agencies as the Board of Investment has begun to bear fruit as many private companies have started to expand their industrial investment and production in various high-tech manufacturing industries such as in the electronic and automotive industries.

Even with the world economic slowdown, foreign investors still have confidence in Thailand as an investment market. A BOI survey (*BOI Investment Review*, December 2002) found that they generally believe that the business climate in the country has actually improved in 2002 compared to previous years, and business will continue to improve over the next few years. The country has no history of ethnic conflict and Thais are generally famous for their gentle and peaceful nature. Thailand also has a good network of roads spanning the country, and a good telecommunications infrastructure. However, in general the skill level of the Thai labour force is quite low and there is still room for improvement in the education system. The domestic financial and capital markets are not fully developed, and communication networks are expensive and not so strong outside of the Bangkok Metropolitan Area. In addition, government bureaucracy and corruption create notable add-on costs in doing business in the country, as observed by Porter (2003).

In 2002, the Thai Prime Minister, Dr Thaksin Shinawatra, announced that the government would pursue export growth with an emphasis on increasing foreign direct investment, while stimulating sustained domestic-led growth by emphasising poverty eradication. He also commented that the government's next set of goals would be reforming the nation's education systems, reforming the public sector bureaucracy and enhancing the country's economic competitiveness in the international arena (*BOI Investment Review*, September 2002).

By encouraging FDI, Thailand hopes not only to import more efficient foreign technologies, but also to generate technological spillages for local firms through several channels such as increased competition and imitation. Thailand's progress in becoming the ASEAN automotive production hub is still fragile, and the potential for growth will be limited until it significantly improves its sophistication and technology. Therefore, Thailand should apply effective resource allocation or mechanisms designed to increase the private sector's abilities to implement its own technological learning, strengthening their own design, engineering and other technological development capabilities or undertake their own innovative activities.

In conclusion, we shall offer some final comments:

Qualified Workforce

A strong technical human resource base is central to the technological capability development in any nation. It is essential to have a skilled workforce that can understand new technology transfers, and in some cases be able to reverse-engineer important technologies.

Government Policy on FDI and R&D

The government should increase its emphasis on promoting technology development in and transfers from MNCs or joint ventures. R&D activities which are commercially focused and commercially efficient should also be encouraged. The government should also provide financial support through more attractive investment incentives, without sacrificing local industries, as the current incentives are in some ways less attractive than those in neighbouring Malaysia and Singapore. In addition, strict local laws and applications that encourage investment and sustained upgrading in R&D, such as disciplined enforcement of intellectual property rights, will be very beneficial to high-tech industry development. Policies that provide incentives to companies that focus on staff development would also help to attract foreign investors.

Organisations which Promote Innovation, Technology Transfer, and R&D

It would be very helpful to the country to have a powerful organisation which oversees the R&D and technology transfer activities and provides funding in the areas is most needed. At present, commercial R&D activities are normally conducted on a micro scale on a company-by-company basis. The government should raise awareness of and induce higher demand for technology services at the firm level. One way is to follow other more developed countries by deploying a network of industrial technology advisers who visit firms, provide initial diagnostic services, prepare a plan of action for productivity and technological upgrading, and establish the right connections among manufacturing firms and specialist service providers and technical consultants. In order to do this, the government will need to recruit, train and retain a large network of industrial technology advisers who can fulfil the above tasks. The employment need not be a life-time commitment on the government's side, but the contract must be sufficiently long term to attract competent and interested professionals.

Establishment of an Active Network or Industry Association

The establishment of an active network or industry association between professionals who work in high-tech fields should be encouraged and supported, together with effective incubators for new businesses in each high-tech industry.

Transparent Government Policy

As Ang et al. (2001) have suggested, questionable contracting procedures and political delays are some of Thailand's weaknesses that need to be overcome in order to portray a cleaner image of the government and attract further investment.

Thailand is at a crossroads with regard to its potential to become a high-tech industrial economy in the near future. Although its past performance has not been impressive in terms of its specific technological achievements, Thailand is regarded as being very successful in opening up its economy to benefit from induced technological progress from the outside. This induced technological progress is likely to have at least two effects: one is to generate a high economic growth rate which may lead to the other effect of generating greater possibilities for more indigenous high-tech industrial development. The success in advanced research in two fields, software development and biotechnology, is made possible by the increase in public resources and genuine public concern to move the country forward on the high-tech industrialisation path. This is not an impossible dream for Thailand.

NOTES

1. 'The Global Information Technology Report 2001–2002', Harvard University and World Economic Forum.
2. As a result of the public sector reforms in 2002, the Ministry of Natural Resources was created, and the jurisdiction on environmental management was transferred from the former Ministry of Science, Technology, and Environment, now renamed Ministry of Science and Technology.
3. Full Report on ICT's role in SME operations, June 2002, Industry Promotion Foundation.
4. This is one of the Thaksin government's policies to encourage each *tambon* (subdistrict) in Thailand to develop at least one product and try to market it locally, nationally and even internationally.
5. McKinsey & Co., Thai Productivity Report 'Prosperity through Productivity', 2002, as cited in the ICT Master Plan, p. 16.
6. Agricultural products which used bio fertilisers, bio insecticides and other biologically controlled (safe and non-chemical) products.

7. *BOI Investment Review*, 30 December 2002.
8. TISTR was first established in the late 1970s as The Applied Scientific Research Corporation of Thailand.

REFERENCES

Ang, D., A. Hun, K.G. Lee, and D.S. Lim (2001), 'Chapter 8: The Future of Thailand's Hi-tech Manufacturing Industry', in John J. Williams and Ian McGovern (eds), *Business Opportunities in Thailand*, Englewood Cliffs, NJ: Prentice-Hall.

APEC (2002), *APEC Outcomes and Outlook*, Asia-Pacific Economic Conference, December 2002.

Arnold, E., J. Bessant and P. Brimble (2000), *Enhancing Policy and Institutional Support for Industrial Technology Development in Thailand: Overall Policy Framework and Development of Industrial Innovation System*, report for the National Science and Technology Development Agency (NSTDA) and the World Bank, Bangkok: NSTDA.

Bangkok Post (2003), 'B2.7 billion R&D facility announced', *Bangkok Post*, 13 June.

Business Week (2003), 'Asia Tech: Special Report', *Business Week Magazine*, 14 April, pp. 52–60.

Einhorn, Bruce (2001), 'Can Thai Tech Go High-Tech?', *Business Week*, 21 May.

Hall, Philip (2002), 'New ICT Minister Sets Out Goals', *Asia Computer Weekly Online*, 18 November.

Hassaya, Chartmontri (2003), 'New Technologies Key to Development', *The Nation*, 24 December.

Heamakarn, Sricharatchanya (2003), 'Growth Potential', *Bangkok Post*, 14 May.

Ho, Andy (2001), 'Maintaining a High-Tech Hub', *The Straits Times*, 1 April.

Karnjana, Karnjanatawe (2003), 'SME Bank to Give Funding to Software Developers Based on IP Assets', *Bangkok Post*, 30 April.

Lim, C.H., C.M. Lee, W.Y. Chua, K.E. Chuah (2001), 'Chapter 7: Agriculture Development in Thailand', in John J. Williams and Ian McGovern (eds), *Business Opportunities in Thailand*, Englewood Cliffs, NJ, Prentice-Hall.

Morakot, Tanticharoen (2002), 'Chapter 7 Thailand', in G.J. Persley and L.R. MacIntyre (eds), *Agricultural Biotechnology: Country Case Studies*, CAB International, pp. 113-125.

NECTEC (2002), *Report on the Demand of IT Professionals in Thailand*, January 2002.

Office of the Board of the Innovation Development Fund (2002), *Thailand's Innovation Technology Policies, January 2002* (in Thai).

Porter, Michael E. (2003), *Presentation on Thailand's Competitiveness: Creating the Foundations for Higher Productivity* (Preliminary Findings), Bangkok, Thailand, 4 May.

Saggi, Kamal (2002), 'Trade, Foreign Direct Investment, and International Technology Transfer: A Survey', *The World Bank Research Observer*, vol. 17, no. 2, pp. 191–235.

Sasin Graduate School of Business (2003), *Study on Thailand's Competitiveness*, Chulalongkorn University Thailand.

Sasiwimon Boonruang (2003), 'ICT Ministry Draws Praise from Industry', *Bangkok Post*, 30 April.

Sukanya Jitpleecheep (2003), 'Room Air-conditioner Base Moved to Thailand', *Bangkok Post*, 1 April.
Sumeth Vongpanitlerd (1992), 'The Development of Thailand's Technological Capability in Industry', in Thailand Development Research Institute (ed.), *Research Monograph* No. 9, December.
Thailand Board of Investment (2002a), *BOI Investment Review*, Vol. 11, No. 3, 30 September.
Thailand Board of Investment (2002b), *BOI Investment Review*, Vol. 11, No. 4, 30 December.
Thailand Board of Investment: http://www.boi.go.th
UNIDO (2002) 'Thailand's Manufacturing Competitiveness: Promoting Technology, Productivity and Linkages', *SME Technical Working Papers*, Series #8, United Nations Industrial Development Organization.
University of Wollongong/The Brooker Group Plc (2002), Presentation Material: NSTDA/World Bank Project on: Analysis of Financial Incentives for Enhancing Thailand's Industrial Technological Capabilities, 5 June.
Wai Chamornmarn (2001), 'Research on the upgrading of Thai Workers' Capability in Obtaining and Retaining Knowledge in the Science and Technology Field in Multinational Businesses', in Supranee Srichatrapimuk (ed.), *The Human Resource Institute*, Vol. 6, April Thammasat University (in Thai).
Waltham, Tony (2002), 'Conference Puts ICT in the Limelight', *20th ASOCIO General Assembly and Symposium*, Chiangmai, Thailand, 4 December.
Wichit Chantanusornsiri (2003), 'Key Thai Industries Face Threat of Being Sidelined', *Bangkok Post*, 3 May.
Wilson, Drew (2002), 'Downturn Rekindles Fizzled Fab', *Semiconductor Magazine*, Vol 3, No. 12, December.

8. China's High-Tech Drive

Erh-Cheng Hwa and Xielin Liu *

1. INTRODUCTION

China, with GDP growth averaging more than 9 per cent per annum, has achieved phenomenal economic growth since the launching of economic reform in the late 1970s. Growth has allowed China to reduce poverty significantly to extent that China has accounted for a significant portion of the decline in world poverty over the last two decades. China's economic modernization drive (dubbed the 'four modernizations' by the late Deng Xiaoping) has given 'high and new technology' ('high tech' or 'HT') prominent attention. Indeed, after two decades of development, especially since the emergence of large inflows of foreign direct investment in the 1990s, the HT industry has grown into a leading sector in the economy. The remainder of this chapter is divided into three sections. Section 2 evaluates the recent (post-1995) performance of the HT sector in the economy including, *inter alia*, its contributions to output, employment, productivity and foreign trade. Section 3 discusses several key institutional underpinnings for China high-tech sector including research and development (R&D) efforts, university- and research institutes- sponsored high-tech companies, as well as the overarching policy/incentive framework for high-tech development. Section 4 concludes with an assessment of the overall performance of China's high-tech industry and offers some recommendations for developing the domestic high-tech industry.

2. THE PERFORMANCE OF THE HIGH-TECH SECTOR

This section assesses the impact of the HT sector on value-added, employment generation, productivity growth in the economy, and exports. Because China has an enormous land size, this section concludes with an analysis of the regional dispersion of the HT industry. The assessment

* The authors would like to thank Shih-Yi Eu for helpful assistance.

only considers the first-round effect, since it does not take into account of the wider economic repercussions after establishing the HT industry.

How do we define the high-tech industry? What industries are considered 'high-tech'? This chapter uses the high-tech data compiled by the National Bureau of Statistics (NBS), China's chief statistical agency. The NBS follows the OECD convention which categorizes the high-tech industry into five manufacturing groups: (1) pharmaceuticals, (2) aeronautics, (3) electronics & communications, (4) computer & office facilities, and (5) medical equipment & apparatus (see Appendix 8A). China's adoption the OECD convention facilitates cross-country comparisons of the high-tech industry performance, but it should be noted that in any single high-tech industrial group, the Chinese technology level is expected to be somewhat lower than the OECD countries.

The HT Industry Has Emerged as the Leading Sector in the Economy

The HT industry grew at more than twice the rate of the growth of GDP over the 1995–2001 period (Tables 8.1 and 8.2). The average annual compounded rates of growth for value added, gross output, profits before tax, and sales were approximately 20 per cent per annum compared with 8.6 per cent growth in nominal GDP. Thus, the share of value-added of the HT industry in GDP nearly doubled in six years. By 2001, with gross output valued at US$148.5 billion, China had the third-largest HT industry in the world, next only to the USA and Japan.[1]

The HT industry was also the most dynamic branch in the manufacturing sector. Tables 8.3–8.5, respectively, highlight the rising role played by the high-tech industry in the manufacturing sector in terms of value added, employment and productivity. In 1996, the high-tech industry accounted for 15.0 per cent of the growth in manufacturing value added and by 2000 this contribution rose to a quarter (25.4 per cent), followed by a slight setback in 2001.[2] As a result, the value-added share of the NT industry in manufacturing rose from 8.8 per cent in 1995 to 13.9 per cent in 2001 (Table 8.3).

Within the high-tech sector, computers and office equipment grew at an average annual growth rate of 29.3 per cent, or by nearly five times during 1995–2001 (Table 8.3-1), which is certainly the fastest growth sub-sector. Computers and office equipment, however, ranked only the third largest sub-sector in terms of value added in 2001 (2.0 per cent). The largest HT sub-sector was electronics and telecommunications equipment by taking up 7.7 per cent of the value-added share in manufacturing, followed by medical and pharmaceutical products (6.1 per cent) in second place (Table 8.3-2).

Table 8.1 Performance of High-tech Industry in China (billion yuan)

	Value added	Gross value	Profit before tax	Sales	Total employment (1000)
1995	108.1	409.8	32.6	391.7	4484.2
1996	127.2	490.9	38.0	449.7	4610.2
1997	154.0	597.2	51.7	561.8	4304.5
1998	178.5	711.1	55.5	658.0	3926.9
1999	210.7	821.7	71.3	782.0	3844.7
2000	275.9	1041.1	103.3	1003.4	3899.8
2001	309.5	1226.3	110.8	1201.5	3983.5

*Source: NBS (*2003)

Table 8.2 Value-added Share of High-tech Industry and GDP (billion yuan)

	Value added of HT industry	GDP	Value added of high-tech/GDP (per cent)
1995	108.1	5847.8	1.8
1996	127.2	6788.4	1.9
1997	154.0	7446.3	2.1
1998	178.5	7834.5	2.3
1999	210.7	8206.8	2.6
2000	275.9	8944.2	3.1
2001	309.5	9593.3	3.2

Source: NBS 2003.

Table 8.3 Value-added Share of High-tech Industry in Manufacturing (billion yuan)

	HT industry	Manufacturing	HT Share in manufacturing (per cent)
1995	108.1	1222.1	8.8
1996	127.2	1359.3	9.4
1997	154.0	1575.6	9.8
1998	178.5	1526.0	11.7
1999	210.7	1685.2	12.5
2000	275.9	1970.1	14.0
2001	309.5	2231.2	13.9

Source: NBS (2003).

Table 8.3-1 Value-added share of High-tech Sub-sectors (billion yuan)

	1995	1996	1997	1998	1999	2000	2001
Medical and Pharmaceutical Products	26.47	35.97	41.15	43.29	51.49	63.39	72.24
Aircraft and Spacecraft	8.00	7.39	9.81	8.74	9.20	10.56	12.41
Electronic and Telecommunication Equipment	54.33	58.83	72.87	86.98	112.22	147.13	162.27
Computers and Office Equipment	9.27	13.68	18.09	26.56	24.14	37.43	43.24
Medical equipment and meters	9.99	11.33	12.07	12.93	13.66	17.37	19.32
High-tech total	108.1	127.2	154.0	178.5	210.7	275.9	309.5

Source: NBS (2003).

Table 8.3-2 Value-added Share of High-Tech Industry in Manufacturing (per cent)

	1995	1996	1997	1998	1999	2000	2001
High-tech industry	8.8	9.4	9.8	11.7	12.5	14	13.9
Of which: Medical & Pharmaceutical Products	2.2	2.6	2.6	2.8	3.1	3.2	3.2
Aircraft & Spacecraft	0.7	0.5	0.6	0.6	0.5	0.5	0.6
Electronic & Telecommunications Equipment	4.4	4.3	4.6	5.7	6.7	7.5	7.3
Computers and Office Equipment	0.8	1.0	1.1	1.7	1.4	1.9	2.0
Medical equipment	0.8	0.8	0.7	0.8	0.8	0.9	0.9
Memo: manufacturing value-added (billion Yuan)	1222.1	1359.3	1575.6	1526.0	1685.2	1970.1	2231.2

*Source: NBS (*2003).

Table 8.4 shows that the employment share of the high-tech sector went up from 6.3 per cent in 1995 to 8.8 per cent in 2001. But the employment in the high-tech sector witnessed a decline, only that the decline, by only half a million (11.1 per cent), was much less than those of the other manufacturing sectors as a whole, 25.0 million (37.7 per cent). Why the non-tech manufacturing had borne the brunt of the employment decline, while one would expect that the opposite should be the case considering that the high-tech sector had a higher labour productivity growth and therefore lower labour absorption rate than the non-tech sector (discussed below)? The main reason was that the high-tech sector was a newly emerging sector in the economy, with many of them being set up by foreign investors recently. On the other hand, traditional SOEs dominated the rest of the manufacturing sector and were going through a period of reform and restructuring, which resulted in a large shedding off their redundant workforce. More generally, HT companies enjoyed better monopoly power in the marketplace than the rest of manufacturing and could cope with economic shocks better than the latter.

Table 8.4 Employment in High-tech Industry (million persons)

	HT industry	Manufacturing	HT share in manufacturing (per cent)
1995	4.48	70.77	6.3
1996	4.61	65.80	7.0
1997	4.30	64.52	6.7
1998	3.93	51.00	7.7
1999	3.84	47.60	8.1
2000	3.90	46.06	8.5
2001	3.98	45.29	8.8

*Source: NBS (*2003).

The high-tech industry enjoyed higher productivity growth than the manufacturing industry as a whole during the 1994-2000 period: the former, 22.5 per cent and the latter, 16.7 per cent (an average annual compound rate of growth). The led to a widening of productivity gap between the high-tech industry and the manufacturing sector as a whole from 1.24 times in 1994 to 1.65 times in 2000. With an average annual growth rate of 27.2 per cent, the computers and office equipment subgroup experienced the highest labour productivity growth, mirroring its output growth (Table 8.5).[3]

As the high-tech industry grew rapidly, its contribution to the nation's exports also rose impressively. Having achieved an annual average rate of growth of 28.0 per cent during the 1998-2001 period, high-tech exports more than doubled in value during this period (Table 8.6), while total merchandise exports rose by 1.45 times in this period (from US$183.7 billion to US$266.2 billion).[4] Thus, high-tech exports rose from 11.1 per cent of total merchandise exports in 1998 to 16.1 per cent in 2001. Equally interesting, the share of high-tech exports to high-tech (gross) output rose from 24.8 to 35.0 per cent in just three years. In 2001, more than a third of high-tech output was earmarked to the foreign market. What is the underlying driver for this rather dynamic growth in high-tech exports? To a significant extent, the increase in high-tech exports can be ascribed to foreign-invested enterprises, whose contribution to high-tech

Table 8.5 Productivity of High-tech Industry (value added per person in 1,000 yuan)

	1994	1995	1996	1997	1998	1999	2000
Manufacturing industry	1.7	1.7	2.1	2.4	3.0	3.6	4.3
High-tech Industry	2.1	2.5	2.9	3.8	4.6	5.6	7.1
Of which: Aircraft and Spacecraft	1.0	1.4	1.0	1.8	1.7	1.9	2.3
Computers and Office Equipment	3.7	6.6	9.4	10.6	12.4	11.6	15.7
Electronics, Telecommunication Equipment	2.5	2.7	3.1	4.0	4.8	6.4	7.7
Medical and Pharma-ceutical Products	1.9	2.2	3.0	3.4	3.8	4.7	6.1
Other Electronic Equipment	1.7	1.7	2.1	2.3	2.9	3.4	4.1

Source : NBS (2003)

Table 8.6 High-tech Exports (billion yuan)

	1998		1999		2000		2001	
		Share		Share		Share		Share
Total	204.11	100	241.30	100	338.84	100	428.20	100
State-owned Enterprises	17.34	8.49	51.64	21.40	70.63	20.85	73.38	17.14
Foreign joint ventures	166.70	81.67	202.43	83.89	288.21	83.28	369.97	86.40

Note: Since foreign-invested enterprises have established joint ventures with state-owned enterprises, the sum of the two could involve double counting and therefore exceed 100.

Source : NBS (2003, p.55).

exports rose steadily, from 81.2 per cent of total exports in 1998 to 86.4 per cent in 2001 (Jiang, 2001). Even the worldwide recession in 2001 sharply slowed down total exports growth, it did not seem to dent high-tech exports that continued to grow briskly at 26.4 per cent in that year. This shows that high-tech exports are far more resilient to changes in global demand than other merchandise exports. This result also supports the strong high-tech employment performance discussed earlier.

The High-tech Industry was Concentrated in the Eastern Region

Much of the rapid growth of the high-tech industry occurred in the eastern region of the country. Table 8.7 bears out this fact. In terms of gross output produced, the eastern region grew at an annual average growth of 26.2 per cent during 1996–2001, while the middle and western regions grew, respectively, at 14.6 and 12.3 per cent, also impressive but significantly less so.[5] Consequently, in terms of output produced, the eastern region's take in the high-tech industry rose from 76.3 per cent in 1996 to 83.5 per cent in 2001. If the trend continues, as is likely, the technological gap between the East and the West is destined to grow wider.

Table 8.7 Output of the High-tech Industry by Region (billion yuan)

	Gross output			Value added		
	East	Middle	West	East	Middle	West
1996	374.6	59.3	57.0	92.4	17.4	17.4
1997	460.3	68.3	68.5	113.4	20.6	20.0
1998	561.7	74.1	75.2	133.8	22.2	22.5
1999	662.5	83.2	75.9	162.6	26.4	21.8
2000	856.7	100.4	84.1	216.6	32.1	27.2
2001	1023.7	110.7	91.9	244.1	36.3	29.1
Share (2001)	83.5%	9.0%	7.5%	78.9%	11.7%	9.4%
Av. growth	26.2%	14.6%	12.3%	22.5%	18.4%	12.4%

Source : *NBS*(2003)

Within the prosperous eastern region, the growth in the high-tech industry has, in turn, concentrated in just four areas: Guangdong province, Jiangsu province, and Shanghai and Beijing municipalities (Table 8.8). Their combined value-added share in the nation's total has increased steadily in recent years and reached 53.0 per cent by 2001, up from 48.3 per cent in 1997. Having benefited from its pioneer status of economic reforms and opening to the outside world, Guangdong enjoys a significant lead over the other three eastern areas, as it attracted the first wave of foreign direct investment in China. Nevertheless, Guandong's lead over Shanghai and the whole Yangtze-river region is being eroded as the latter has risen up quickly to challenge the former since the early 1990s. Shanghai and Beijing also house many of the nation's leading research institutions and universities. They have a more favourable R&D environment that is conducive to the development of high-tech industry. In contrast, in the less-developed middle and western regions, only Xi'an and Wuhuan have a significant R&D presence.

China's High-tech Sector Ranks Low in International Comparison

The world economy at large has certainly taken notice of the rapid development of China's high-tech industry. Although it is presumptuous to argue, as some analysts have, that China is on its way to becoming the factory of the world, it is nevertheless true that goods rolling out of

Table 8.8 Value-added Share of High-Tech Industry in Eastern Regions (billion yuan)

	1997		1998		1999		2000		2001	
		% of State		% of State		% of State		% of State		% of State
Guangdong	34.41	22.3	42.46	23.8	50.47	24.0	67.78	24.6	75.98	24.6
Jiangsu	16.60	10.8	17.92	10.0	23.40	11.1	30.20	10.9	35.78	11.6
Shanghai	12.56	8.2	15.25	8.5	19.35	9.2	23.53	8.5	28.76	9.3
Beijing	10.73	7.0	15.65	8.8	19.02	9.0	24.68	8.9	23.62	7.6
Total	74.30	48.3	91.28	51.1	112.24	53.3	146.19	52.9	164.14	53.1

Source: NBS (2003)

Table 8.9 Patents granted by US Patent and Trademark Office

Rank	Economy	All patents					
		Items		Share		Growth rate	
		1990	2001	1999	2001	2001	1990-2001 compound average
1	USA	52,977	98,663	53.4	53.6	1.7	5.8
2	Japan	20,743	34,890	20.9	19.0	6.0	4.8
3	Germany	7,862	11,894	7.9	6.5	9.9	3.8
4	Taiwan	861	6,545	0.87	3.6	12.7	20.3
5	France	3,093	4,456	3.1	2.4	6.8	3.4
6	UK	3,017	4,356	3.0	2.4	6.5	3.4
7	Canada	2,087	4,063	2.1	2.2	3.5	6.2
8	Korea	290	3,763	0.29	2.0	8.4	26.2
9	Italy	1,498	1,978	1.5	1.1	0.56	2.6
10	Sweden	885	1,935	0.89	1.1	11.3	7.4
14	Israel	311	1,031	0.31	0.56	11.3	11.5
18	Hong Kong	151	620	0.15	0.34	11.3	13.7
21	Singapore	16	304	0.02	0.17	11.3	30.7
23	China	48	266	0.05	0.14	11.3	16.8
		Utility patents					
Rank	Economy	Items		Share		Growth rate	
		1990	2001	1999	2001	2001	1990-2001 compound average
1	USA	47,391	87,607	52.4	52.8	3.0	5.7
2	Japan	19,525	33,233	21.6	20.0	6.2	5.0
3	Germany	7,614	11,260	8.4	6.8	10.0	3.6
4	Taiwan	732	5,371	0.81	3.2	15.1	19.9
5	France	2,866	4,041	3.2	2.4	5.8	3.2
6	UK	2,789	3,965	3.1	2.4	8.1	3.3
7	Canada	1,859	3,606	2.1	2.2	5.5	6.2
8	Korea	225	3,538	0.25	2.1	6.8	28.5
9	Sweden	768	1,743	0.85	1.1	10.5	7.7
10	Italy	1,259	1,709	1.4	1.0	- 0.29	2.8
13	Israel	299	970	0.31	0.56	0.33	0.58
19	Hong Kong	12	296	0.15	0.34	0.01	0.18
22	Singapore	52	237	0.02	0.17	0.06	0.14
24	China	47	195	0.05	0.14	0.05	0.12

Source: 'The Dynamics of Industrial Technology Innovation and its Comparative Advantages among Asia-Pacific Economies', conference proceedings, Taiwan Institute of Economic Research, 15-16, August 2002, pp.6-9.

Table 8.10 High-tech Exports by Country (2000, US$ million)

Country	Value	Share in manufacturing (%)
USA	197,033	30.31
Japan	127,368	28.32
South Korea	53,950	34.89
India	1,245	4.29
Malaysia	39,964	49.16
Singapore	73,643	62.39
China	40,387	18.57

Source: IMD (2002).

Chinese factory mills are penetrating the world markets at rapid pace. Undergoing a brisk economic development and transition notwithstanding, China's technological prowess is still no match to more advanced economies. A look at the patents granted by the patent office of the United States (Table 8.9) shows that China's technology and innovative capacity still lags far behind. The number of patents held by every 100,000 people ranks 36th in the world (1998). China's R&D accounts for about 0.7 per cent of GDP (Table 8.11) that is also low by international standards (IMD, 2002).

In 2000, China's high-tech exports of US$40,387 million were about the same as Malaysia's and were lightly below Korea's (Table 8.10). But, in spite of the recent surge in high-tech exports as analyzed above, China's high-tech exports share in manufacturing, 18.6 per cent, was still significantly lower than Malaysia's 49.2 per cent and Korea's 34.9 per cent. This may suggest that high-tech played a lesser role in the Chinese manufacturing than in the case of Malaysia and Korea, even allowing for the fact that China likely exported a smaller portion of high-tech output than Malaysia and Korea because of her greater domestic market.

3. DRIVERS TO INDIGENOUS HIGH-TECH DEVELOPMENT

China's ultimate objective of high-tech development is to nourish a worldly competitive industrial base rooted in indigenous technology capacity. To achieve this end, China has followed a two-pronged approach: elevating the overall indigenous technology/innovation capability to nourish domestic high-tech companies while actively attracting foreign high-tech firms through improving the policy environment and offering them generous financial, tax and other incentives. This section reviews the pillars of China's innovation system for nourishing indigenous technology. These include:

1. R&D policy and efforts;
2. high-tech enterprises sprung from the universities, research institutions, and returning overseas Chinese high-tech manpower;
3. high-tech industrial parks; and
4. several high-tech policy initiatives and programmes: '863' Programme, commercialisation of technology, preferential policies for integrated circuit (IC) and software products, and the venture-capital fund.

R&D Spending and Funding Patterns

Governments worldwide adopt preferential policies to promote high-tech enterprises, justified not merely by their direct benefits on the economy, but more importantly for their 'externalities'. High-tech enterprises are

Table 8.11 China: R&D Expenditure (US$ 100 million yuan)

	Amount	Real annual growth (%)	Share in GDP (%)
1990	125.43	5.78	0.71
1991	150.79	20.2	0.70
1992	209.84	39.2	0.79
1993	256.19	22.1	0.74
1994	309.10	20.7	0.66
1995	348.69	12.8	0.60
1996	404.48	16.0	0.60
1997	481.47	19.0	0.64
1998	551.12	14.5	0.69

Sources : China Science and Technology Indicators (1998) and *China Technology Statistical Yearbook 1999*.

the conduit for innovation, since they typically invest much more than an average company on R&D. Thus, R&D spending is commonly employed to assess a country's indigenous technology sophistication. Developing economies generally do not spend as much on R&D as developed economies for two reasons. First, they have fewer financial resources and less intellectual capacity to engage in R&D than richer countries. Second, since most of their needed technologies may already exist, it is by far cheaper and less risky to import them from abroad than engaging their own R&D. This is the latecomer's advantage.

Table 8.11 shows that China's R&D spending in the 1990s hovered around 0.7 per cent of GDP, which is significantly less than what developing countries generally spend on R&D in the range of 1-2 per cent of GDP and even far less than what developed countries typically spend on R&D – 3 per cent of GDP.

China's R&D is conducted, in order of magnitude, by (large and medium-sized) enterprises, R&D institutions and universities (Table 8.12). In the early central planning era, corporate R&D was paltry since SOEs were mandated to execute production plans and had no R&D function at all (Liu and White, 2001). This compelled universities and research institutions to play an exceptionally important role in China's R&D functions.

What are the funding sources for R&D spending in China? The two major sources of funding are the government and self-raised funds. But the exact answer varies according to the institutions executing R&D. In 2000, corporate R&D was almost entirely funded by enterprises themselves (Table 8.13). At the other extreme, research institutes relied heavily on the government for R&D. But this was only natural since they were government-administered institutions. Recently they have taken

Table 8.12 National R&D Expenditure by Source (percentage)

	Enterprises	R&D institutions	Higher education	Others
1999	49.6	38.5	9.3	2.6
2000	60.3	28.8	8.6	2.3
2001	60.4	27.7	9.8	2.1

Source: Ministry of Science of Technology, *China Science and Technology Statistics*, 2000 - 2002.

Table 8.13 National R&D Expenditure by Sources of Funding (year 2000, percent)

	Total	R&D institutions	Enterprises	Higher education	Others
Government	33.4	81.4	6.9	58.5	30.9
Enterprise	57.6	5.2	86.4	32.3	57.2
Overseas funds	2.7	0.7	4.0	1.2	0.4
Others	6.3	12.7	2.8	8.0	11.4
Total	100.0	100.0	100.0	100.0	100.0

Source: Ministry of Science of Technology, *China Science and Technology Statistics*, 2000 - 2002.

Table 8.14 R&D Personnel in Different Institutions, 2001

Total	Independent research institutes	Large and medium-sized enterprises	Higher education	others
956,482	204,806 (21.4%)	379,336 (39.6%)	171,126 (17.9%)	201,214 (23.8%)

Source: Ministry of Science of Technology, *China Statistical Yearbook of Science and Technology, 2002*, China Statistics Press, 2002.

steps to become independent and market-oriented entities. Higher-education institutions like universities also relied on the government for the largest chunk of their R&D budget but, interestingly enough, they funded nearly one-third of their own R&D spending through the high-tech enterprises in which they invested. This subject will be discussed in the next section.

In the case of R&D personnel, independent R&D institutions and

Table 8.15 Domestic Patents Granted by Institutions (1998-2001)

	Granted			
	1998	1999	2000	2001
Total	954	1685	2824	2614
Universities & colleges	243	425	652	579 (22.1%)
Scientific research institutions	337	543	910	800 (30.6%)
Enterprises	182	462	1016	1089 (41.7%)
Govt & social organizations	192	255	246	146 (5.6%)

Source: Ministry of Science of Technology, *China Statistical Yearbook of Science and Technology, 2002*, China Statistics Press, 2002.

universities together have about the same number of researchers as large and medium-sized enterprises (Table 8.14). However, most leading scientists have come from research institutions and universities rather than from enterprises.

Corporate R&D was less efficiently executed than that by universities and other R&D institutions. Universities and research institutions combined spent less on R&D than the corporates, but managed to acquire far more domestic patent authorizations (Table 8.15). This singles out the prominent role played by universities and research institutions in fostering China's innovation system. This is the subject of discussion for the next section.

University-affiliated High-tech Enterprises

China launched the reform of the science and technology system in 1985. Since then Chinese universities, under recurrent financial pressure, have begun to think how to turn their accumulated research into commercial use and in the process gradually became an important force in the nation's high-tech industrial development.

The development of university-sponsored enterprises has undergone two stages. The first spanned from early 1980s to 1990 during which time China began to implement the economic reform and opening policy.

In 1985, the CPC Central Committee issued '*Resolutions of the CPC Central Committee on the Reform of the Education System*', which encouraged educational institutions to participate in economic and social development. Many universities have started to provide services to the general public since then, mainly in the area of technological transfer and development and consultation services (MOST, 1999a).

The second stage of university-sponsored enterprises has lasted from 1991 to present. In 1991, the State Council approved the proposal submitted by the State Education Commission and the State Commission of Science and Technology which gave a green light to university-sponsored enterprises and provided guidelines for the management of such enterprises. The development pace of university-sponsored enterprises has accelerated since then and achieved a phenomenal scale. The annual sales of such enterprises grew 27-fold in ten years, rising from just 1.8 billion in 1991 to 48.4 billion in 2000, for an average annual rate of growth of 39.0 per cent (Table 8.16). The Asian financial crisis slowed sales and profits growth but rapid growth resumed in the aftermath.

In 2000, university-sponsored enterprises totalled 5,451 nationwide according to the statistics compiled by concerned education departments. Table 8.17 profiles the universities with the top ten sales in 2000. With

Table 8.16 Performance of University-affiliated Enterprises (billion yuan)

Year	Sales	Profit	Tax	Revenue of universities
1997	29.55	2.72	1.23	1.58
1998	31.56 (6.8%)	2.59 (- 5.6%)	1.35 (9.7%)	1.50 (- 5.1%)
1999	37.90 (20.1%)	3.05 (18.0%)	1.66 (18.6%)	1.59 (6.0%)
2000	48.46 (27.9%)	4.56 (49.5%)	2.54 (53.3%)	n.a.

Notes: Figures in parenthesis are percentage changes over the previous year.

Source : Lan (2002).

almost no exceptions, all these universities are located in either big cities or major provincial capitals. Well-known university-sponsored enterprises include the Beijing University Founder Group and the Tsinghua Tongfang Group.

The Tsinghua Tongfang Group, a Tsinghua University sponsored enterprise, typifies how a university-affiliated enterprise can speed up the commercialization of university-based research. Tsinghua Tongfang developed container detectors and successfully marketed them to China's Customs department for monitoring smuggling. However, this outcome was not what was originally intended. Professors of the Department of Engineering Physics in the University did the original research on the

Table 8.17 Sales of University-affiliated Enterprises of Top Ten Universities (US$2000 million yuan)

Sales	University	Sales	Location
1	Beijing University	11,006.71	Beijing
2	Qinghua University	6,287.89	Beijing
3	Shanghai Jiaotong University	1,728.72	Shanghai
4	Harbin Industrial University	1,654.20	Harbin
5	Dongbei University	1,324.84	Shenyang
6	China Petroleum University	1,239.74	Beijing
7	Nankai University	1,200.06	Tianjing
8	Fudan University	1,157.02	Shanghai
9	Xian Jiaotong University	1,128.05	Xian
10	Zhejiang University	1,120.96	Hanzhou

Source: Lan (2002).

container detectors in late 1990s when they saw the potential market for the products, since smuggling was rampant. However, the researchers could neither introduce market-compatible products, nor find any enterprise willing to commercialize the technology and make up for the shortfall of resources. The Tsinghua Tongfang Group was able to over the problem by tapping into the resources of the Department of Engineering Physics and came up with the products that was acceptable to the market (Lan, 2002). This success rests on the exploitation of the mutual comparative advantage between university-sponsored enterprises and the universities themselves. The enterprises being closer to the market are more sensitive to the market needs than university researchers, but the universities are better equipped with R&D capability and ingenuity than the enterprises.

Research Institute-affiliated Enterprises

Decisions on the Reform of the Science & Technology System promulgated by the central government in 1985 had a far-fetching influence over China's science and technology system and was also a major impetus for institution-sponsored enterprises.

The subsidiaries of the Chinese Academy of Sciences founded their own enterprises even earlier than the universities. The pioneers of Beijing's Zhongguancun Electronics Street, of whom the best known was Liu Chuanzhi—founder of the Legend Group, were mostly researchers from the Chinese Academy of Sciences. Legend Group's emergence from an institution-sponsored enterprise to a leading PC producer in Asia provides a showcase example of the role played by institution-sponsored enterprises in China's high-tech development. The impetus for institution-sponsored enterprises is similar to that for university-sponsored ones; namely, the reform of the science and technology system stepped up the financial pressure. Fortunately, thanks to their long involvement in R&D, institution-sponsored enterprises enjoyed technological advantages over SOEs enabling an easier entry into the high-tech industry.

Lu (2000), a leading authority on Legend and other similar companies, noted that all of the leading Chinese PC companies have followed a new path of technological development. They would initiate an indigenous innovation based upon the refinement of certain existing products. Once convinced that the market had accepted their new or improved products, they would make constant innovation in product design so as to expand the market share until they could exploit their comparative advantage up to a certain scale economy. Then, they would invest to build up their own production capacity. Lu would view this development path as a 'Top to Bottom' model in technology learning.

Reviewing various enterprises located in China's high-tech parks provides a general picture on the development of institution-sponsored enterprises. In 1999, institution-sponsored enterprises registered an output value 10 times larger than that of university-sponsored enterprises, partly because many national research institutions were restructured into companies and grew substantially afterwards (Table 8.18).

Enterprises Established by Returned Overseas Scholars

China's reform and opening and economic policies have improved the environment for returned overseas scholars to establish their own businesses. All levels of government have offered preferential policies and various services to them. Some regions have specially designed incubators, like the so-called 'pioneer parks', to help returned scholars to start their high-tech businesses. Generally speaking, pioneer parks provide four kinds of services: business services; administration; incubation; and recruitment of high-tech experts (MOST, 2001.) Nationwide, a total of 53 pioneer parks were established by the end of year 2000, which hosted over 1,300 enterprises set the scholars, according to preliminary statistics. In total, about 10,000 of them established more than 4,000 high-tech enterprises by the end of year 2000, which generated an annual output value of over 10 billion yuan. Shanghai alone had about 1,000 such enterprises (Zhang Youshui and others, 2001).

The enterprises established by returned scholars contributed to domestic high-tech industry in the following ways. First, since most are high-tech companies, they enhance domestic high-tech development. Second, they introduce advanced management practices to the domestic economy. Third, they augment domestic equity capital. Even if returned overseas scholars did not have much capital themselves, some were able to attract capital from big foreign companies or venture-capital companies. Fourth, they serve as a vital link to the pool of overseas scholars and markets. They would continue to recruit overseas scholars needed by their enterprises. They would also help to track global high-tech development and foster international cooperation due to their stronger connections abroad.

Table 8.18 Profile of System-affiliated Enterprises in the High-tech Zone

		1995	1996	1997	1998	1999
Independent research institute	Number of companies	1727	1661	1513	1743	1453
	Employment (thousand)	137	110	117	138	135
	Gross output value (billion yuan)	13.2	13.8	16.3	35.0	30.2
University	Number of companies	743	774	679	526	466
	Employment (thousand)	23	33	40	33	32
	Gross output value (billion yuan)	3.3	5.3	9.7	2.7	3.0
Enterprise	number of companies	2468	2207	2454	2925	2619
	Employment (thousand)	267	249	293	397	513
	Gross output value (billion yuan)	30.9	45.7	58.1	94.2	181.9
Governmental agency	Number of companies	909	1371	1328	1799	1471
	Employment (thousand)	160	357	425	461	522
	Gross output value (billion yuan)	27.1	62.8	91.4	108.9	136.4
High-tech zone affiliated	Number of companies	4274	2215	2210	2268	2523
	Employment (thousand)	273	191	224	277	329
	Gross output value (billion yuan)	47.2	34.0	58.6	67.9	81.4
Military-affiliated	Number of companies	103	96	101	52	32
	Employment (thousand)	10	12	20	9	8
	Gross output value (billion yuan)	0.8	1.5	2.3	0.9	1.3
Other (including foreign company)	Number of companies	2713	5398	5396	6622	8934
	Employment (thousand)	121	339	356	487	672
	Gross output value (billion yuan)	15.7	51.1	74.5	99.8	160.2

Source: MOST (2001).

High-tech Industrial Parks

China's high-tech park development has benefited from organizational innovation and most significantly from the Silicon Valley model. Taiwan's successful Hsin-Zhu Science-Based Park has also provided useful experience. Currently, the State Council has sanctioned a total of 53 high-tech parks. The central governments has introduced a variety of preferential policies to help develop such parks, including the several tax incentives granted to the high-tech enterprises in the park:

1. exemption from product tax and value-added tax (VAT) on goods and equipment related to imports and exports;
2. a preferential income tax rate of 15 per cent; and
3. exemption from income tax for two years and 50 per cent reduction of income tax for 3–5 years to foreign-invested high-tech enterprises (MOST, 1999b).

Since high-tech parks, under the auspices of local governments, enjoy a certain degree of autonomy, they sometimes offer additional preferential treatment to enterprises in the park.

High-tech parks play three important roles in China's high-tech development. First, through incubation, they are an important location to commercialize high-tech R&D achievements and to develop new high-tech enterprises. In 2000, China had 465 incubators for various high-tech enterprises, which was the second-highest number, next only to the USA. Occupying a land area of 768 m^2 institutions, incubators provided 788 agency services, hosted 15,449 enterprises that generated an annual sales of 42.36 billion yuan and employed 292,000 workers, invested 13.35 billion yuan in construction, and attracted venture-capital funds totalling 2.27 billion yuan. So far 3,887 enterprises have graduated from incubators, of which 32 have gone public initial public offering. Incubators have transferred 7,930 items of provincial-level or above high-tech achievements for industrial use, developed 22,046 new products, and obtained 6,650 patents (MOST, 2003b).

Second, high-tech parks provide an important site for mutual learning between enterprises (Saxenian, 1996) and for industrial clustering to realize externalities. Finally, high-tech parks offer the fresh opportunity to instil modern management practices and to break away from the ingrained dogmatism and low-efficiency associated with the government bureaucracy.

In a decade, high-tech parks have achieved a sustained rapid growth through constantly improving the investment environment and the transfer and commercialization of high-tech products. In 2000, the 53 national high-tech parks had 20,796 enterprises which realized a total output value of 794.2 billion yuan and a pre-tax profit of 105.7 billion yuan, respectively, 112 times and 89 times greater than the corresponding figures in 1991 when the parks were first established. Industrial value added reached 197.9 billion yuan, an increase of 34 per cent from the previous year and accounting for 24 per cent of total industrial valued-added. The annual labour productivity averaged 316,000 yuan (US$38,257), far greater than the average labour productivity of industrial enterprises. Thus, high-tech parks have become an economic growth pull for China (MOST, 2001).

High-Tech Policy Initiatives

'863' programme

With nearly three decades of centrally planned economy as the background, the Chinese government surely has had rich economic planning experience. It is nature that it chose to accelerate high-tech development through government-sponsored programmes. The most important is the *State High-tech R&D Programme* that is the '863' programme. Its basic guidelines and aims are as follows: to limit the objectives while focusing on priorities; and to achieve a breakthrough in fields where China enjoys advantages in order to narrow the technology gap between China and developed countries in order to enhance the country's overall economic strength. In 1993, the programme shifted its aim from 'catching up the leading technology in the world' to a more balanced one of 'achieving high-tech heights while promoting high-tech industrialization'.

The '863' programme targeted six major fields with an investment expenditure of 11 billion in 15 years. Table 8.19 outlines the allocation of funds in the six fields.

After 15 years of implementation, the '863' programme has produced significant achievements. It established several high-tech R&D bases, nourished a new generation of high-tech researchers, achieved breakthroughs in a number of key technologies, hybrid rice for instance, and finally established footholds in international high-tech fields. The programme also contributed to the development of supercomputers.

Commercialization of technology

As noted earlier, the rate of commercialization of high-tech research is low in China. The commercialization of high-tech achievements has to be carried out by science and technology experts. Therefore, effective

Table 8.19 Fund Allocation of the '863' Programme (percentage)

	1999	2000
Biology	25%	25%
Communications	27%	27%
Automation	14%	15%
Energy	12%	16%
New materials	11%	10%
Ocean science	11%	7%

Source: Ministry of Science and Technology, *Annual Report of the State Programmes of Science and Technology Development*, 2000-2001.

incentive policies should be enacted to motivate inventors and those who put inventions into industrial use. In the last two decades, the government has reformed on the incentive policy. Incentives were redirected to individuals away from their work units; and rewards of property rights over innovation substituted for material rewards or rewards in kind. But the progress on the ground was slow. The real breakthrough did not occur until a series of documents were issued following the 1999 National Technology Innovation Conference to acknowledge that individuals' property rights over technology could be construed as a form of capital. For example, the *Basic Rules on the Commercialization of High-Tech Achievements,* issued by the Ministry of Science and Technology and several other ministries and commissions in March 1999, either confirmed or reiterated the following (MOST):

1. Technology know-how can account for as much as 35 per cent of the registered capital for limited liability incorporated companies or non-corporate entities.
2. Research institutions and universities should reward the staff who transferred or commercialized high-tech research. Such rewards should be no less than 20 per cent of the net income derived from the technology transfer or commercialization; or at least 5 per cent of the retained profits for 3–5 consecutive years. Personnel with technology shares in joint-stock companies should be awarded equity shares worth no less than 20 per cent of the value of their technology shares, in addition to sharing company proceeds according to their equity shares.
3. Science and technology personnel can engage in R&D and high-tech commercialization in other work units provided they have completed the assignment in their own work unit (MOST, 2000).

China passed the *Patent Law of the People's Republic of China* as early as 1984 to protect the intellectual property rights (IPR) of technology inventors. The law has greatly motivated enterprises and technological talent to invent and innovate. Nevertheless, the government's long-standing policy stipulating that the state owns the IPR derived from government-sponsored research programmes has dampened the enthusiasm of research institutions and universities to commercialize the technology know-how. Hence, upon entering the 21st century, the government enacted regulations emphasizing that research institutions own the usage and transfer rights of IPR derived from government-sponsored research programmes. However, the widespread piracy in IPR has substantially subdued the enthusiasm for innovation by high-tech enterprises and, in particular, software enterprises. On the other hand, the government continues to strengthen IPR protection measures.

Preferential policies for integrated circuits and software

In recent years, the Chinese government has promulgated a number of preferential policies to support the domestic high-tech industry. And the policy focus has shifted from direct intervention to the creation of favourable policy environments. The best-known cases are those encouraging the development of the software and IC industries. Document No. 18, *Policies Promoting the Development of the Software Industry and the Integrated Circuit Industry* issued by the State Council in June 2000, is a case in point. This document attracted significant foreign and domestic investments to the risky and capital-intensive IC industry. In 2002, the Ministry of Finance (MOF) and the State Administration of Taxation (SAT) issued the *Notice on Tax Policies to Encourage the Development of the Software Industry and the Integrated Circuit Industry* (MOF-SAT Document No. 70, 2002) to implement Document No. 18. This document stipulated:

1. from 1 January 2002 to the end of 2010, after the collection a VAT tax of 17 per cent on IC products, the tax exceeding 6 per cent is immediately rebated to the company;
2. producers of IC 0.8 or fewer micron products will enjoy 'tax exemption for two years after making profits and 50 per cent tax reduction in the next three years; and
3. from 1 January 2002 to the end of 2010, investors in IC or packaging enterprises who re-invest their after-tax profits in other high-tech businesses or open new manufacturing or packaging enterprises will enjoy a 40 per cent rebate of income tax paid for re-investment, if their enterprises have been in operation for at least five years. The

rebate rate rose to 80 per cent in western China (*China Business Post*, 2003).

The aforesaid policies are widely welcomed by domestic enterprises such as the Shanghai-based Semiconductor Manufacturing International Corporation (SMIC). However, foreign companies, including the US Semiconductor Industry Association, have claimed that these policies are at odds with WTO principles (*China Business*, 2003). Finally, in July 2004, China and the US signed a memorandum of understanding to end the VAT rebate for domestic IC firms.

Government venture-capital fund

The Chinese government has attached great importance to promoting high-tech development through venture capital (VC) since 1998. At present, the domestic VC industry is still in its infancy. In 2001, government VC totalled 9.3 billion yuan while domestic corporate VC totalled 8.8 billion yuan. Venture capital from overseas contributed 3.4 billion yuan. The IDG Technology Venture Investment Fund, the WI Harper Group, Softbank, Intel, Goldman Sachs, and the National Science & Technology Board of Singapore were among the most active overseas VCs that together had invested over US$200 million in China's software industry and internet-related fields. Beijing received the lion's share of VC, with Shenzhen being the distant second (Table 8.20).

In May 1999, the central government set up the Innovation Fund for Small Technology-based Firms (Innovation Fund) to support innovations of small technology-based firms that are routinely suffered from capital shortages. Since its inception, the Innovation Fund has offered grants and low interest-rate loans to small technology enterprises, but only an insignificant amount of equity capital. In 2000, the Innovation Fund allocated 659 million yuan to high-tech companies averaging 756,000 yuan for each project. Electronic information, biology & pharmaceuticals, materials, optical, mechanical and electronic integration, new energy and energy conservation, natural resources and environment were the beneficiaries of the Innovation Fund.

Meanwhile, local governments have attached a great deal of importance to the Innovation Fund and have provided counterpart funding and other complementary measures to cultivate a sound investment-financing environment for small technology-based firms in their respective localities. A survey of 1,383 projects financed by the Innovation Fund shows that out of a total of 13.436 billion yuan raised by the enterprises, the Innovation Fund contributed 694 million yuan (5.1 per cent); local governments, 414 million yuan (3.0 per cent); bank loans

Table 8.20 Venture Capital in Main Regions of China (2001, million yuan)

	Government		Domestic company		Foreign company	
Region		Percentage (%)		Percentage (%)		Percentage (%)
Beijing	1,936.81	20.79	3,646.58	41.32	2,768.38	79.53
Shenzhen	1,316.68	14.13	2,216.44	25.11	451.26	12.96
Shanghai	1,634.67	17.55	3,74.20	4.24	14.50	1.19
Other regions	4,427.14	47.53	2,588.22	29.33	219.58	6.31
Total	9,315.30	100	8,825.44	100	3,480.72	100

Source: China Venture Capital Development Report (2002), Department of Financing and Infrastructure of MOST, 2002. Central Government funding consists mainly of grants and subsidized loans and as such is not strictly venture capital. To a lesser extent, this applies also to local government funding.

5.495 billion yuan (40.7 per cent); and enterprises' own capital 6.128 billion yuan (45.6 per cent). This may indicate that the Innovation Fund leveraged 18.4 times its own financial contribution from all other funding sources including enterprises' own capital (MOST and MOF, 2000).

4. CONCLUDING REMARKS – AN ASSESSMENT OF CHINA'S HIGH-TECH DEVELOPMENT

Although China began to produce some high-tech products as early as the 1950s, it was not until the 1980s that China began to see the budding of a high-tech industry. In as little as two decades, China's high-tech industry had grown from almost nothing to become the driver to nation's economic growth. A significant number of pillar enterprises with substantial R&D capacity and competitive power have emerged in the following industrial sectors: aeronautics, nuclear power, electronics, computers and communications, bioengineering, optics, mechanics and electrics, and oceanic resources. Nevertheless, significant issues remain that need to be further studied and dealt with properly in due course in order to sustain high-tech growth.

Is the High Dependence on Foreign Technology and Foreign Funded Enterprises (FFEs) an Issue?

As noted in Table 8.6, FFEs contribute over 80 percent of China's high-tech exports and the share is still growing. China's high-tech dependence on FFEs reflects to a large extent its chosen development strategy that can best be reflected by the pervasive tax and other incentives (such as land concessions) offered by the government to them (Hwa, 1999). FFEs are also attracted to China by its competitively priced costs of labour and land, good infrastructure and, above all, a steadily improved investment climate as a result of a deepening in market reform and an expansion in market access. In return, FFEs has increased China's technology capacity by establishing firms with more advanced technology than indigenous ones (Jiang, 2002) and research centres and labs, often in collaboration with local universities (Liu and Zhao, 2003). [6] Still, one wonders whether this can be a sustainable strategy for developing China's high-tech industry and power China's growth, considering the immense size of the Chinese economy. One could equally wonder whether the 'footloose' multinationals would 'abandon' China in favour of other localities, if and when the opportunities present themselves. Would the corporate interests of multinationals unduly influence China's high-tech industry development? Indeed, in the global supply chain, China is targeted by multinationals mainly in the lower valued-added portion. Zhu's (2003) survey of 67 multinationals in China's three most prosperous cities in Jiangsu province: Suzhou, Wuxi, and Zhangzhou, finds that multinationals have generated few backward linkages to the Chinese economy. The high value-added inputs required to sustain their Chinese operations ranging from technology to raw materials and capital goods are predominantly imported from abroad (Tables 8.21 and 8.22). The finding seems to be consistent with the fact that the bulk of high-tech foreign trade undertaken by FFE consists of processing trade, as noted earlier.

But these worries may be exaggerated. The low level of backward linkages can be traced to a variety of interlocking factors. On the policy level, up until its recent accession to the WTO, China has encouraged FFEs to produce predominantly for the external market. This is the main reason behind the high share of processing trade in total trade. And most of the technology, capital and essential materials engaged in

Table 8.21 Sources of Key Technologies Employed by Multinationals in China

Cities	Sample	Sources of technology					
		China mainland	Share (%)	Overseas	Share (%)	Parent company	Share (%)
Suzhou	22	0	0.0	3	13.64	19	86.36
Wuxi	29	0	0.0	4	13.79	25	86.21
Zhangzhou	16	0	0.0	2	12.50	15	87.50
	67	0	0.0	9	14.43	58	86.57

Source: Zhu (2003)..

Table 8.22 Sources of Raw Materials Used by Multinationals in China

Cities	Sample	From					
		Local	Share (%)	Other domestic regions	Share (%)	Parent company	Share (%)
Suzhou	22	1	4.55	2	9.09	19	86.36
Wuxi	29	0	0.00	3	10.34	26	86.66
Zhangzhou	16	2	12.50	1	6.25	13	81.25
	67	3	4.48	6	8.95	58	86.57

Source: Zhu (2003).

production for export would need to be imported from abroad because they are either not available or not in ample supply in the domestic economy. For the future, however, the trend is likely to reverse itself for the following reasons. To begin with, FFEs will shift from their China strategy from viewing China as the springboard for the global market to one aiming at China's vast domestic market. Their shift of the China strategy will be prompted by China's adjustment of its own development strategy away from the single obsess with the external market to the development of its own domestic market. For instance, China's accession to WTO has for the first time allowed FFEs to enter the distribution chain, financial and other services catering mainly to the domestic market. China will also soon equalize the fiscal and other incentives between producing for the domestic market and the foreign market. These and the other like measures will open new opportunities in China's large domestic market, which will induce FFEs to orient more their production to the domestic market and to 'localize' their operations as a companion step. Increased local sourcing R&D by multinationals is a case in point and the trend is likely to continue or even accelerate provided that China continues to open its vast and rapidly growing domestic market to foreign ventures as well as continue to upgrade its knowledge economy (World Bank, 2001). Multinationals are also likely to increase their domestic sourcing of other inputs to sustain their operations in China as the local industrial base is steadily strengthened. Over the long haul, the more likely scenario is for multinationals to be fully integrated into the Chinese economy.

Actions Required to Develop High-tech Enterprises

It has not been a zero-sum game between FFEs and domestic high-tech companies. New products introduced by foreign high-tech firms have stimulated local firms to participate in the same product markets. The electronics and telecommunications equipment industries offer the best examples. Starting from a near-zero base, Chinese companies are able to command a formidable presence in the mobile phone and electronics markets within a short period of time. The Huawei Company in the telecom equipment market is an example. While domestic firms still produce for the relatively lower-end markets for now, with the rapid emergence of private enterprises in the high-tech industry, the catch-up is expected to accelerate in the future. China's high-tech industrial policy should continue to 'walk on two legs': in addition to wooing foreign high-tech firms, China should also improve the overall technology policy framework. The following actions are worthy of consideration.

Raise R&D expenditure and efficiency

Section 2.3 showed that China's R&D spending ranks low in international comparison although China has probably the world's largest R&D staff. Even more critical, China needs to raise the R&D efficiency. This can best be done by improving the work incentives of high-quality research and technical staff, upgrading their skills, as well as stepping up efforts to reverse the 'brain-drain'. At present, China's relatively low-cost and abundant high-tech personnel resources are not sufficiently motivated to fully exploit their comparative advantage. The state-dominated science and technology apparatus that tends to base reward on seniority instead of merit, as well as on non-material means, is chiefly to blame. While corporate reform is taking place, significant progress still lies ahead. Creating the right incentive system and environment can also help high-tech firms and research institutes to tap into overseas high-quality manpower. The potential is large since only one-third of the more than 300,000 Chinese who have gone abroad in pursuit of advanced education since 1978 have returned to China thus far. Moreover, Chinese firms will face ever more competition from multinationals for China's limited high-level manpower in the aftermath of its accession to the WTO.

Raise the rate of commercialization of technology

The supply of research output has far outstripped the market demand for it, resulting in a low rate of commercialization of technology research. It is estimated that fewer than 30 per cent of the sponsored and completed research subjects could be commercialized and only 5 per cent in the high-tech field; both are much lower that developed nation standards. Economics of supply and demand can explain this phenomenon. Consider supply: research institutions and universities have undertaken much high-tech research but the output has had only limited potential. Now the demand: the unclear ownership right for most Chinese enterprises, ostensibly state-owned enterprises, diminishes the motivation for pursuing technological improvement and hence the demand for new technology (see next subsection), not to mention the operational and financial risks entailed by acquiring a new technology. A promising development is that the government is withdrawing financial support to poor-performing government-sponsored research institutions to re-orient their research towards the market.

Reform the corporate governance

Establishing high-tech companies involves high financial risks that can only be motivated by equally high financial returns derived from such an undertaking. That is, risk-taking behaviour should be rewarded by returns commensurate with the risk undertaken. Generally speaking, managers in a state-owned company are unlikely to be rewarded sufficiently to take the high risk entailed by adopting a new technology. Therefore, they would refrain from undertaking R&D and developing new products, thus jeopardizing their company's long-term survival. Therefore, it is essential to continue to reform the governance of state-owned enterprises. In addition, the property rights of private enterprises should be protected, for only private companies are capable of taking the high risks that accompany high-tech investments.

Develop the equity market and the venture-capital industry

Venture capital in the US provides the best example of how the risk capital can support high-tech companies and even establish a whole high-tech industry. Venture capital essentially serves as the vital link between technology development in the research lab and the market. The US experience also suggests that the success and failure of venture investments is intimately tied to the vagary of the stock market prices.

China does not as yet has effective private equity markets in which high-tech enterprises can raise either start-up capital or late-stage expansion capital. That is the reason why the central government launched the Innovation Fund to fill the funding vacuum. The Innovation Fund, however, is anything but a VC fund. The dearth of private equity capital is due to the lack of exits for venture investment; neither the mergers and acquisitions option nor the IPO option is operative in China. This is, in turn, due to the lack of an equity market for pricing and trading shares of high-tech companies.

China's capital market is also in an early stage of development. No doubt, it will gradually develop as economic development progresses, but it will also take a long time to mature. In 2004, after intensive preparations, a stock exchange for small and medium enterprises was finally established in Shenzhen. It is not quite a high-tech stock exchange yet, but it is a step in that direction.

Since high-tech start-ups are typically cash-stricken, allowing them to use company shares to exchange for technology know-how or high-tech personnel can substantially raise the odds of their very survival. Company Law requires modification to accommodate high-tech companies.

Protect intellectual property rights

China has made great progress in IPR protection, but the protection remains far behind the standards of developed market economies. Enterprises are still barely aware of IPR and show little respect for those of others'. IPR disputes are commonplace. It is estimated that software piracy alone costs the domestic software industry 40 billion yuan annually. The IPR issue, most evident in pharmacy, light industry, garments, and machinery manufacturing is also a vexing issue perplexing many multinationals. While the situation has been improved in recent years, the problem remains unsolved (*China Business Times,* 2003).

APPENDIX 8A: DEFINITION OF HIGH-TECH INDUSTRY

The 2002 China Statistics Yearbook on the HT Industry categorizes the HT industry into five groups; pharmacy, aeronautics, electronics & communications, computer & office facilities, and medical equipment & apparatus as in Table 8A.1. This classification is close to OCED definition and is used in this Chapter.

Table 8A.1 Definition of High-tech Industry in China

Industrial classification (code)	Name of the industry	Code
Manufacture of Medical and Pharmaceutical Products (27)	Chemical Pharmaceutical Products	271+272
	Processing of Traditional Chinese	274
	Medicine	276
	Biology Products	
Manufacture of Aircraft and Spacecraft (376)	Manufacture and Repair of Aircraft	3761
	Manufacture of Spacecraft	3762
	Other aircraft products	3769
Electronic and Telecommunications Equipment (40—404)	Telecommunication Equipment	40—404
	Telecommunication	4011
	Transmission Unit	4012
	Telecommunication Exchange	4013
	Unit	4014
	Telecommunication Terminal	402
	Unit	403
	Mobile & terminal unit	405
	Radar and Peripheral Equipment	4051
	Broadcast and Television Equipment	4052
	Electronic Apparatus	4053
	Electronic Vacuum Apparatus	4059
	Semiconductor Separated Parts	406
	Integrated Circuits	407
	Photo-electronics and other	409
	parts	
	Electronic Components	
	Household Audiovisual Equipment	
	Other Electronic Equipment	

Industrial classification (code)	Name of the industry	Code
Computers and Office Equipment (404+4154+4155)	Computers	4041
	Network Equipment	4042
	Peripheral Computer Equipment	4043
	Office Equipment	4154+4155
Medical Equipment and Monitors (368+411+412+4141+419)	Medical Equipments and Instruments and Monitors	368 411+412+4141+419

NOTES

1. The US industry was worth about US$ 810 billion and Japan's, about US$500 billion (OECD, 2003).
2. The calculation is based upon nominal magnitudes due to the lack of appropriate deflators for value added.
3. It should be noted that the labour productivity growth estimates have not been adjusted for the effect of inflation.
4. Total merchandise exports were US$1,837 billion in 1998, US$1,949 in 1999, US$2,492 in 2000 and US$2,662 in 2001 (Chinese customs data).
5. Calculations are based upon nominal values; real magnitudes would have been preferable but were not possible due to the lack of appropriate deflators.
6. It is common for a developing economy to use fiscal incentives to woo foreign technology and capital in order to stimulate growth although this gives rise to tax competition reducing their effectiveness.

REFERENCES

China Business (2003), 'USA Interfered with China's Chip Industry with an Eye on Document No. 18', *China Business*, 6 June.

Chinese Business Times (2003), 'Multinationals Adjusted Technological Strategy, Six Major Changes in China Attracted Broad Attention', *Chinese Business Times*, 8 February.

China Business Post (2003), 'Document No. 18 of the State Council Exerted the Biggest Impact Over IC Development', 9 March.

IMD (International Institute for Management Development)(2002), The World Competitiveness Yearbook, 2002.

Jiang, Xiaojuan 'et al' (2001), 'Significance, Current Situation and Policy Suggestions of Foreign Capital in High-tech Industry', *China Industrial Economy*, no.2, 55-63.

Jiang, Xiaojuan (2002), 'Analysis of Foreign Investment in China in 2002 and Prospect for 2002', *Management World*, no.1, P29-30.

Hwa, Erh-Cheng (1998), "China: Fiscal Policy and Structure Adjustment", December 15, 1998, mimeo, The World Bank, Washington, D.C.

Lan, Xue (2002), *Links Between Universities and Market, Research Report on*

China's Scientific and Technological Development, Economic Management Press, Beijing, China.

Liu, Xielin and Steven White (2001), 'Comparing Innovation Systems: A Framework and Application to China's Transitional Context', *Research Policy*, Vol. 30 (7), 1091-1114.

Liu, Xielin and Jue Zhao (2003), 'Analysis of R&D Activities of 19 Multination.als in Beijing' in *the Forum on Science and Technology in China*, pp. 65-69, Beijing, China.

Lu, Qiwen (2000), *China's Leap Into Information Age*, the Oxford University Press.

MOST (Ministry of Science and technology) (1999a), *Report on China's High-tech Industrialization*, Science Press, and Beijing, China.

MOST (Ministry of Science and technology) (1999b), *Selected Laws and Regulations on Technological Innovation*, China Science and Technology Press, Beijing, China.

MOST (Ministry of Science and Technology) (2000), *Policies for Technological Innovation in China*, Science and Technology Documents Publishing House, Beijing, China.

MOST (Ministry of Science and Technology) (2001), 'High-tech Development Centre of the Torch Program', Statistic Report on Ten-year Development of National High-tech Parks 1999-2000, Science and Technology Documents Publishing House, Beijing, China.

MOST (Ministry of Science and technology) (2003a), *China Scientific and Technological Indices (2002)*, Science and technology Documents Publishing House, Beijing, China.

MOST (Ministry of Science and Technology) (2003b), www.most.gov.cn.

MOST (Ministry of Science and Technology) and MOF (Ministry of Finance) (2000), *Annual Report on Innovation Funds*, 4-5, Beijing, China.

National Bureau of Statistics (2003), *2002 China Statistics Yearbook on High-technology Industry*, China Statistics Press, Beijing, China.

OECD (2003), Structural Analysis, online databank, 2003, Paris.

Saxenian, A (1996), *Regional Advantage*, Cambridge, MA: Harvard University Press.

World Bank (2001), *China and the Knowledge Economy: Seizing the 21st Century*, Dahlman, C.J., and Aubert, J.E., Washington, D.C. US.

Zhu, Huagui (2003), 'Research on Technological Externalities of Branches of Multinationals in China', *Scientific Research Management*, Vol. 24, (2), 138-144, Beijing, China.

9. Indonesian Industrial Policies: Before and After the Crisis

Kelly Bird and Hal Hill*

1. INTRODUCTION

Indonesia, the world's fourth most populous nation, has an unusual industrialization history, certainly in comparative East Asian context. Apart from a brief growth episode in the 1930s, it experienced little industrialization until the late 1960s. Then, during the Soeharto period, there were three decades of rapid industrial growth and catch-up, which saw the emergence of a large and increasingly broad-based factory sector. However, during the recent economic crisis, industrial output fell sharply, and has only recently recovered to pre-crisis levels.

Indonesia therefore emerges as something of an 'East Asian outlier', whose experience is more akin to the Philippines than to the rest of the region. Indonesia is an industrial late-starter. Its relative resource endowments also dictate a different industrial path. It shared 30 years of rapid industrialization with most of its neighbours. But even pre-crisis it was some way off having to face the industrial upgrading challenges necessitated by the loss of comparative advantage, as experienced in many of the regional economies. Since the crisis, these differences have been further accentuated, including for example the continuing negative foreign direct investment (FDI) flows. Consequently, the 'high–tech' focus of this volume necessarily receives less emphasis than is the case for neighbouring high-growth economies on the verge of major industrial restructuring and upgrading. However, the general lessons to emerge from Indonesia's experience are arguably more relevant to much of the developing world. The country's inclusion in this comparative project therefore broadens the analytical and policy relevance of these case studies.

In this chapter, our organization is as follows. We first provide an

* For helpful comments on earlier versions of this chapter, we are grateful to Yun-Peng Chu and Thee Kian Wie.

overview of the parameters which have shaped Indonesia's industrialization and industrial policy. In Section 3 we provide a summary picture of Indonesia's changing industrial structure since the 1970s. Next, by way of historical context, we examine the record of industrial policy making and reform during the Soeharto era, as a backdrop to our analysis of contemporary policy issues. In Section 5, we address what we argue are likely to be the key determinants of industrial growth for the next decade and beyond. These include the quality of governance and the legal system, the incentives regime, and some key supply-side issues. The discussion here underlines both the continuities and changes in the country's pre- and post-crisis commercial environment. A concluding section summarizes our major arguments.

2. AN OVERVIEW[1]

The pace and pattern of industrialization is invariably determined by general economic conditions. Thus, there was little industrial investment in Indonesia prior to 1966 owing to the highly uncertain commercial environment, including the widespread expropriation of foreign property after 1958 and progressive disengagement from the international economy, including the international development institutions. Industrial output then grew very rapidly from 1967 owing to the adoption of 'orthodox' economic policies. The macro economy was stabilized, trade and investment were opened up, there were major investments in infrastructure, and the commercial environment suddenly became broadly predictable and business friendly.

During the oil boom decade from 1973, Indonesia's industrialization was further propelled by large investments in the state enterprise sector, and by some resort to protection. These positive effects were more significant than the negative 'Dutch disease' impacts of the squeeze on non-oil tradable goods activities. As oil prices began to soften in the early 1980s, then fall sharply in the middle of the decade, the government responded with an increasingly bold and effective reform strategy. There were tax and customs reforms, trade liberalization, and a more open policy towards FDI. These microeconomic reforms were underpinned by large exchange rate depreciations which, in the context of prudent macroeconomic management, delivered a major boost to competitiveness. The supply response was swift. Manufactured exports began to grow rapidly. For the first time in its industrial history, Indonesia became progressively more 'East Asian', with labour-intensive, exported-oriented industrialization emerging as a major driver of growth.

This rapid growth continued until 1997, albeit with a weakening of the reform momentum during the 1990s. Indonesia then experienced a deep economic crisis. GDP contracted by almost 14 per cent in 1998, the largest decline among the crisis-affected East Asian economies in that year. Unlike its neighbours, it had 'twin crises', both economic and political, the latter involving the fall of Soeharto after 32 years of authoritarian rule. Economic recovery since 1998 has also been the slowest among these economies, reflecting the depth of the financial collapse, continuing political uncertainty, and the fragility of institutional reform.

How does one position Indonesia in comparative East Asian development context? There are both broad similarities and marked differences. The following features would appear to be important in any discussion of comparative industrialization patterns:

1. *Size*: With its large population and land area, there has always been a temptation to look inward. Indonesia shares *more in common with* China and India in this respect, and less with the newly industrializing economies (NIEs) and Malaysia.
2. *Resource endowments*: Indonesia is a moderately resource-rich economy as compared to the NIEs. External shocks therefore affect it differently, and the bases of its comparative advantage are also different.
3. *Weak institutions*: Indonesia generally ranks low in international comparisons of institutional quality. Its civil service is poorly paid and vulnerable to capture (though there are pockets of quality). It especially ranks very poorly in international comparisons of corruption, while its legal system is weak and offers little commercial protection.
4. *A latecomer in human capital*: Reflecting colonial neglect and faltering economic progress during its first two decades of independence, Indonesia's human capital base is underdeveloped, certainly compared to the NIEs and the Philippines. Quantitative educational indicators improved significantly during the Soeharto period. But quality remains problematic, especially at the tertiary level and in the hard sciences.
5. *Ownership complexities*: Indonesia's industrial policy and reform are complicated by its history. There continues to be widespread reservation towards high levels of foreign ownership. This in part reflects the country's colonial history, the protracted struggle for independence, and the xenophobia exploited during the Sukarno era through to 1966. The crisis of 1997–98, and the sometimes heavy-handed intervention of the international donor community,

exacerbated these sentiments more recently. In addition, the so-called non-*pribumi* community (that is, non-'indigenous', mainly ethnic Chinese) have traditionally dominated much of the modern sector commerce and industry, even though they constitute around 2 per cent of the population. Governments therefore have to move cautiously on reform, lest it be seen that foreigners and non-*pribumi* groups benefit disproportionately. Poorly performing state enterprises are also more likely to be tolerated for this reason.

6. *Absence of a coherent industrial policy*: Industrial policy in Indonesia has always been a contested field, with the result that it has lacked analytical coherence and consistency. The 'technocrats' have broadly favoured economic orthodoxy, while 'nationalists' have sought to promote a state-led and protected strategy. There has also been an active 'high-tech' group. Policy directions have been influenced by whichever group has been in the ascendancy.

As we shall see below, there has been a focus on 'high-tech' industrialization in Indonesia. For two decades, Dr B.J. Habibie was the Minister for Research and Technology, and briefly Vice President and President in 1998–99. With a German education and industrial experience, Habibie articulated an ambitious industrialization strategy. President Soeharto lent his support for this programme, and during periods of fiscal abundance – the oil boom era to 1983 and export-led growth from the late 1980s – his grandiose projects received major government support. During the boom years of the 1990s, Habibie and his supporters became more influential in general economic policy making. With the economic crisis and fiscal austerity of 1997–98, however, these projects mostly collapsed, with little to show for the $3 billion or more investments, apart from some industrial learning and human capital investment.

Of particular relevance to this volume, notions of a high-tech industrialization drive therefore rank low in Indonesia's current policy priorities, for at least three reasons. First, the immediate concern is post-crisis recovery, and the restoration of high growth to ensure broad-based improvements in living standards. This entails adoption of the usual post-crisis macro- and microeconomic recovery strategies. Investment collapsed in the wake of the crisis and remains well below that of the high-growth era. FDI has been negative for most of the period since 1998. The challenge is to establish a secure and predictable business-friendly environment. Moreover, the government is fiscally constrained, with debt service absorbing about one-third of expenditure and large public debt maturities on the near horizon. Hence, its capacity to introduce sector-specific promotional measures is severely constrained.

A second reason why high-tech issues are not a high priority is that Indonesia remains some way off the turning point where it will begin to lose comparative advantage in labour- and resource-intensive activities. It will remain a 'labour surplus' economy for some time yet. The slower the recovery from economic crisis the further that turning point recedes into the future. In this respect it differs fundamentally from Malaysia and also Thailand, let alone the NIEs.

A third factor is that Indonesia's policy approach to upgrading and high-tech issues will be shaped by differences in the factors adumbrated above. For example, its educational foundations need considerable strengthening before any concerted technology drive is contemplated. An additional consideration is institutional weaknesses, in particular the danger that a discretionary and interventionist policy framework could easily be derailed by large investments in uneconomic projects or captured by vested interests.

Finally in this overview, while much of this chapter will draw attention to continuities in Indonesia's industrialization, it is important also to emphasize the fundamental discontinuity introduced by the crisis of 1997–98. Consistently high growth has been replaced by sluggish expansion. The industrial sector is no longer a principal engine of growth. In fact a form of 'de-industrialization' is occurring, at least temporarily, in the wake of the shock to the modern financial sector, as tradable goods lose competitiveness. The commercial 'rules of the game' under Soeharto were largely predictable and stable, albeit in the context of an authoritarian and corrupt administration. Now they are uncertain and insecure, not only as a consequence of regime change and political turbulence, but also of the major decentralization programme introduced in January 2001. The earlier confidence has given way to uncertainty: over whether Indonesia can compete with China, over whether the exodus of foreign investors will continue, and over whether the country can adjust to challenges ranging from the full implementation of the ASEAN (Association of South East Asian Nations) Free Trade Agreement (AFTA) to the removal of MFA (Multi fibre Agreement) quotas. In some cases, as we shall see, the magnitude of these challenges may be overstated. But they nevertheless account for the general air of pessimism that has permeated much of Indonesia's commercial climate – and particularly foreign perceptions of it – since 1997.

3. CHANGING INDUSTRIAL STRUCTURE

This section presents a summary descriptive/analytic account of Indonesia's changing industrial structure over the past quarter-century or so, focusing on growth, structural change, ownership patterns and exports.

Growth and Structural Change

Indonesia experienced very rapid industrial growth and structural change through to 1997. Tables 9.1–3 provide a summary picture. Indonesia recorded industrial growth of at least 9 per cent in all but two of the 27 years, 1970–96 (Table 9.1). Non-oil manufacturing, separately estimated from 1984, grew at double-digit rates each year 1984–96.[2] Initially, growth was driven by catch-up, by the indirect effects of the 1970s oil boom, and by some state-instigated investment projects. From the mid-1980s, labour-intensive exports became a significant engine of growth. There was just a very short period of adjustment from what may be broadly termed the import-substitution phase to a more export-oriented strategy, indicating how swiftly and effectively the reform process was executed and produced results. This growth came to an abrupt halt with the crisis of 1997–98. The contraction in the manufacturing sector was about the same as for the economy as a whole. Thereafter, positive growth has been recorded from 1999, but at much lower rates than pre-crisis.

Accompanying this growth has been rapid structural change. The process of modern industrialization did not commence in Indonesia until the late 1960s. From 1965 to 1997, the share of the industrial sector in

Table 9.1 Growth Rates by Major Economic Sector (%, av. per year)

Economic sector		Agriculture	Industry	(Non-oil manufacturing)	Services	GDP
	1970–80	3.9	12.0	na	9.7	8.0
Pre-reform	1980–85	3.4	4.2	na	6.9	5.0
Reform	1985–90	3.0	7.0	12.0	7.3	6.3
High growth	1990–97	2.5	9.1	11.5	6.9	7.0
Crisis	1998	−1.3	−14.0	−13.1	−16.5	−13.1
Recovery	1999	2.2	2.0	3.5	−1.0	0.8
	2000	1.9	5.9	7.0	5.2	4.9
	2001	4.1	2.7	4.9	5.0	3.8
	2002	2.8	4.4	6.4	4.7	4.3
	2003*	3.1	3.6	5.4	6.2	4.5

Notes: * Industry includes mining, manufacturing, utilities and construction.

Source: National Income Accounts, Badan Pusat Statistik, 1970–2003.

GDP more than trebled (Table 9.2). The rising share of industry was initially concealed by the price effects of the oil boom through until 1980. The share of manufacturing then doubled in the 1980s, and continued rising through to 1997. Non-oil manufacturing accounted for the entire increase in the share of manufacturing from its first year of separate enumeration (1983) to 1997. Since the crisis, the share of manufacturing has remained broadly stable, with a slight dip in 1998–99 when resources returned to agriculture as a crisis-survival strategy.

Within manufacturing, structural change has been equally rapid. Since 1975, when the industrial structure began to be measured on a regular and reasonably accurate basis, there has been a shift towards a more diversified industrial structure (Table 9.3).[3] The major labour-intensive and footloose industries in ISIC 32 grew rapidly during the export phase. Wood products expanded fast in response to the prohibition on the export of unprocessed timber, before encountering environmental constraints in the 1990s. Heavy industry (ISIC 35 and 37) grew quickly through to the mid-1980s in response to protection and major state investments. Within machinery and equipment, the auto industry grew rapidly under the impetus of prohibitive protection, but collapsed in 1998–99. Electronics has become increasingly important and export oriented, but never as prominent as in neighbouring East Asian economies.

Trends in some major industries highlight the diversity of the manufacturing sector. By way of illustration, we refer here briefly to the

Table 9.2 Share of GDP by Major Economic Sectors, 1965–2003 (%)

	Agriculture	Industry	(Non-oil manufacturing)	Services
1965*	52.4	14.1	na	33.5
1970**	45.5	21.7	na	32.8
1980**	30.7	30.9	9.9	38.4
1990***	20.1	37.9	17.3	42.0
1997***	14.9	43.2	22.4	42.0
1998***	16.9	42.8	22.4	40.3
2002+	15.4	45.5	24.6	39.1
2003+	15.2	45.1	24.8	39.7

Notes:
* Based on constant 1960 prices.
** Based on constant 1973 prices.
*** Based on constant 1993 prices.
\+ Based on constant 2000 prices.

Source: National Income Accounts, Badan Pusat Statistik, 1965–2002.

Table 9.3 Share of MVA by Broad Industry Group, 1975–2002 (%)

	1975	1985	1995	2000	2002
31. Food and beverages	46.7	32.2	22.3	21.0	23.2
32. TCF	10.8	14.1	17.8	16.1	11.6
33. Wood products	2.5	8.8	8.2	6.7	10.7
34. Paper products	4.6	3.5	4.8	6.3	6.0
35. Chemicals	20.0	16.1	13.1	15.0	14.2
36. Non-met products	2.6	5.1	3.6	3.4	6.2
37. Iron and steel	0.7	7.3	7.5	3.3	6.2
38. Machinery, autos etc.	12.1	12.8	21.9	27.4	21.4
39. Other	0.1	0.3	0.7	0.8	0.4
All Industry	100.0	100.0	100.0	100.0	100.0

Source: Annual Manufacturing Survey, BADAN PUSAT STATISTIK, 1975–2002.

experience of three key but different industries: electronics, garments, and the automotive industry.

In *electronics*, although exports began to grow quickly from the early 1990s, Indonesia has never been a major player in export-oriented segments, unlike the NIEs and its more industrialized Southeast Asian neighbours (Hobday, 2001). The principal explanation is that, while its labour costs were competitive, Indonesia was slower to offer the two major commercial policy features sought by the major multinational enterprise (MNE) players in the industry. The first was highly efficient export–import procedures, to facilitate vertically integrated, multi-plant 'global factory' operations. Only in Batam has Indonesia been able to offer an environment broadly competitive with that of Malaysia and the Philippines, for example, and it is no coincidence that this region has generated most of the growth in electronics exports since the mid-1990s. The second major requirement is that these MNEs be permitted to operate with 100 per cent foreign equity, or at least majority foreign shares and minimal pressure for divestment. In this respect, also, Indonesia has been slow to reform, except in a few regions (Batam again), and in the post-crisis period when in any case other aspects of the business environment negated this positive concession.

In the immediate post-crisis period, electronics exports grew rapidly, almost doubling to $6 billion in 2000, fuelled by both the global boom in the industry and Indonesia's increased competitiveness. Since 2001, however, the industry's fortunes have waned. The global industry slumped in 2001 and 2002. Moreover, at the end of that year, the high-profile relocation of Sony from Indonesia to Malaysia occurred. Malaysia was said to be chosen as part of the firm's global restructuring, and owing to the country's more stable labour relations system, its smoother customs procedures, a more conducive investment climate, and a more developed electronics industry, including a flourishing components sector. An additional problem for the appliance subsector has been rampant smuggling in Indonesia, due in part to high luxury taxes and corruption in customs. These luxury taxes were eventually reduced in early 2003, and the industry has began to recover, with more recent players such as Samsung taking market share from the more established Japanese firms.

The *automotive* industry grew rapidly from the early 1970s (Aswicahyono et al., 2000). This was a classic case of very high – in some cases prohibitive – protection spawning an inefficient and fragmented industry. Lacking any substantial export base until very recently, the industry's growth was essentially determined by domestic demand conditions and the scope for backward integration. An ambitious local content and enforced subcontracting scheme was drawn up in the late 1970s, and implemented erratically for almost 20 years. Lacking both scale economies and industrial capabilities, the industry was the least developed among the more industrialized Southeast Asian economies. The slower pace of liberalization also resulted in Thailand effectively positioning itself as the region's major automotive producer from the early 1990s. Such a 'back-to-front' emphasis on the assembly subsector therefore stymied the growth of the components sector, where over time Indonesia could have built competitive strengths based on scale and labour cost advantages.

The crisis hit the Indonesian industry particularly hard, with annual production slumping from almost 400,000 vehicles to about 60,000, similar to that of the mid-1970s. In the process, however, there has been a painful but necessary process of structural adjustment. Protection has declined, and there has been a significant consolidation among both assemblers and suppliers. There has been a period of industrial learning, albeit inefficient, since the 1970s. Foreign firms have bought distressed assets, and injected new technologies and market opportunities. In some sectors, the industry is approaching international competitiveness: for example, certain utility vehicles (notably the 'Kijang' minibus, which accounts for about half of this segment of the industry), motor cycles, and a few labour-intensive components. Exports of the latter now total about $700 million. Industry sales have recovered to pre-crisis levels. Provided that the reform process continues, there are reasonably good prospects that an efficient, more specialized and less fragmented industry will evolve.

The *garment* industry led Indonesia's footloose, labour-intensive, export-oriented industrialization drive. Exports grew rapidly, from $100 million in 1980 to $1.6 billion in 1990, doubling again in 1990–95. Several factors accounted for its success. Initially, Indonesia enjoyed largely unhindered export access, as it had ample scope for growth in quota-constrained MFA markets. Firms successfully used this learning experience as the basis for later penetrating the more competitive non-MFA markets. The major customs reforms of 1985 greatly simplified import procedures, enabling garment exporters to source their cloth internationally from least-cost suppliers. The devaluations of 1983 and 1986 boosted competitiveness. Fortuitously, these changes occurred just as the NIEs were losing comparative advantage in labour-intensive activities. Indonesian exporters were thus able to tap into international markets, often facilitated by NIE connections, principally as buyers and strategic advisors rather than investors, owing to the mature, standardized technology prevailing in the industry.

As with the export-oriented electronics industry, the 1997–98 crisis offered garment exporters major opportunities. However, the record since 1998 has been mixed (see James et al., 2003). Garment exports reached record levels of $4.7 billion in 2000, but then declined sharply to $3.7 billion by 2002. Labour problems – the rapid increase in wage costs 2000–02 and general labour relations difficulties – have resulted in some major international buyers shifting to China and Vietnam. Uncertain supply deliveries are just as important a deterrent as rising costs, given tight international delivery schedules and the importance of a nimble response to unfilled export quota opportunities. Moreover, since the mid-1990s, Indonesia's garment exporters have not been able to reduce their reliance on MFA markets. Consequently, the imminent phase-out of the MFA arrangements has induced some buyers to anticipate a major relocation of international supply networks, particularly centred on China.

Export Performance

Indonesia emerged as a significant industrial exporter only in the mid-1980s (Table 9.4). Initially these exports were concentrated in resource-based activities, especially plywood, reflecting the country's natural resource endowments and the prohibition of unprocessed commodities (of which logs and rattan were the most important). Its industrial export base began to widen significantly post-reform, with textiles, garments, footwear, electronics, furniture, sporting goods and toys also registering rapid growth.

The share of labour-intensive products in total manufactured exports increased in the wake of the 1980s reforms, from about 45 per cent in the mid-1980s to 61 per cent by 1996. Initially, much of the increase came from typical 'early stage' manufactures as noted above. But exports

Table 9.4 Exports of Manufactured Products According to Factor Intensity, 1980–2002 ($ million or per cent)

Description	1980	1990	1993	1996	2000	2002
Natural resource intensive (NRI)	119	3,850	5,359	5,052	3,697	3,381
% of total manufactures	24	34.2	27.2	19.3	10.0	10.6
Major items:						
Wood and cork	68	3,586	5,129	4,843	3,260	2,853
Labour intensive (ULI)	287	5,188	10,918	14,918	23,132	19,023
% of total manufactures	57	46.1	55.4	56.9	62.8	59.8
Major items:						
Textiles	46	1,470	2,637	2,834	3,505	2,896
Furniture	na	338	676	952	1,518	1,512
Clothing	98	2,001	3,502	3,591	4,734	3,945
Footwear	na	694	1,661	2,195	1,672	1,148
Perfumes	na	184	133	199	318	369
Electronics + components	na	na	136	760	3,660	2,185
Telecommunications/electrical machinery	na	243	1,367	3,124	5,954	5,685
Capital intensive	97	2,222	3,445	6,218	9,979	9,399
% of total manufacture	19	19.7	17.5	23.8	27.1	29.6
Major items:						
Steel products	na	820	605	1,000	1,491	1,388
Organic chemicals	35	100	244	505	1,140	1,124
Other machinery	na	39	141	349	911	938
Rubber products	na	85	106	299	371	455
Paper & paper products	na	182	494	942	2,261	2,074
Road vehicles	na	46	334	348	489	561
Plastics	na	55	68	313	665	494
Total	501	11,260	19,723	26,177	36,808	31,804

Notes:
na. Refers to value less than US$10 million.
Resource intensive = SITC items 61, 63, 66 (excl. 664–6), 671.
Labour intensive = SITC items 54, 55, 65, 664–6, 695–7, 76–7, 81–5, 89.
Capital intensive = SITC items 5 (excl. 54 and 55), 62, 64, 67 (excl. 671), 69 (excl. 695–7), 7 (excl. 76–8), 86–8.

Source: Trade Statistics, Badan Pusat Statistik.

gradually began to broaden to other products, especially electronics. Export growth then tapered off in the post-reform period, mainly due to slower demand in Indonesia's major export destinations. Increased competition in export markets, slower productivity growth and the real rupiah appreciation around the mid-1990s were also contributing factors (Aswicahyono and Pangestu, 2000). During the post-crisis period, Indonesia's manufactured exports rebounded strongly as a result of the large exchange rate depreciation. However, this growth ran out of steam after 2000 in response to slowing global demand and emerging domestic problems, to be discussed shortly. The exports of capital-intensive manufactures, which generally have a significant natural resource component, displayed a similar pattern.

Notwithstanding these regional similarities, Indonesia's export structure has always differed considerably from that of its neighbours. First, as would be expected, resource-based manufactures are relatively more important. This is so even though their share has been declining in recent years owing to the relaxation of the prohibition on the export of unprocessed goods, and the mismanagement of that resource base (especially forests) resulting in its rapid depletion. Second, the scale of its manufactured exports is much less than its more established and consistently more outward-oriented neighbours. For example, Malaysia's manufactured exports are about 2.5 times those of Indonesia, while Thailand's are about 50 per cent higher. Third, as noted, electronics are much less important in Indonesia's exports as compared to most of its neighbours. The consequence is that the country is less connected to one of the fastest-growing segments on world trade. As a corollary, of course, Indonesia's export pattern is more diversified, and it is somewhat more insulated from the global electronics cycle.

Ownership Patterns

Indonesia's industrial ownership patterns reflect the interplay of policy and industrial organization factors. As is the case in most countries, domestic firms are the major players in Indonesian industry. They account for more than 50 per cent of manufacturing value added in all 2-digit industries except ISIC 38 (fabricated metals, dominated by electronics) and the relatively minor ISIC 39 (miscellaneous manufacturing). They also employ most of the industrial workforce in aggregate (79 per cent), including at least one-half in all 2-digit industries (Table 9.5).[4] Among domestic firms, state-owned enterprises (SOEs) are important in certain 'strategic' industries, such as fertilizers, steel and cement, together with some firms which were inherited from the pre-1966 nationalizations (e.g., sugar processing) and never subsequently relinquished. During the recent crisis, the SOE sector in general contracted, most notably the prestige projects which were heavily dependent on direct government support, such as former president Habibie's flagship aircraft factory in Bandung. Other firms have been accidentally nationalized as part of the financial crisis, but the government has indicated that it intends to divest most of them.

By the mid-1960s, there was no foreign investment in Indonesian manufacturing, apart from a few de facto joint ventures with firms from the former Soviet bloc under the guise of 'production sharing agreements'. A law enacted in 1967 then suddenly ushered in a liberal policy, which was maintained for almost a decade, when greater restrictions were placed on MNE entry. A gradually more open regime was introduced from the mid-1980s, and has been more or less maintained ever since. As shown in Table 9.5, foreign ownership of Indonesian manufacturing has risen steadily since: about 23 per cent of non-oil manufacturing value added in 1975; slightly less in 1985 (owing to a decade of restrictions, plus the growth of the SOE sector over this period); rising to 29 per cent in 1995 in the wake of more liberalized entry and growing export orientation. The share rose higher still in the wake of the crisis, in response to further policy liberalizations and the opportunity for foreign firms to buy distressed local assets. Moreover, foreign firms have been better able to endure the crisis as compared to domestic firms, owing to the former's 'deeper pockets', and their access to international financial

Table 9.5 The Share of Foreign Firms in Manufacturing, 1980–2002 (per cent of industry value added)

Sector	1980	1985	1990	1995	2000	2002
All sectors	30	22	22	29	39	34
31. Food Products	21	12	8	11	16	13
32. Textiles	22	29	17	24	35	32
33. Wood	12	13	10	12	9	16
34. Paper	13	9	30	32	12	49
35. Chemicals	42	27	34	41	46	30
36. Non-metallic minerals	55	41	21	25	38	31
37. Basic metal industry	32	19	24	43	39	25
38. Fabricated metals	44	29	46	44	66	65
39. Other manufacturing	60	41	20	62	48	15

Source: Annual Manufacturing Survey of Medium and Large-Scale Establishments, Badan Pusat Statistik, 1980 to 2001. Note: industry classifications changed in 2001 so comparison with earlier years may not be exact.

and export markets.

When entry is unimpeded, foreign firms tend to concentrate mainly in industries in which MNEs possess some sort of competitive advantage. That is, they are usually important in capital and R&D-intensive industries, such as petrochemicals, steel and automobiles. They are also prominent in internationally vertically integrated industries, such as electronics and autos. In newer industries, their dominance may arise from superior knowledge of international market channels, for example footwear. A fourth category of traditional dominance is where international brand names are important, for example in cigarettes, commercial drink products and pharmaceuticals. Some of these patterns are evident in the 2-digit ownership shares shown in Table 9.5. Thus, for example, MNEs are important in ISIC 38, dominated by electronics and the automotive industry. They are also important in base metals (principally steel and related products), the chemical industries, and some more labour-intensive activities (in textiles, garments, footwear and miscellaneous manufactures) where knowledge of export markets is important.

The balance of payments (BOP) estimates of foreign investment suggest that FDI inflows have been minimal since 1997, and frequently negative. Investment approvals data also suggest a much reduced foreign interest over this period (Table 9.6). In contrast to the 1997 approvals of $33.8 billion, most subsequent years have been well under one-third of this figure. Moreover, the proportion of FDI taking the form of 'greenfield' investment has declined, with both expansions and mergers and acquisitions (M&As) rising. These approvals data are approximate, and bear no necessary relation to the BOP-based estimates of realized flows. The approvals data include FDI from all sources, including

Table 9.6 Foreign Direct Investment Approvals, 1990 to 2003

	Total approved FDI (US$ billions)						
	1990	1996	1997	1998	1999	2002	2003*
Approvals	9.6	29.9	33.8	13.6	10.9	9.8	4.4
	Percentage by category						
New FDI	–	61	50	63	61	44	32
Expansions	–	19	32	18	23	38	14
A&M	–	20	18	20	17	18	55

Notes: * Total for first semester of 2003.

Source: Indonesian Investment Coordinating Board, Jakarta.

investment funded from domestic resources (e.g., reinvested profits). They also include borrowed funds. Being approvals, they are at best a broad indication of investor interest in the country. Moreover, in recent years it is widely believed that some of the domestic capital which exited the country in 1997 may have returned under the guise of foreign investment, to take advantage of the greater protection which foreign ownership is presumed to confer. The fact that Singapore has recently emerged as the largest source of FDI in Indonesia further confirms such a suspicion.

4. INDUSTRIAL POLICY AND REFORM: SOME LESSONS FROM RECENT HISTORY

In this section we examine various aspects of industrial policy and reform during the Soeharto era. These provide a clue to likely future policy directions as Indonesia achieves a broad-based recovery from the effects of the crisis. Indonesia's political economy architecture has of course changed since 1998, in some instances radically. But the continuities are arguably just as important. We focus on three major and contentious areas of industry policy: trade reform, small and medium-sized enterprise (SME) development and technology strategies.

Trade Reform[5]

There were major changes in trade policy during the Soeharto era. As noted, the political turbulence of 1965–66 initially ushered in a radical shift towards economic orthodoxy, including an open commercial policy. This period of liberalism was short-lived, however. By the early 1970s, there was a nationalist resurgence, prompted in 1973–74 by the sudden increase in international oil prices, and vigorous anti-Japanese protests. Tariffs were increased, but more importantly the government embarked on an ambitious programme of heavy industrialization underpinned by increased resort to non-tariff barriers (NTBs). Such a strategy persisted for about a decade. It was only in the early 1980s, and more decisively in the mid-1980s as oil prices continued to fall, that key economic policy makers were able to arrest the trend, and then embark on a series of major trade reforms.6 Since then, progress has waned, but at least there has been no significant regress. These trade liberalizations were part of a general and coherent package of reforms, including two major currency depreciations; prudent fiscal policy; comprehensive tax reform; and sweeping (and in hindsight mismanaged) financial deregulation (see Hill, 2000, chapter 6). Of course, the protection estimates do not take account

Table 9.7 Indonesian Trade Policy, 1966–95: Major Contending Groups and Policy Outcomes

Period	1) Technocrats: policy stance	Role in policy making	2) Economic nationalists: policy stance	Role in policy making	3) Rent seekers[a)] stance policy	Role in policy making	4) Foreign institutions: policy stance	Role in policy making	5) Trade policy outcomes
1966–72	Pro-market	Strong	Protectionist	Moderate	Protectionist	Strong	Pro-market	Strong	Relatively liberal
1973–82	Adopted ISI, although still relatively pro-market	Declining	Protectionist	Strong	Protectionist	Strong	Pro-market	Declining	Increasing protection
1982–85	Began to support economic liberalization	Increasing	Protectionist	Declining, although still quite strong	Protectionist	Strong	Pro-market	Moderate	Ambivalence
1985–90	Pro-market and support for trade reform	Strong	Protectionist	Declining	Tended to be in favour of protectionist policy	Strong	Pro-market	Increasing	Relatively liberal

1990–95	Pro-market and support for trade reform	Declining	Protectionist	Increasing	Tended to be in favour of protectionist policy	Strong	Pro-market	Relatively strong [b)]	Relatively liberal

Notes:
ISI: import-substitution strategy.
a) Scattered around military, Soeharto and bureaucracy in the 1970s, and concentrated around Soeharto from the 1980s.
b) Increasing due to the important role of trade agreements such as APEC and AFTA.

Source: Basri and Hill (2004).

of much crony-related and generally firm-specific protection which began to proliferate from 1990 (Schwarz, 1994).

Given the country's industrial policy history, it may appear surprising that such major reforms were introduced, and have broadly endured since then. The key is to identify the major actors in economic policy, and to understand the absolutely dominant role of Soeharto as the ultimate arbiter of economic policy during this period. Table 9.7 provides a summary and highly simplified picture of four main groups with influence over economic policy, their broad policy predisposition, and their influence at particular points of time.

The drift towards increased protection in the 1970s occurred because both the technocrats and foreign influences on policy were on the wane, at least in the realm of microeconomic policy. They were less needed during these 'good times', and there was less imperative to follow their policy orthodoxy. Moreover, neither group was completely united on core trade policy issues. Economic nationalists were becoming increasingly powerful over this period, and were able to build opportunistic alliances with various rent seekers.

By contrast, in the mid-1980s, the opposition to trade liberalization was on the wane. The technocrats were united and stronger in their resolve to reform and, at that time of looming crisis, they had the ear of Soeharto. Foreign influences were becoming more clearly pro-reform, and they had more weight. Indonesia needed funding from the international financial organizations to help it adjust to lower oil prices. Foreigner investors were becoming more interested in the country as a low-cost export platform or as part of internationally integrated manufacturing operations, rather than as a relatively small and protected domestic market. Neighbouring countries, most especially China, were liberalizing and growing rapidly, in the process constituting a powerful competitive spur for reform. Finally, the idea of coordinated, open, region-wide liberalization, in the form of both the Asia-Pacific Economic Conference (APEC) and AFTA, was beginning to take root in elite government and business circles.

There was of course opposition to trade policy reforms, but it could be swept aside or bought off on a case-by-case basis. An example of the former was the Ministry of Trade and Industry which, accustomed to the traditional approach of regulations and licensing, became increasingly irrelevant in the new debate about growth and efficiency. Similarly, the large SOE sector was heavily government dependent, and therefore had little scope for autonomous action. Finally, the business empires of the Soeharto children were still in their infancy, and did not yet have the reach which would a decade later begin to stymie the reform process.

As we shall see shortly, this framework in modified form can also shed useful light on the dynamics of trade reform in the post-Soeharto era.

SMEs and Industry Policy

The promotion of SMEs has been an article of faith for successive Indonesian governments. These firms are seen as an important counter-weight to the dominant foreign and non-*pribumi* corporate sector, and as a means of distributing the benefits of growth. Since the crisis, there has also been a perception that these firms have been more resilient in the face of adversity.

A bewildering variety of programmes have been introduced to assist SMEs. However, while there are some success stories, most studies conclude that it has been the government's general economic policies which have facilitated SME growth rather than micro interventions. We draw here on a variety of case studies, together with a more general literature on SME dynamics before and after the crisis.[7]

The Bali garment industry, which grew spectacularly in the 1980s and is almost exclusively based on small firms, was practically an 'accidental' case of industrialization (Cole, 1998). Foreign tourists, mainly surfers wishing to support a recreational life-style, saw commercial opportunities in Balinese garments and its indigenous design capacity. They were able to act as marketing intermediaries, connecting local producers with retail outlets abroad, in the process dispensing important information on designs and production techniques. Later, as the island's fame spread, these links developed quickly, and the industry mushroomed from its seasonal, cottage origins to larger production units and some local design capacity. It was also a magnet for creative designers and entrepreneurs from other parts of Indonesia, who sought to emulate Bali's successes in their own regions. By contrast, export-oriented SME furniture manufacturers in Jepara, in northern Central Java, had their origins further back, but here too exports began to grow quickly in the 1980s. The industry lacked the tourism connection, but it did have a good local skills base together with access to raw materials, and foreigners quickly saw the opportunities for profitable export as deregulation proceeded.

These studies suggest a model of successful and innovative SME development in which the following ingredients appear to be important. First, there was some basic industrial competence in a particular field of activity (e.g., sewing and wood-working in these cases). Second, there was generally a conducive macroeconomic environment (except at the peak of the 1997–98 crisis), including especially a competitive exchange rate. Third, there was reasonably good physical infrastructure, including proximity to adequately functioning import and export facilities. Finally, of importance also have been injections of technical, design and marketing expertise which link small producers to new ideas and major markets.

With the partial exception of the first ingredient, all four elements are directly amenable to public policy These case studies also have important implications for government policy. Neither resulted from any deliberate government promotional measures. The government did play an important role in providing a supportive macroeconomic environment and in the provision of a rapidly improving infrastructure. In Bali, the local government generally adopted a fairly open policy towards the presence of foreign entrepreneurs, and export procedures were not unduly burdensome most of the time. These of course hardly constitute 'contributions' from government, except in the negative sense of avoiding a harshly restrictive regulatory regime.[8]

It is possible to extend these conclusions to industry policy more generally. That is, it is difficult to discern instances where micro-level interventions in Indonesia have been effective. Studies which have examined the relationship between inter-industry variations in government assistance and subsequent (lagged) performance, according to a variety of measures, have found little evidence of causality. For example, an analysis of selective policy instruments introduced or extended in the 1970s and early 1980s – protection, credit subsidies, state enterprises – detected very little evidence of such a strategy having 'worked' according to a range of subsequent performance criteria (Hill, 1996). Yet Indonesia achieved rapid economic growth for the three decades up to 1997. The conclusion is therefore inescapable that it was the general policy environment (macroeconomic management, openness, investments in infrastructure and people, and a broadly predictable business climate) which explained this rapid growth.

This conclusion is moreover confirmed by the nature of the various industry-specific promotional interventions. Invariably, these have been prone to abuse, implementation has been sporadic and often short-lived, and there has been little systematic attempt to prescribe conditionality, in the sense of linking incentives to tightly defined performance criteria. Protection has generally been motivated by rent seeking rather than efficiency enhancement, that is, some notion of infant industry development (Basri, 2001). The nature and instruments of patronage have changed during Indonesia's rapid transition to democracy, but the underlying principles have not.

Technology Policy

Like most countries, Indonesia has never had a formal 'technology policy'. Rather, there has been a range of interventions with implications for the development of technological capabilities. Some have been intended, while others have been quite incidental. It is possible to identify some quite effective programmes, alongside those which have been costly. In general, our argument is that most pre-crisis interventions were misguided. In the post-crisis environment, the major programmes have been discontinued or so starved of resources as to be ineffectual. It is unlikely that the government will accord a high priority to technology programmes in at least the medium-term future.

In the 20 years to 1997, Indonesia's Research and Technology Minister, B.J. Habibie completely dominated Indonesian technology policy.[9] Indonesia had ambitious plans for technological development over this period. For example, in its Second Long-Term Development Plan, 1994–2019, drawn up well before the crisis, expenditure on R&D was projected to rise sharply, from around 0.2 per cent of GDP to 2 per cent. Government expenditure on R&D pre-crisis was quite modest, about $400 million, but was dwarfed by the huge investments in the nation's 'showcase' aircraft factory, IPTN, which totalled approximately $3 billion. Reflecting in part the background of the key personalities involved, programme objectives also emphasized 'technological self-sufficiency', through a four-stage evolution from basic manufacturing capacity to the mastery of leading-edge technologies as embodied in advanced manufacturing.[10]

According to all the standard indicators, Indonesia is very much a technological latecomer. It is only since the late 1960s that it has been open to international technology markets and has had a government commitment to the development of any sort of scientific base. The government has been the dominant actor in formal R&D activity, both as a founder and as an 'implementer'. It has provided about 80 per cent of the resources, and directly carries out over 60 per cent of the activities. International comparisons suggest that the Indonesian government's role is unusually high. Other indicators also underline Indonesia's latecomer status. Owing to its size, it has the largest stock of scientific personnel in ASEAN, but when adjusted for quality of training, hardware support and active scientific activity, its R&D effort is again much closer to that of the Philippines and Thailand. Its manufactured exports rank low in terms of their technology intensity.[11]

Over the 1967–97 period, the government's major contributions to technological progress were clearly in providing a reasonably open and enabling commercial environment. As we have argued elsewhere, it achieved macroeconomic stability, it opened the economy to foreign trade and investment, it connected Indonesia to the global economy through large infrastructure investments, and it promoted near-universal education at primary and junior secondary levels. However, with the partial exception of some programmes in agricultural extension, scholarships for advanced training abroad, and activities within the Indonesian Institute of Sciences (LIPI), it would be difficult to sustain an argument that the government's microeconomic interventions in this field were effective. The show-case high-tech aircraft manufacturing project collapsed as soon as financial support was discontinued in the wake of the economic crisis.[12] A similar fate has beset the other Habibie-sponsored activities.

The problem is that many of these interventions were based on a fundamental misunderstanding of the process of technological innovation. They constituted a case of 'back-to-front' and 'big-bang' industrialization. Instead of developing the supply-side competence of industrial skills through focused training and extension programmes, the government promoted the most complex final stage processes. These projects lacked the necessary support base of SME suppliers and engineering skills. Moreover, they were undertaken against a general backdrop of bureaucratic weakness. Programme implementation was always a major challenge. Programmes often lacked coordination across ministries. Corruption has been a continuing and serious problem. The state enterprises which were nominally tasked to support these projects have had a poor financial record, and achieved few of their supposed 'developmental objectives'. Meanwhile, education at all levels (including, of particular relevance to this technological mission, technical and higher education) and public R&D institutes have been starved of resources. Even in the case of the high-level scientific training abroad, the skills acquired have tended to be highly specialized and of limited use outside the aircraft industry.

5. CONTEMPORARY ISSUES AND CHALLENGES

An Overview

Post-crisis, Indonesia faces two key, interrelated challenges. The immediate one is to restore economic growth to the levels it enjoyed up to 1997. In the manufacturing sector, this means especially the promotion of efficient, export-oriented, labour-intensive industries. It also entails the development of institutional arrangements which assure investors that the commercial environment is reasonably predictable and stable, in addition to immediate issues associated with debt and financial restructuring.

The process of recovery and reform is of course particularly complex for a country experiencing deep economic and financial crises. In this respect, Indonesia currently resembles the Philippines in the late 1980s, Russia in the mid-1990s, and Argentina in the early 21st century. Indonesia's recovery has been complicated by the fact that the authorities are grappling with complex technical issues against a backdrop of political uncertainty and institutional decay (Thee, 2003). Alongside these challenges is the importance of developing a longer-term policy framework which encourages industrial growth and innovation. As the literature on East Asia emphasizes, although there is no single path to rapid industrialization, a number of 'core' factors have been present:[13]

1. *Openness and incentives*: This is important at a general level since more open economies grow more quickly. It is also central to industrial innovation among borrowers and latecomers since almost all the world's R&D is undertaken in OECD economies. There are, of course, myriad transmission mechanisms. FDI is probably still the major channel of technology flows. Technology is also embodied in the trade of goods, and in the cross-border movement of non-factor services. Labour migration and associated flows of human capital are increasingly significant.
2. *Human capital*: This is a critical factor at all levels of development, in creating the ability to absorb, assimilate and diffuse imported technology, and to innovate on the basis of these imports. At early stages of development, mass literacy and universal primary education deserve priority, while upgrading strategies at middle incomes require more attention to technical and higher education, to quality, and to the composition of educational output among disciplines.
3. *Infrastructure and institutions*: This encompasses a diverse set of factors. An efficient business operation able to connect to the global economy requires high-quality telecommunications and physical infrastructure. There needs to be a commercial climate which is conducive to long-term planning horizons and investment decisions. Property rights and an adequately functioning legal system are also important. Financial institutions, domestic and foreign, need to be present. Macroeconomic stability is crucial.
4. *Commercial environment*: Competition is crucial to innovation, as a spur to raise productivity. Monopolies invariably prefer a 'quiet life', and should therefore be a cause for concern, unless they are 'Schumpeterian' in nature, or operate in an open international

environment. Whether industrial structure more generally (e.g., the level of foreign or state ownership, the size distribution of firms) matters is a more complex question with no clear answers.

5. *R&D institutions*: These constitute the formal component of most governments' technology and innovation policy. Important questions here concern the scale of these institutions, their funding, and their interaction with enterprises and the education system. For a latecomer country like Indonesia, they are of less immediate importance, especially during the crisis-recovery period.

Our discussion in this section is informed by this analytical framework. In particular, we focus on openness, incentives and infrastructure. As argued above, the establishment of R&D institutes is a longer-term policy priority, and the development of human capital is not a short-term policy lever. Table 9.8 provides a summary of major policy initiatives pre- and post-crisis.

Openness and Incentives

We discuss here various elements related to openness and the incentives regime, including the exchange rate, protection, some partial trade reforms (principally a range of export schemes), labour market policies and the foreign investment climate. The Indonesian economy has remained broadly open in the wake of the crisis, and this has been a factor facilitating the partial recovery. However, the picture is a complex and varied one.

The exchange rate

Indonesia experienced a major boost to its competitiveness as its exchange rate collapsed in 1997–98. Its nominal exchange rate depreciated far more than any of the other crisis economies during this period. The result was some improvement in its export performance, though for various reasons the increase in competitiveness was not as great as might have been expected. First, it was the only crisis economy to experience serious inflation, which reached about 80 per cent on an annualized basis during the first half of 1998. Consequently, as Figure 9.1 demonstrates, as compared to its lower-income ASEAN neighbours, most of the initial boost to competitiveness was eroded by the end of 1998. That is, the real exchange rate appreciated quickly in the second half of that year. Moreover, even though macro stability was restored by mid-1998, inflation has been consistently higher than the country's major trading partners, resulting in a gradual appreciation for most years.

Table 9.8 Shifts in Investment and Trade Policy in Indonesia since the 1970s

Period	Growth performance	Policy regime	Selected policies
1971–82	Oil boom-led high economic growth (averaged 7.5% p.a.)	Import substitution	High import tariffs and numerous NTBs Substantial public investments in infrastructure and agriculture development Increasing restrictions on FDI
1982–86	Slow growth period (about 4% p.a.)	Increasing interventions	Proliferation of import controls and investment restrictions
1986–97	High growth period (7.3% p.a.)	Trade and investment reforms	See below
	1987–91 – export led	– major reforms	Major reforms included: substantial cuts in import tariffs and elimination of many NTBs opening up sectors to foreign investment; relaxation of FDI equity restrictions for exporters
	1992–97 – increasing diversification	– reform fatigue	Slowdown in the pace of reforms. But: in 1995 a major tariff reform package setting out a schedule for further tariff reductions major relaxation of equity and divestment restrictions on FDI outside the export sector

Period	Growth performance	Policy regime	Selected policies
1998–2003	Economic crisis and recession in 1998 (–13%) Recovery period (1% in 1999; 4% from 2000–03)	GOI/IMF crisis-induced reform programme, 1998–2003	Major reforms covering: reduction in tariffs especially on agricultural commodities elimination of Bulog's monopoly over agricultural commodities, except in rice elimination of government-sanctioned cartels and marketing restrictions deregulation of the domestic retail and wholesale sectors enactment of several economic laws covering competition, bankruptcy bank and financial sector rehabilitation and reforms corporate debt restructuring initiatives Recent emergence of some protection from 2002 on: increase in import licensing across sectors NTBs on sugar new manpower law increases 'restrictiveness' of labour regulations related to dismissals, severance payments and restricts temporary work and outsourcing of jobs

A second factor has been uncertainty about future exchange rate movements, owing principally to the volatility introduced by political turbulence. Although the floating rate regime has worked quite well – arguably better than most had expected – capital flight during periods of political turmoil has resulted in considerable fluctuation. Hence, economic agents will tend to discount any short-term exchange rate advantages. There have also been supply-side constraints limiting the export response to these favourable price movements. At the peak of the crisis, the sudden import compression resulted in a shortage of international transport capacity. There is typically a lag in the production of agricultural crops. Transport and financial bottlenecks have also been present.

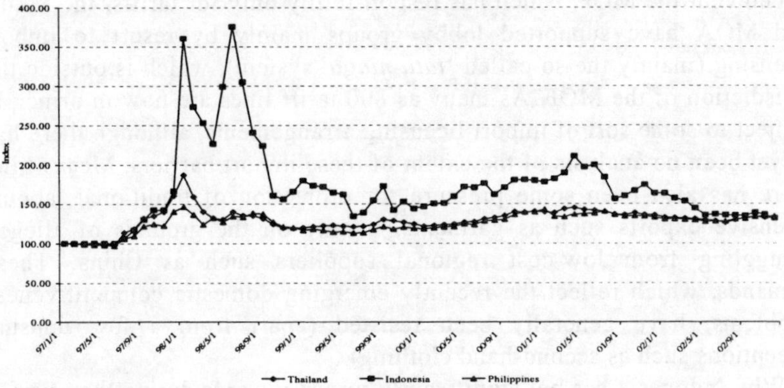

Figure 9.1 Real Effective Exchange Rates

On balance, exchange rate movements have been a positive factor in Indonesia's recovery, and in underpinning its export growth. The real rate is below that of the pre-crisis period and also that of its major non-crisis competitors such as China and Vietnam.

Protection

Here the picture is mixed. There has been no concerted attempt to turn inward in the wake of the crisis, and Indonesia therefore remains a fairly open economy. There have been no major studies of the trade regime since Fane and Condon (1996), but their broad conclusions reported above remain generally valid. There was some liberalization as part of the International Monetary Fund (IMF) packages of 1997–98.[14] With weakened central government authority, smuggling has become more extensive, in effect rendering redundant much of the very high protection.[15]

Nevertheless, there has been persistent pressure for protection as the immediate impacts of the crisis passed, much of it emanating from the ministries of Industry and Trade (MOIT), and Agriculture (MOA). This pressure has often come from the pre-crisis beneficiaries of protection, whose industries were deregulated in 1998 as part of the IMF rescue package. Examples include the flour miller Bogasari, heavy truck manufacturing, and commodities like soybean, cloves and cocoa. There have been some tariff increases, including for rice, sugar and the flour miller Bogasari,[16] but these have been limited. The principal resistance to increased tariffs has been the Ministry of Finance (MOF), especially during the tenure of Dr Boediono in the Megawati administration. To circumvent the MOF, which has responsibility only for tariffs, the MOIT and MOA have supported lobby groups mainly by resort to import licensing (mainly the so-called '*tata niaga*' system), which is outside the jurisdiction of the MOF. As many as 800 tariff lines are now in principle subject to some sort of import licensing arrangements, although there has as yet been no analysis of the extent of these import barriers. Meanwhile, there has also been some pressure for protection of traditional labour-intensive exports such as garments, mainly on the grounds of alleged smuggling from low-cost regional suppliers such as China. These demands, which reflect the recently emerging domestic competitiveness problems, have generally been resisted (apart from a few unusual exceptions such as second-hand clothing).

Why Indonesia has not significantly turned inwards during the crisis is an important issue, but one which is tangential to this chapter. Briefly, the key point is that several factors have constrained protectionist pressures since 1997 (Basri and Hill, 2004). First, as noted there has been a very large decline in the real exchange rate (RER), principally as a result of the collapse in the nominal exchange rate in 1997-98. Thus tradable goods industries have received very large 'exchange rate protection'. Second, the international financial institutions, particularly the IMF, have become much more powerful, and have been able to impose a range of conditionality measures on a weakened Indonesian government. Third, as the domestic market has shrunk, while competitiveness has been enhanced, Indonesia's main attraction to foreign investors is its export-oriented commodity and manufacturing sectors. Thus, there is little protectionist pressure emanating from foreign investors. Fourth, in an era of fiscal austerity, economic nationalists, heavily reliant on government subsidies, have little scope to promote their favoured projects, even though some of these initiatives still command significant community support. Finally, and more speculatively, in the wake of democratization and decentralization, it is possible that the increased political clout of the more export-oriented regions off-Java may constitute a check on any protectionist pressures emanating from Jakarta.

Nevertheless, there is no strong *institutionalized* advocate of openness in Indonesia's current political constellations. Reforms introduced simply to placate the IMF may lack durability, and may be unpopular precisely because they were 'dictated from abroad'. Moreover, suddenly more powerful parliamentarians require campaign funds, and thus there is a greatly increased danger of opportunistic pressures for protectionism. In addition, these vested interests are also able to play on the increased community mistrust of globalization, in the wake of the financial crisis and the perceived heavy-handed foreign intrusion in domestic policy affairs. This appears to be the case over the period 2002–04. The government, and especially the Minister for Trade and Industry in the Megawati administration, appears to have been captured by groups lobbying for protection, barriers to entry or export controls in sugar, rice, textiles, soybeans, cocoa, rubber, leather products, fertilizers, steel, and modern-sector retailing. As with the Soeharto regime, the division of administrative responsibilities inhibits the process of trade liberalization. That is, the more outward-oriented Department of Finance controls tariffs, while the more protectionist-inclined Departments of Trade and Industry, and Agriculture, retain the power to set NTBs. Moreover, further complicating the picture, the rapid shift towards regional fiscal autonomy has resulted in a proliferation of subnational (and illegal) trade barriers, mainly in the form of informal exactions on goods crossing provincial boundaries.

Export support schemes

In addition to remaining broadly open, exporters have also been somewhat insulated through a variety of export support schemes. Indonesia has never adopted export processing zones on the scale evident in most of its ASEAN neighbours, but various schemes enable exporters to secure access to imported inputs on a duty free basis, and to minimize the blatant corruption and administrative complexities associated with the customs administration.

Three types of export programmes operate. One is bonded factories (known in Indonesia as *Kawasan Berikat* (KB), which operate as a physically separate customs zone.[17] Another is a duty rebate system known as '*Bintek*', under which exporters are reimbursed for duty paid on imported inputs which are subsequently exported. This scheme is not widely used since there are long delays (12 months or more) for reimbursement and 'facilitation payments' are generally required. A final option is the special case of the Singapore-centred Batam/Bintan zone. (There are in addition some smaller zones in other parts of the country.) The KBs and Batam have been the major source of Indonesia's manufactured exports since the early 1990s. For example, exports from these zones in 1992 totalled about $1.3 billion, compared to non-oil manufactured exports of $12.2 billion. By 2000, the former had risen to $15.7 billion out of a total of $30.2 billion. That is, the zones accounted for 80 per cent of the increment to non-oil manufactured exports over this period. Batam alone accounts for almost 30 per cent of the zones' exports. It has also been a growing source of employment, especially of young females. The numbers increased from 32,000 in 1992 to 136,000 in 1997. Moreover, during the crisis, when aggregate modern sector employment declined slightly, job growth in Batam remained strong, with the total rising to 156,000 in 2000.

These zones therefore figure very prominently in Indonesia's global engagement, and they generally offer firms, both foreign and domestic, a more congenial commercial environment in which to operate. They are of course suboptimal, second-best policy instruments. A first-best solution would be to simplify administrative and customs procedures, and enable all firms to operate on a free-trade footing. There are also concerns about the possible erosion of the government's fiscal base, especially in Batam where both producers and consumers are exempt from paying customs duties, value-added tax (VAT) and excise taxes on most goods and services consumed locally. (Companies and individuals are subject to income tax, however.) The schemes also have the effect of inhibiting the development of backward linkages from exporters to local suppliers, notwithstanding government statements to the contrary emphasizing their importance. This is so because bonded factories are only permitted to sell imported inputs to other such factories. There are also complications introduced by the decentralization programme, with recently established local governments intent on establishing zones in an ad hoc, uncoordinated manner.[18]

Labour regulations

There has been a rapid change in Indonesia's labour market policies in the wake of the crisis. During the Soeharto era, labour market outcomes more or less accorded with 'East Asian norms' (Manning, 1998). Rapid economic growth generated rising real wages, with a lag. Much of the increase was associated with the transfer of workers from the informal and agricultural sectors to formal, non-agricultural employment. Trade unions existed, but were heavily managed. Minimum wages were prescribed but they were generally below market levels in the formal sector, and were not enforced systematically.

With the crisis and the collapse of the Soeharto regime, there were two major labour market effects. The first and immediate one was that, given the relatively unregulated nature of the market, real wages fell sharply in 1998, by more than in any other crisis-affected economy. This meant that the major labour market impacts were on price (i.e., real wages) rather than quantity (i.e., unemployment which, in a formal sense, was little affected). Such an outcome was an important factor in ameliorating the adverse social consequences of the crisis (Manning, 2000).

As the worst of the economic contraction passed, however, and as a shaky and populist democracy evolved, powerful pro-labour pressures emerged. The constraints on trade unions were largely removed, and competing labour interests began to vie for worker support. Under successive ministers of manpower, the government strongly supported worker entitlements and wage claims. Minimum wages, responsibility for which was handed over to newly established second-level (sub-provincial) governments, began to increase rapidly. Several new regulations also resulted in significantly increased labour market rigidities. In this respect, Indonesia differs from, for example, Korea, where labour market controls and democratization were introduced during a period of economic prosperity and growth. Rather, Indonesia currently resembles the post-Marcos Philippines, where a somewhat similar set of events has bequeathed an indifferent record of formal sector employment since the late 1980s.

Figure 9.2 documents these trends in several real-wage series over the 1996–2002 period, that is, just prior to the crisis, through the crisis, and the sluggish recovery period. Real wages fell sharply from late 1997 for about a year. The magnitudes vary across the series, from about 30 to 80 per cent, with the smallest decline in the mandated minimum wages series. From 1999, as the process of economic contraction reversed, and labour market populism commenced, real wages begin to increase quickly. The regulated minimum wage series increased by over 90 per cent in the three years March 1999–2002, to the point where these real wages in dollar terms now exceeded pre-crisis levels. Wages in the manufacturing sector have followed, especially among larger firms, with a lag. It is notable that wages in the unregulated domestic service and agricultural sector increased very little over this period. The latter series constitute a more accurate indication of supply and demand conditions for unskilled workers. From 2002, the pressure for wage increases has abated,

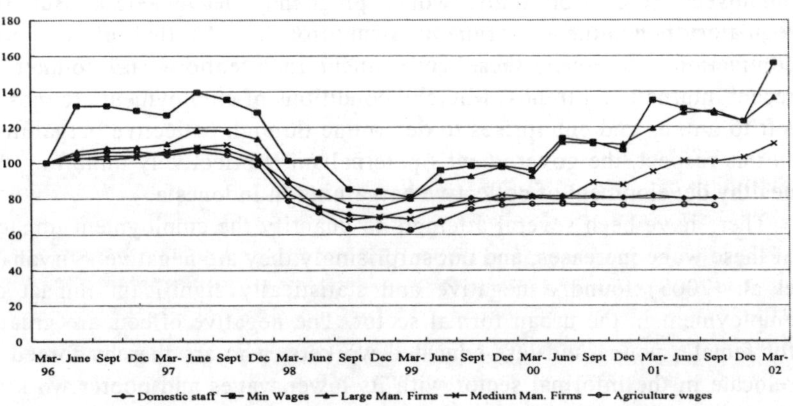

Figure 9.2 Trends in Real Wages (indexed to 100 in 1996)

especially in the wake of the October Bali bombings. However, there has been no decline in mandated minimum wages. In fact, the simple average of real minimum wages across all regions rose by about 8 per cent in 2003 (a 13 per cent nominal increase in the context of an underlying inflation rate of 5 per cent).

These real-wage increases have also greatly increased business uncertainty, and encouraged workers to expect continuing increases. In addition, the regulatory environment has introduced rigidities into hiring processes which discourage firms from taking on additional labour. For example, severance pay entitlements have also been increased, and are now among the highest in East Asia. There has been pressure to convert contract workers into permanent employees, which poses particular problems for industries that have traditionally employed workers on a contract basis, where work is seasonal or fluctuates considerably. This includes several major labour-intensive industries, such as garments, toys and electronics. The mandatory conversion of workers into permanent status will mean that employers incur high compliance costs of hiring and dismissing workers, which will in turn adversely affect employment growth.

For some years, employers were required to pay striking workers until both parties entered mediation or unless workers do not follow certain procedures. This provision has recently been watered down. Multiple unions vying for worker membership has been a problem in some cases, although new laws have clarified union representation in the collective bargaining process. It should be emphasized, of course, that workers deserve protection from abuse, are entitled to safe and sanitary work conditions, and that there needs to be special provisions for female employees (e.g., for night work, pregnancy leave, etc.). But the regulatory pendulum has arguably swung too far, at a time of economic contraction. Moreover, these government interventions run counter to recent international trends, whereby conditions of employment are mostly left to unions and enterprises to determine through collective bargaining. In this respect, the government's paternalism is effectively inhibiting the healthy development of collective bargaining in Indonesia.

There have been several attempts to quantify the employment impacts of these wage increases, and not surprisingly they are negative. Suryahadi et al. (2003) found a negative and statistically significant impact on employment in the urban formal sector. The negative effects are greater for female, young and less-educated workers, who are thereby forced to relocate in the informal sector with its lower wages and poorer working conditions. By contrast, employment prospects for white-collar workers are enhanced, resulting in increased wage inequality.

Bird and Manning (2002) examined the impact of minimum wage increases on the relocation of jobs between the formal and informal sectors. They found that the effects had been to expand employment, and to depress the earnings of some workers, in the informal sector. The main effect occurred through labour displacement and slower employment growth in the formal sector. The negative effects are observed after the crisis of 1997–98, when real minimum wages were pushed up quickly, compliance was tightened, and labour was suddenly free to organize and enforce its rights, all in the context of low economic growth. Consistent with Suryahadi et al. (2003), Bird and Manning also found that the effects were most pronounced for the vulnerable groups of workers – particularly females and younger workers – the alleged beneficiaries of the policy changes.

Foreign investment

Indonesia therefore presents a rather unattractive location for footloose manufactured exports, and FDI more generally. The competitive boost from the sharp exchange rate depreciation has been quickly eroded by subsequent nominal appreciations, inflation rates are higher than major trading partners, and until recently there has been rapid real-wage growth. In addition, there is increased policy and political uncertainty, and the complicated process of corporate recovery.

As noted, Indonesia was a very large recipient of FDI through to 1997. For the first half of the Soeharto era, most of this FDI went into the resource sector and import-substituting manufacturing. From the early 1980s, following the decline in commodity prices and the major reforms, export-oriented manufacturing and services predominated. Since 1997, however, the picture has been a gloomy one (Bird, 2004). There have been persistently negative FDI inflows, unlike the other crisis-affected economies, where FDI either held up or in some cases actually increased. Several major investing countries, including Japan[19] and Korea, have reported negative sentiment towards Indonesia from their business communities. There have also been plenty of well-documented 'horror stories' in the press. These have included Cemex's (the huge Mexican cement company) difficulties acquiring Semen Gresik; the protracted saga of the Canadian insurance company Manulife; Standard Chartered Bank's aborted takeover of Bank Bali; footwear exporters moving to Vietnam; Sony electronics moving to Malaysia; and much else.

These trends are not surprising, for reasons documented elsewhere in this chapter. The extent to which FDI can play a role in the recovery process depends substantially on how congenial is the domestic commercial environment. It is not uncommon for FDI to increase during a crisis, while short-term capital flees the country. This in essence is the Krugman 'fire-sale FDI' phenomenon, of foreigners purchasing distressed assets cheapened by financial and exchange rate collapses, and facilitated by more crisis-driven liberalized entry. However, this prognostication only works if the commercial rules of the game remain broadly predictable, as they did in the other crisis economies. In the post-Soeharto era, the abrupt changes in the political and institutional rules of the game, and much diminished regime credibility in economic management, continue to deter most investors (Thee, 2003).

It may be that the worst is over for Indonesia in this respect. Foreigner investors now appear to have a better understanding of the new political economy realities. There is a clearer working relationship between the presidency, the legislature and the bureaucracy. Capital appears to be returning. The country's stock exchange has been performing well since late 2002, even in the wake of the serious Marriott Hotel bombing. While much of this capital is short term and speculative, at least some finds its way into longer-term investments. Several case studies (e.g., Lindblad, 2003) have shown that MNEs with deep pockets and longer time horizons have availed themselves of opportunities to buy cheapened assets in Indonesia. Debt–equity swaps have also taken place. There have also been some (necessarily) quiet success stories. For example, the two major cement producers, Indocement and Semen Cibinong, are now foreign controlled. Unilever has bought several local brands, though not the factories, in the consumer market (e.g., the Sari Wangi tea bag). The modern retail sector has been opened to FDI, with the French firm Carrefour opening a dozen outlets since late 1998.

There are no quick fixes in attracting foreign capital back. Thus far, most of the capital flows appear to have been short term in nature. In all likelihood, much of it is probably domestic capital which fled in 1997 and was 'parked' abroad. It could leave again with renewed political turbulence, for FDI involves a long-term commercial calculus. Superficial regulatory reform (e.g., of procedures within the investment board, BKPM) is unlikely to have much impact. The mooted return of fiscal incentives, in the absence of far-reaching and durable reform, could be costly and counter-productive. As Indonesia illustrated with its 1984 tax reform, the abolition of these incentives in the context of general tax and other growth-oriented reforms, is actually more attractive to foreign investors and more cost-effective for governments.

Institutions and Infrastructure

During the Soeharto era, there was a massive expansion in all forms of physical infrastructure throughout the archipelago: roads, ports, airports, irrigation and telecommunications. Institutions – the civil service and the legal system – remained quite weak (with pockets of excellence in the former), but from a commercial perspective nonetheless performed adequately. That is, both the civil service and the legal system were highly corrupt, but more or less 'predictable'. There was very limited recourse to the commercial courts, for example, with firms preferring to enlist the support of powerful backers, drawn mainly from the senior echelons of military through to around the mid-1980s, and from among the Soeharto family in its last decade of power. This was, in McLeod's (2000) terminology, a 'better class of corruption'. Businesses paid a 'tax surcharge', in exchange for protection and an accommodating commercial environment, in the context of rapid growth.[20]

In various respects, Indonesia's commercial environment is now much less attractive. In addition to labour issues discussed above, we briefly illustrate this proposition with reference to physical infrastructure and the legal system.

Public investment in physical infrastructure has declined since the crisis. Moreover, some of the responsibilities have been handed over to the regional governments since 2001, with ill-defined coordination arrangements and divisions of authority. The country can for a period 'live off' its large investments built up over 30 years. But seven years after the onset of the crisis, there are indications that serious supply-side constraints are emerging, especially in power generation, much as in the Philippines in the early 1990s when growth resumed there.

Foreign investors might be able to play a role in the infrastructure sector, but the government appears reluctant to establish a conducive investment climate for these long-term projects. Reform of the complex electricity tariff is one issue. Another is a consequence of the breakdown in relations between the government and power sector suppliers during the crisis. This resulted in a $200 million settlement being paid to the US EXIM Bank, and a subsequent reluctance on the part of the authorities to provide any guarantees to investments in this sector.

Indonesian ports also harm exporter competitiveness. Indonesia's average berth productivity is low relative to its neighbours.[21] For example, at the country's largest (and probably most efficient) port, Jakarta's Tanjung Priok, the number of containers lifted per hour, the usual measure of port productivity, is about 35. This compares to 75–80 for the better-performing ports in Korea, Singapore, Taiwan and Thailand. Nor is this lower physical productivity compensated by lower charges. In fact, Tanjung Priok's lift costs (per container) are the highest among the eight major Asian ports for which Ray (2003) has assembled data. In addition,

Indonesia's problems are compounded by the lack of competition among port operators, which until recently have been state owned and tightly regulated.

With regard to the legal system, foreign investors and creditors have traditionally had little faith in resolving disputes through formal legal mechanisms in Indonesia.[22] For example, since the crisis, domestic parties in dispute with their foreign partners or creditors were able to use the legal system to thwart the latter's contractual claims. In some cases, this resulted in the temporary freezing of foreigners' assets in Indonesia, including even the temporary imprisonment of the local representative of a foreign company. This was, for example, the experience of Tempo Pharmaceutical (a joint venture between a well-connected Soeharto-era business and Roche) and the protracted Manulife insurance case.

In many developing countries, foreign investors typically prefer to resolve disputes with their local partner or the host government through domestic or international arbitration, as this process is regarded as fairer, quicker, and less vulnerable to corruption. As a result, many investment agreements in Indonesia and elsewhere include provisions for some sort of arbitration. However, since the crisis, Indonesian partners have often refused to go to arbitration, and rather used local courts to over-rule the arbitration provisions (as in the Tempo Pharmaceuticals example). There are also cases of local partners persuading the courts to over-rule the arbitration award in the belief that they will receive more favourable treatment from local courts and judges. This is so in spite of the general assessment among legal experts that Indonesia's Arbitration Law of 1999 is a sound one, and explicitly upholds arbitration provisions and awards. More recently, there have been some reforms which have improved the operation of the system. One example has been the introduction of ad hoc judges to the Supreme Court.

However, legal reforms remain slow. Part of the problem is that judges are 'career judges', whereas in most modern legal jurisdictions – especially those which use the common law, such as the UK – judges are appointed from the legal profession, based on experience and reputation, and are adequately remunerated. Meanwhile, the commercial courts, which were initially regarded as an opportunity to overcome corruption and incompetence in bankruptcy cases, have proved to be disappointing. Very few international investors resort to these courts (Schroeder-van Waes and Sidharta, 2004).

In addition to general problems in the legal system, exporters in resource-based industries, in which Indonesia has a significant potential comparative advantage, are constrained by leasing arrangements on government-owned land. Leases must be renewed every 30 years, which is relatively short for modern agricultural activities requiring large investments. Lessees therefore have little incentive to invest in land-care. The renewal system is also widely abused by the relevant regulatory agencies.

6. CONCLUSION

Our major arguments may be summarized as follows. First, Indonesia differs significantly from its higher-income East Asian neighbours, and thus a different analytical and policy framework is required to understand its industrialization record and policy challenges. In particular, it is still some way off the 'turning point' where it loses comparative advantage in labour-intensive activities, and has to manage the process of upgrading to more skill and capital-intensive industries. Thus, the concern with notions of a 'high-tech drive' evident in other chapters of this volume is a much lower policy priority.

Second, the economic crisis of 1997–98 has introduced a major discontinuity in Indonesia's development trajectory. The country has in effect 'lost' almost a decade of growth, and only returned to pre-crisis per capita income levels in 2004. And, while broad-based economic recovery appears to be evident, growth is 2–4 percentage points lower than in the past. Moreover, several features of the new institutional and policy environment may continue to retard growth. Examples referred to in this chapter include activist labour market interventions, the slow pace of financial restructuring and debt work-outs, reduced infrastructure expenditure, more complicated export–import procedures, and increased commercial uncertainty.

Our third general argument concerns the industrial policy framework. Here we argue that, consistent with theory and empirical evidence, the keys to growth include openness (broadly defined to include all factors which impede the flow of goods and services across boundaries), macroeconomic stability, investments in human capital and infrastructure, a predictable commercial policy regime, and clearly defined property rights. As we have argued, there has been substantial progress in most of these areas since 1998. But the unfinished reform agenda remains daunting, and the more slowly these impediments to growth are removed, the longer will Indonesia's industrialization continue to lag behind its high-growth neighbours.

NOTES

1. See Hill (1997) for a set of essays on various aspects of Indonesian industrialization in the pre-crisis period.
2. Indonesian industrial statistics conventionally distinguish between 'non-oil' and total manufacturing, in recognition of the large oil and gas processing sector. At its peak, the latter accounted for about one-third of manufacturing value added. Since the mid-1980s, however, its importance has declined, as oil and gas have become relatively less important and as footloose export-oriented manufacturing has grown rapidly.
3. Note that the data in Table 9.3 refer to firms with a workforce of at least 20 workers and exclude the oil and gas processing sector. The former omission is understandable given the difficulties of accurately obtaining data on small and cottage enterprises. The omission of the latter is a continuing puzzle.
4. Note again that these statistics refer only to firms with a workforce of at least 20 employees. Including the smaller firms, almost entirely domestically owned, would increase the domestic shares significantly.
5. This subsection draws heavily on Basri and Hill (2004).
6. The clearest evidence of the reforms is provided by Fane and Condon's (1996) study of effective protection. They concluded that the weighted average rate of effective protection for manufacturing (excluding the special case of oil and gas processing) declined from 59 per cent to 16 per cent over the 1987–95 period, while the dispersion (standard deviation) fell from 102 to 39. The coverage of NTBs fell even faster: the percentage of non-oil manufacturing value added affected by NTBs declined from 77 per cent to 17 per cent.
7. Important studies include Schiller and Martin-Schiller (1997), Cole (1998), Berry and Levy (1999), Sandee et al (2000), Berry et al. (2001), and Sandee and van Diermen (2004).
8. As Cole (1998 pp. 277) puts it, '[b]eyond these points, the role the government played seems more positive in its absence than in its actions'.
9. See the contributions to Hill and Thee (eds, 1998) for detailed analyses of the pre-crisis situation. The best early study of the Habibie mega projects is McKendrick (1992).
10. Dr Habibie's disdain for labour-intensive industries was also well known. He frequently singled out the garments industry arguing that, in spite of its prominence in early stage East Asian industrialization, Indonesia should not aspire to be just 'tailors for the world'.
11. Though, as argued elsewhere in this chapter, these data can be misleading when exports are dominated by low-end electronics.
12. Since the crisis, the aircraft factory has been attempting to retool its operations. One recent initiative has been a programme to manufacture a modern version of the ubiquitous motorized three-wheeler bicycles, known as '*bajaj*', including an attempt to force the authorities in major cities to phase out the cheaper Indian-manufactured units.
13. See, for example, Kim and Nelson (eds, 2000), Wong and Ng (eds, 2001), Yusuf et al. (2003). Yusuf et al. (pp. 370 - 72) conclude their volume with '10 policy messages': prudent debt management, flexible exchange rate management, regional coordination, financial reform, deregulation and incentives for innovation in services, legal reform, resources for innovation, networking and clusters, competition policies, and proactive public policies for innovation. The framework adumbrated in this chapter, developed in more detail in Hill (2004), is consistent with this list, as modified for Indonesia's special circumstances.
14. These were particularly focused on the egregious crony-related, firm-specific protection, such as the national car project, a large petrochemical plant, and the

clove monopoly, and therefore had little impact on the overall trade regime.

15. Smuggling has been particularly rampant in cases where high luxury taxes have been imposed. These include luxury cars, expensive electronics and high-end footwear. According to one major foreign electronics producer, their domestic market for audio equipment has practically disappeared.
16. The latter received a 5 per cent increase, despite a requested 20+ per cent increase fully supported by the MOIT.
17. Factories in these zones are exempted from paying customs duties and value-added tax on imported material, and VAT on locally produced material used for export.
18. In this as in several other respects, there are striking similarities between Indonesia since 1998 and the Philippines after its mid-1980s economic crisis. In the latter, for example, since 1990 almost all the export growth and most of the industrial dynamism has come from MNE electronics firms in the export zone enclaves (see Hill, 2003).
19. See Kinoshita (2003) for a recent and knowledgeable Japanese perspective on Indonesia's post-crisis FDI regime.
20. Of course, this environment was often brutally unfair to the weak and disenfranchised, and there were frequent 'petty' disputes, particularly concerning land and access to natural resources.
21. This paragraph draws on Ray (2003, pp. 262–7).
22. See Lindsey (2004) for a review of the evolution of the post-crisis Indonesian legal system.

REFERENCES

Aswicahyono, H., M.C. Basri and H. Hill (2000), 'How Not to Industrialize? Indonesia's Automotive Industry', *Bulletin of Indonesian Economic Studies*, 36 (1), pp. 209–41.

Aswicahyono, H., K. Bird and H. Hill (1996), 'What Happens to Industrial Structure when Countries Liberalize? Indonesia Since the mid 1980s', *Journal of Development Studies*, 32 (3), pp. 340–63.

Aswicahyono, H. and M. Pangestu (2000), 'Indonesia's Recovery: Exports and Regaining Competitiveness', *Developing Economies*, 38 (4), pp. 454–89.

Basri, M.C. (2001), *'The Political Economy of Manufacturing Protection in Indonesia, 1975–1995'*, unpublished doctoral dissertation, Australian National University, Canberra.

Basri, M.C. and P. van der Eng (eds) (2004), *Business in Indonesia: New Challenges, Old Problems*, Singapore: Institute of Southeast Asian Studies.

Basri, M.C. and H. Hill (2004), 'Ideas, Interests and Oil Prices: The Political Economy of Trade Reform During Soeharto's Indonesia', *World Economy*, 27 (5), pp. 633–56.

Berry, A. and B. Levy (1999), 'Technical, Marketing and Financial Support for Indonesia's Small and Medium Industrial Exporters', in B. Levy, A. Berry and J.B. Nugent (eds), *Fulfilling the Export Potential of Small and Medium Firms*, Dordrecht: Kluwer Academic Publishers, pp. 31–72.

Berry, A., E. Rodriguez and H. Sandee (2001), 'Small and Medium Enterprise Dynamics in Indonesia', *Bulletin of Indonesian Economic Studies*, 37 (3), pp. 363–84.

Bird, K. (2004), 'Recent Trends in Foreign Direct Investment', in Basri and van der Eng (eds), pp. 93–107.

Bird, K. and C. Manning (2002), 'The Impact of Minimum Wage Policy on Employment in the Formal and Informal Sectors', Australian National

University, Canberra, unpublished paper.

Cole, W. (1998), 'Bali's Garment Export Boom', in Hill and Thee (eds), pp. 255–78.

Fane, G. and T. Condon (1996), 'Trade Reform in Indonesia, 1987–1995', *Bulletin of Indonesian Economic Studies*, 32 (3), pp. 33–54.

Hill, H. (1996), 'Indonesia's Industrial Policy and Performance: "Orthodoxy" Vindicated', *Economic Development and Cultural Change*, 45 (1), pp. 147–74.

Hill, H. (1997), *Indonesia's Industrial Transformation*, Singapore: Institute of Southeast Asian Studies.

Hill, H. (2000), *The Indonesian Economy*, 2nd ed., Cambridge University Press, Cambridge.

Hill, H. (2003), 'Industry', in A.M. Balisacan and H. Hill (eds), *The Philippine Economy: Development, Policies and Challenges*, Oxford University Press, New York, pp. 219–53.

Hill, H. (2004), 'Technology and Innovation in Developing East Asia', in S. Yusuf, M. A. Altaf and K. Nabeshima (eds), *Global Production Networking and Technological Change in East Asia*, Oxford University Press/World Bank, Washington DC, pp. 353–93.

Hill, H. and Thee Kian Wie (eds) (1998), *Indonesia's Technological Challenge*, Institute of Southeast Asian Studies, Singapore.

Hobday, M. (2001), 'The Electronics Industries of Pacific Asia', *Asian-Pacific Economic Literature*, 15 (1), pp. 13–29.

James, W.E., D.J. Ray and P.J. Minor (2003), 'Indonesia's Textiles and Apparel: The Challenges Ahead', *Bulletin of Indonesian Economic Studies*, 39 (1), pp. 93–103.

Kim Linsu and R.R. Nelson (eds) (2000), *Technology, Learning, and Innovation: Experiences of Newly Industrializing Countries*, Cambridge University Press, Cambridge.

Kinoshita T. (2003), 'Indonesia's Investment Climate', unpublished paper, Waseda University, Tokyo.

Lindblad, J.T. (2003), 'Business Responses to Crisis in Indonesia: The 1930s and the 1990s', *Australian Economic History Review*, 43 (2), 169–82.

Lindsey, T. (2004), 'Legal Infrastructure and Governance Reform in Post-Crisis Asia: The Case of Indonesia', *Asian-Pacific Economic Literature*, 18 (1), pp. 12–40.

McKendrick, D, (1992) 'Obstacles to "Catch-up": The Case of the Indonesian Aircraft Industry', *Bulletin of Indonesian Economic Studies*, 28 (1), pp. 39–66.

McLeod, R. (2000), 'Soeharto's Indonesia: A Better Class of Corruption', *Agenda*, 7 (2), pp. 99–112.

Manning, C. (2000), 'Labour Market Adjustment to Indonesia's Economic Crisis: Context, Trends and Implications', *Bulletin of Indonesian Economic Studies*, 36 (1), pp. 105–36.

Manning, C. (1998), *Indonesian Labour in Transition: An East Asian Success Story?*, Cambridge University Press, Cambridge.

Ray, D. (2003) 'Survey of Recent Developments', *Bulletin of Indonesian Economic Studies*, 39 (3), pp. 245–70.

Sandee, H., R.K. Andadari and S. Sulandjari (2000), 'Small Firm Development during Good Times and Bad: The Jepara Furniture Industry', in C. Manning and P. van Diermen (eds), *Indonesia in Transition: Social Aspects of Reformasi and Crisis*, Institute of Southeast Asian Studies, Singapore, pp. 184–98.

Sandee, H. and P. van Diermen (2004), 'Exports by Small and Medium-Sized Enterprises in Indonesia', in Basri and van der Eng (eds), pp. 108–21.

Schiller, J. and B. Martin-Schiller (1997), 'Market, Culture and State in the Emergence of an Indonesian Export Furniture Industry', *Journal of Asian Business*, 13 (1), pp. 1–23.

Schroeder-van Waes, M.C. and K.O. Sidharta (2004), 'Upholding Indonesian Bankruptcy Legislation', in Basri and van der Eng (eds), pp. 191–203.

Schwarz, A. (1994), *A Nation in Waiting: Indonesia in the 1990s*, Allen & Unwin, Sydney.

Suryahadi, A., W. Widyanti, D. Perwira and S. Sumarto (2003), 'Minimum Wage Policy and its Impact on Employment in the Urban Formal Sector', *Bulletin of Indonesian Economic Studies*, 39 (1), pp. 29–50.

Thee K.W. (2003), 'The Indonesian Economic Crisis of 1997–98 and the Long Road to Recovery', *Australian Economic History Review*, 43 (2), pp. 183–96.

Wong Poh Kam and Ng Chee Yuen (eds) (2001), *Industrial Policy, Innovation and Economic Growth: The Experience of Japan and the Asian NIEs*, Singapore University Press, Singapore.

Yusuf, S. et al (2003), *Innovative East Asia: The Future of Growth*, Oxford University Press, New York, for the World Bank.

Index